Smart Sensors, Measurement and Instrumentation

Volume 13

Series editor

Subhas Chandra Mukhopadhyay
School of Engineering and Advanced Technology (SEAT)
Massey University (Manawatu)
Palmerston North
New Zealand
E-mail: S.C.Mukhopadhyay@massey.ac.nz

More information about this series at http://www.springer.com/series/10617

Henry Leung · Subhas Chandra Mukhopadhyay
Editors

Intelligent Environmental Sensing

 Springer

Editors
Henry Leung
Department of Electrical and Computer
    Engineering
University of Calgary
Calgary Alberta
Canada

Subhas Chandra Mukhopadhyay
School of Engineering and Advanced Techn.
Massey University (Turitea Campus)
Palmerston North
New Zealand

ISSN 2194-8402   ISSN 2194-8410   (electronic)
Smart Sensors, Measurement and Instrumentation
ISBN 978-3-319-12891-7   ISBN 978-3-319-12892-4   (eBook)
DOI 10.1007/978-3-319-12892-4

Library of Congress Control Number: 2014953596

Springer Cham Heidelberg New York Dordrecht London

Printed on acid-free paper

Springer International Publishing AG Switzerland is part of Springer Science+Business Media
(www.springer.com)

Preface

Environmental issues are always on the policy making agenda and industries have to manage their environmental impact. Developing environmental sensing and monitoring technologies become essential especially for industries that may cause severe contamination to the eco-systems. According to industrial analysts, the market for environmental sensing and monitoring technologies is projected to reach US$17 billion by 2020. Currently there are three main approaches to improve environmental sensing: developing novel environmental sensors, designing more effective sensing algorithms to enhance detection performance and using multiple sensors to form environmental sensor networks. These three approaches are not exclusive but complimentary for an improved environmental monitoring.

This book is written by experts using one or more of these three approaches of intelligent environmental sensing in their own applications. The book gives a snapshot of the current state of the art in environmental sensor technology, sensory signal processing and wireless sensor networks for environmental monitoring. It starts with a review of sensing technologies and different environmental monitoring applications such as greenhouse monitoring, food quality monitoring, water monitoring and wildlife monitoring. The other ten chapters are dedicated to current researches on these three approaches to environmental sensing.

Chapters 2 to 5 describe sensor technologies for environmental monitoring. Chapter 2 presents a novel millimeter sized sensors called micro motes and their deployment in situations without radio and Global Positioning System such as underground oil reservoirs. Chapter 3 considers the volcanic ash monitoring problem by using the sensor network approach. In particular, a novel low cost smart multi-sensor node is developed to estimate flow rates, classify granulometry, and discriminate volcanic ash from other types of sediments. Chapter 4 reports a special designed sensor for ocean monitoring - a portable high frequency surface wave radar. With advanced signal processing techniques, this portable radar system can provide long range monitoring of sea currents, waves and winds. Chapter 5 reports a motion sensor system to sense the tilt for landslide monitoring. This system is shown to be able to detect small displacements during a typhoon event.

Chapters 6 to 8 put more focus on the second approach, that is, advanced algorithms, in particular, all three chapters consider using sensor fusion to enhance the sensing performance. Chapter 6 considers the problem of water quality monitoring by using an intelligent water monitoring system. This system uses sensor fusion to combine different sensors including camera, Global Positioning System, temperature sensor, PH sensor, conductivity sensor, dissolved oxygen sensor, turbidity sensor and a special designed planar electrode sensor for water quality monitoring. As remote sensing is widely used in environmental sensing, Chapter 7 presents an effective image fusion approach to combine dissimilar imagery data. The fuzzy integral is used to perform an optimal fusion and the method can learn model parameters adaptively by using Kalman filter and compressed sensing. Chapter 8 considers remote sensing for greenhouse precision cultivation. The proposed novel system combines RFID, multi-spectral imaging and plant-oriented sensing algorithm and develops a variable spraying system for precision irrigation.

Interesting applications and detailed description of the wireless communication aspects on wireless sensor networks for environmental sensing are presented in Chapters 9 to 11. Chapter 9 gives a clear description on how to use the IEEE 1451 standard to develop a wireless sensor network for environmental sensing. The authors use indoor air quality monitoring as a demonstration. Chapter 10 uses wireless sensor network technology for an industrial monitoring problem. It considers different wireless standards such as ZigBee, WirelessHART and then develops a wireless sensor network system to monitor torque, speed and efficiency of induction motors. Chapter 11 considers a unique monitoring problem – debris flow. It introduces different types of debris flow monitoring systems in Taiwan and the performance of the geological monitoring system on providing debris flow warnings.

We would like to whole-heartedly thank all the authors for their contributions to this book.

Henry Leung
Subhas Chandra Mukhopadhyay

Contents

Contents

About the Editors

Dr. Henry Leung is a professor of the Department of Electrical and Computer Engineering of the University of Calgary. Before joining U of C, he was with the Department of National Defence (DND) of Canada as a defence scientist. His main duty there was to conduct research and development of automated surveillance systems, which can perform detection, tracking, identification and data fusion automatically as a decision aid for military operators. His current research interests include adaptive systems, computational intelligence, data mining, information fusion, robotics, sensor networks, signal processing and wireless communications. He has published extensively in the open literature on these topics. He has published over 190 journal papers. Dr. Leung has been the associate editor of various journals such as the International Journal on Information Fusion, IEEE Signal Processing Letters, IEEE Trans. Circuits and Systems, International Journal of Advanced Robotic Systems. He was the chair of the Nonlinear Circuits and Systems of the IEEE Circuit and System Society and has served on the program committee, organizing committee, track chairs for various conferences such as the SPIE Conference on Sensor Fusion, IEEE ISCAS and FUSION. He has also served as guest editors for various journals such as "Intelligent Transportation Systems" for the International Journal on Information Fusion and "Cognitive Sensor Networks" for the IEEE Sensor Journal.

Dr. Subhas Chandra Mukhopadhyay graduated from the Department of Electrical Engineering, Jadavpur University, Calcutta, India in 1987 with a Gold medal and received the Master of Electrical Engineering degree from Indian Institute of Science, Bangalore, India in 1989. He obtained the PhD (Eng.) degree from Jadavpur University, India in 1994 and Doctor of Engineering degree from Kanazawa University, Japan in 2000.

Currently, he is working as a Professor of Sensing Technology with the School of Engineering and Advanced Technology, Massey University, Palmerston North, New Zealand. His fields of interest include Smart Sensors and Sensing Technology, Wireless Sensors Network, Electromagnetics, control, electrical machines and numerical field calculation etc.

He has authored/co-authored over 300 papers in different international journals and conferences, edited eleven conference proceedings. He has also edited ten special issues of international journals and twenty books with Springer-Verlag as guest editor. He is currently the Series editor for the Smart Sensing, Measurements and Instrumentation of Springer-Verlag.

He is a Fellow of IEEE, a Fellow of IET (UK), a Topical editor of IEEE Sensors journal. He is also an Associate Editor for IEEE Transactions on Instrumentation and Measurements and a Technical Editor of IEEE Transactions on Mechatronics. He was a Distinguished Lecturer of IEEE Sensors council. He is Chair of the Technical Committee 18, Environmental Measurements of the IEEE Instrumentation and Measurements Society. He is in the editorial board of many international journals. He has organized many international conferences either as a General Chair or Technical Programme Chair.

Chapter 1
Sensing Technologies for Intelligent Environments: A Review

Hemant Ghayvat, Subhas C. Mukhopadhyay, and X. Gui

Abstract. Sensors are fundamental components for making any environment intelligent. Depending on the applications, different sensors are required to implement specific objectives. This chapter will review different applications and consequently the requirements for different sensors and sensing technologies used in intelligent environment with a special emphasis on smart homes.

1.1 Introduction

The sensing technologies offer a great solution in our daily activity from commercial buildings to daily household, from motorbike to big aeroplane, and from digital thermometer to biosensors for medical applications. Sensors are important in monitoring, data collection and high speed computing systems and based on data different control activities can be carried out. Depending on applications there is a need of different types of sensors. The selection of sensors for a specific application depends on many factors. There are vast ranges of sensors available for one particular application, so it is up to designers find the trade-off with respect to physical dimensions, performance figures, measurement method and cost. It is a difficult task to pick the right sensor, especially as the design can be sometimes for next few decades depending on applications.

There are some important sensor criteria which must be considered before selecting the sensor for any application. Accuracy is one of the most important performance parameters for any sensors and should be as close to 100% of the actual value as possible. In some critical applications it may be more important

Hemant Ghayvat · Subhas C. Mukhopadhyay · X. Gui
School of Engineering and Advanced Technology
Massey University, Palmerston North, New Zealand
e-mail: S.C.Mukhopadhyay@massey.ac.nz

© Springer International Publishing Switzerland 2015
H. Leung and S.C. Mukhopadhyay (eds.), *Intelligent Environmental Sensing*,
Smart Sensors, Measurement and Instrumentation 13, DOI: 10.1007/978-3-319-12892-4_1

than others. The range of measurement of sensors depends on the principle based on which the sensors have been fabricated and it is very much dependent on the materials of the sensors. Sometime it is extremely difficult to obtain a very wide range of sensing and it may be advisable to discretise the ranges by more than one sensor to obtain the highest available measurement accuracy. The sensors are designed with certain limitations that are mostly related to environmental parameters such as temperature, pressure, humidity and other physical inputs; this environmental limitation prevents the sensor from damage. It is sometimes due to technology, and material used in designing. For the same application, we get different sensors whose environmental limits are different.

The sensitivity is the parameter which shows how the sensor responds but it is degraded with time (age) due to the factors such as wear and tear, hysteresis, fabrication material and design technology, etc. It is always good to select the sensor which offers better response and reflect with respect to desired input and show very negligible or ideally zero to other undesired ambient additives. The cost of sensor depends on the complexity of manufacturing technologies and the features it offers, such as: accuracy, precision etc. Sometimes the cost of sensor itself is small, but the cost of instrumentation and signal conditioning circuits which are responsible for signal processing and conversion is high.

For the micro observation and analysis of sensor data, it is always good to have a high-precision sensor. The sensor with high repeatability allows us to take few readings for the particular application and the output sample shows the very small variance for the same condition. In most of the applications, the absolute accuracy may be important but the stability and repeatability play major role on the performance. The performance of the overall system is limited by improper calibration (calibrating environments may be different), due to this sometimes the same sensing system shows different outcome values for the same input stimuli at the same operating environments.

In many big machines and electronic appliances the sensors are just a small part as compared to other circuit elements but sensor's long-lasting accurate function matters a lot for that whole machinery, this comes from the stability of a sensor. The response time of sensor varies from application to application. For some applications such as home monitoring or intelligent buildings, a sensor such as smoke detector must have very fast response time.

Availability of sensors is one important parameter which sometimes was not considered properly. Unnecessary waiting time to avail the sensors may make the whole development process longer and also replacement of sensors delay the system.

1.2 Monitoring of Environments

The market of sensing technologies on monitoring of different environment has already occupied a significant portion and increases at a very fast pace with time. The environmental monitoring is not limited to home for electronic appliances and measuring devices, it is everywhere such as volcanic eruption, global warming, landslide and so on. Sensing technology provides excellent interface between human and nature for monitoring, this monitoring data helps us to prepare any eventuality or prevents us from any danger. In general, everything around us can be defined as environmental but from sensors perspective we divide the chapter based on different applications:

1.2.1 Wireless Systems

The wireless sensors have communication facility to transit measured data towards receiving station wirelessly. Usually, the end device wireless sensor nodes take power from battery while the sensors are installed at outside environment. While the sensors have provisions to take power from electric supply also behave as wireless sensors and communicate wirelessly. There are many wireless protocols available for wireless communication. In recent times, wireless sensors and networks are extended to web-based database server and data delivered by sensors end devices are uploaded into internet by coordinator at home gateway [10, 14, 19, 26, 34, 36, 37, 41, 49, 54, 56, 62] . A typical representation of wireless system is shown in figure 1 [62].

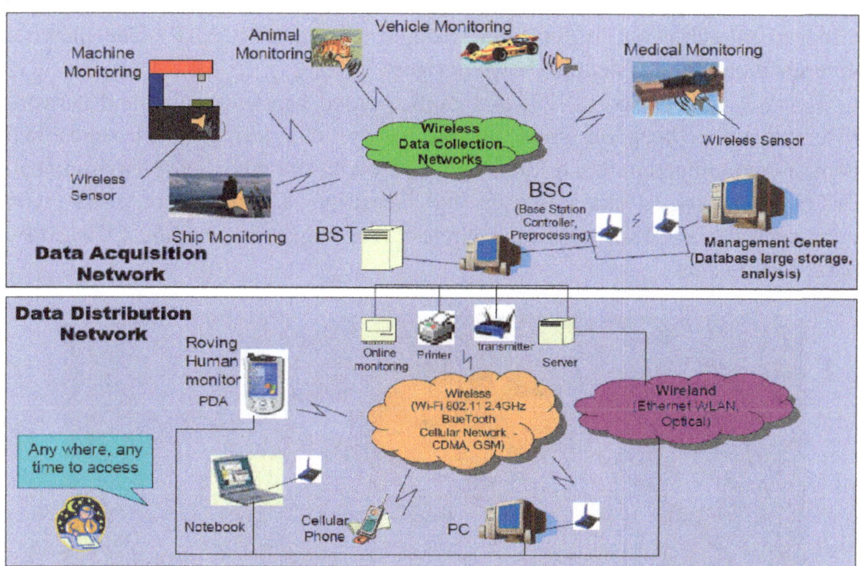

Fig. 1 Layout diagram representation of wireless sensor networks [62]

The wireless area is especially suitable in smart home for eldercare to monitor someone from a remote distance. The body area network based on wireless ANT3 module (ANT11TS33M4IB) is designed to get the static position of the person by tri-axial accelerometer (Accelerometer ADXL345) and angular velocity by gyroscope (ITG3200). The sensor data is fed to microcontroller MSP430F2619 and transmitted to mobile phone by RF transceiver (nRF24AP2-8CH) and coin cell battery CR2032 is used, the block diagram representation is shown in figure 2 [5].

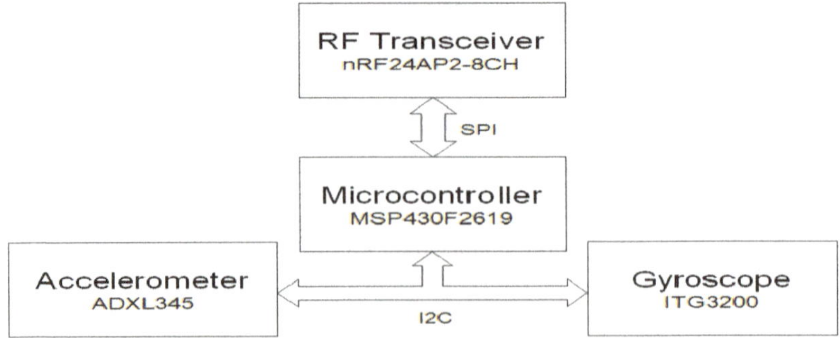

Fig. 2 Block diagram representation of sensors and controller for monitoring and real-time classification of functional activity [5]

The implementation of wireless sensors and network-based monitoring of toxic gasses has been reported in [9]. The web-based system monitors the concentration of greenhouse gasses such as CO_2 and CH_4 at the landfill site. The IR sensors (TDSOO76 by Dynament) have been used for gas measurement. The sensor data is logged to webserver by the GSM module (Siemens MC35iT). The pictorial representation of the system is shown in figure 3.

The study as reported in [27] is a ZigBee-based wireless sensor and network system that has been designed for greenhouse monitoring. It measures and monitors the temperature and humidity of plants in the greenhouse environment. CMOS technology based sensor by Sensirion Company is used with microcontroller PIC18LF4620 and wireless communication module CC2420 by Texas Instruments.

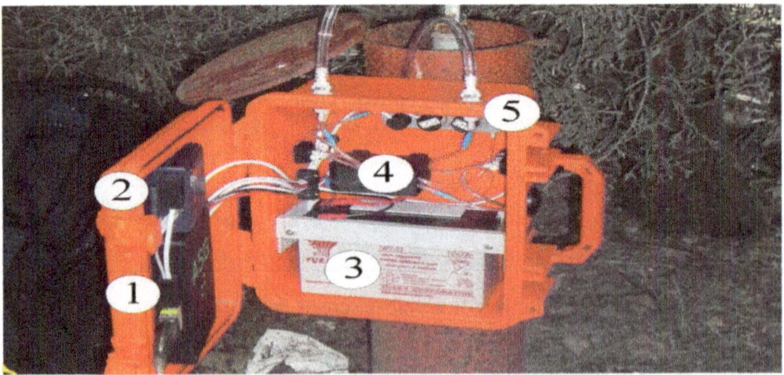

Fig. 3 Block diagram representation of autonomous gas sensing platforms on landfill sites [9]. {Annotation (1) represents the control board, (2) GSM module, (3) Battery backup, (4) extraction pump and (5) sample chamber with infrared gas sensor}

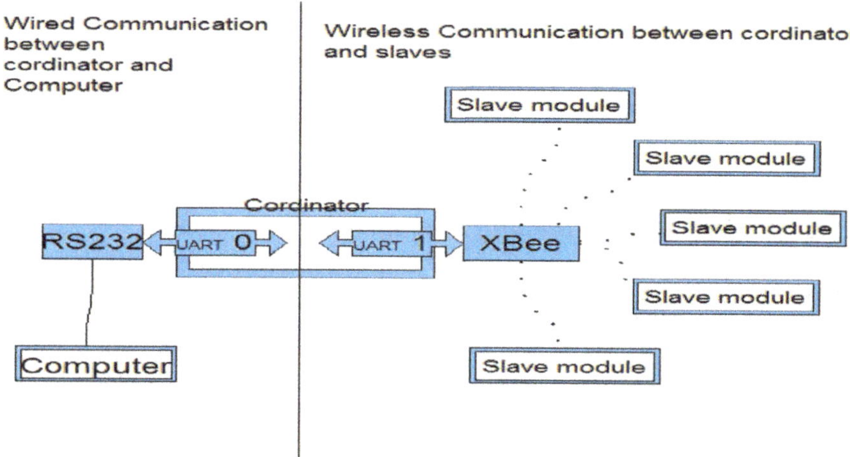

Fig. 4 Block diagram representation of wireless connection between coordinator and end device [29]

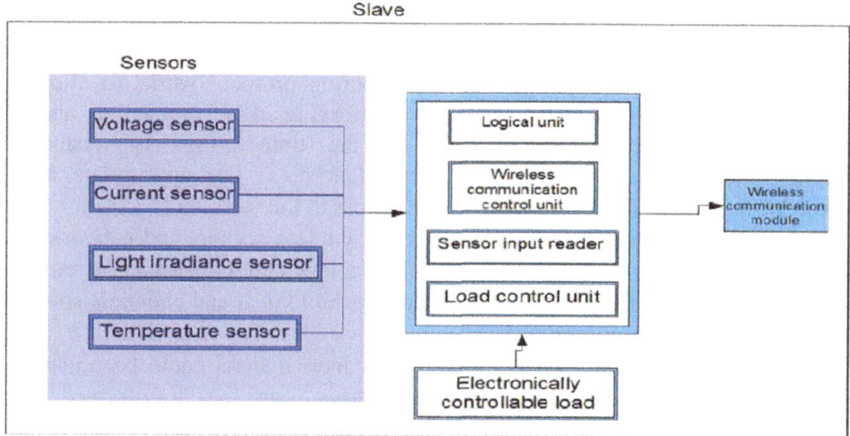

Fig. 5 Block diagram representation of end device module [29]

The solar panel is the good option as a renewable energy source but its cost is not cheap, additionally the maintenance and repair of installed solar panel become difficult due to the fact that there are not many feasible solar panel monitoring options available. The system designed by [29] offers great solution for this: the panel performance is analysed by current, voltage, photodiode (photodiode PDB-C139 as a sensor to analyse the irradiance) and thermistor (thermistor B57164K472J is used for solar temperature measurement) sensors; and the repairing as well as replacement of a particular solar panel block is subject to the wireless data logged in the local computer. The block diagram representation of the system is shown in figures 4 and 5. In a similar study of wireless interface, the Bluetooth-based system is proposed in [52].

In [30] the monitoring system is developed with XBee device which works on ZigBee wireless communication protocol, the system uses the heat, light and intensity detection sensors. The system proposed in [58] for wireless compost monitoring uses the Tmote Sky integrated platform which contains temperature, humidity and light sensors.

1.2.2   Energy Harvesting and Management

The energy harvesting is all about collecting energy for wireless sensor node operation from ambient sources. The effective use of energy with respective protocol is energy management, so energy management and harvesting are two essential considerations for wireless sensor and networks. The wireless sensor node in home environment is heterogeneous most of the time and distributed in such a fashion where we cannot frequently replace the battery easily always, so we have to make self-sustainable wireless sensor node and when we find no essential communication between master (coordinator) and slave (end-device) the system should be either in hibernate mode, sleep mode or any protocol developed efficient mode, so that the power would be saved for long time [90, 91,100]. We are aware that the energy generated from various energy harvesting device is limited, energy efficient protocol scheme at different layers of system operation is highly recommended. The energy efficient routing protocol works to allot the optimal path for the data delivery. Power conserving MAC protocol is another important consideration because most of the time power is wasted in communication of undesired and less important data. The appropriate MAC protocol can prohibit this energy-inefficient access to the medium [23,102].

The common energy harvesting sources for wireless sensors and networks are mechanical vibration, thermal energy and photovoltaic cell. Other than these, there are more sources of energy such as body energy, biological and chemical sources, but they are not very common.

When the device is under vibration state its inertial mass could be utilised to produce the movement, this vibrating energy can be transformed into electricity energy. In the paper [101] the electromagnetic energy harvesting scheme is realised by using a composite magneto electric (ME) transducer and a power management circuit. In the transducer the magnetostrictive Terfenol-D plate is subjected under dynamic magnetic field with the ultrasonic horn to produce vibrations and these vibrating waves are converted into electricity by piezoelectric element.

The energy harvesting is capturing miniature quantity of energy from natural sources, i.e., environment renewable sources, storing and using them effectively whenever desired. The recent advancement in technology offers the efficient energy harvesting devices to capture, accumulate and store in maintenance free as batteries at very economical cost. The problem with wireless sensor nodes is that they are deployed most of the time in environments that is difficult to access and charging of batteries is always an issue [7, 40]. The system proposed in [4] scavenges ambient RF energy power for the wireless sensors and networks.

The conventional BAN sensors are powered by batteries. Usually, the available batteries have disadvantages such as limited lifetime, bulky size, a finite amount of energy and chemicals that could cause a hazard. The concept of battery-less wireless sensor is quite good for body area network, in which the energy for system-on-chip physiological sensor operation is generated by body heat [99]. A study to scavenge power from human activity has been reported in [28]. In this system the wristwatch extracts necessary power from human body for its functions. The maximum force of body exerts on our foot, so deployment of magnetic generator as well as electrostatic generator in the footwear can produce energy which could charge a battery. The block diagram representations of scavenge power from human activities are shown in figures 6, 7 and 8 respectively.

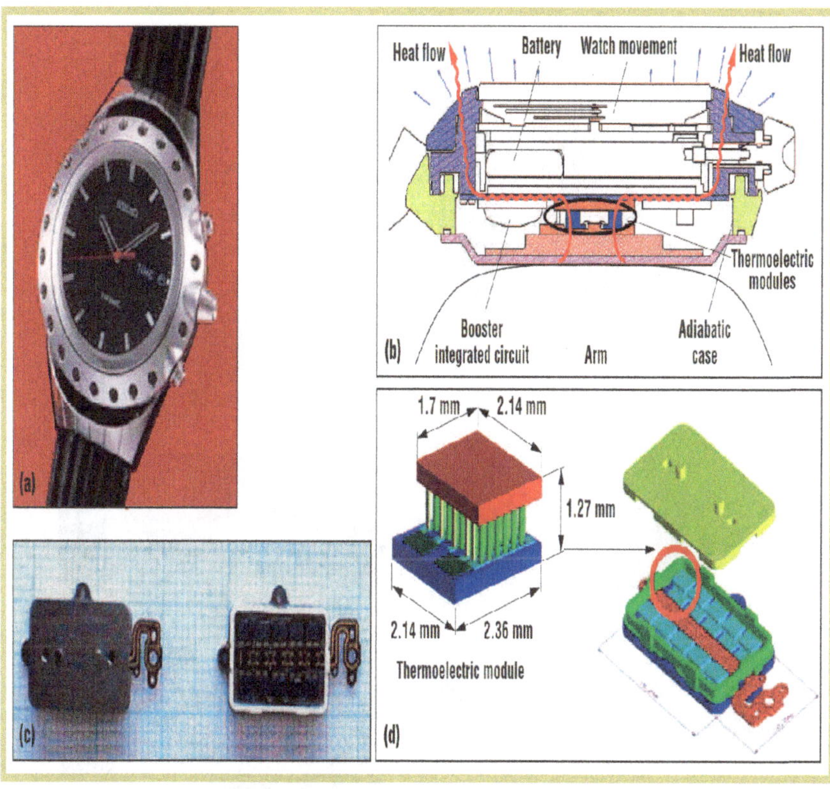

Fig. 6 Block diagram representation of The Seiko Thermic wristwatch: (a) the product; (b) a cross-sectional diagram; (c) thermoelectric modules; (d) a thermopile Array [28]

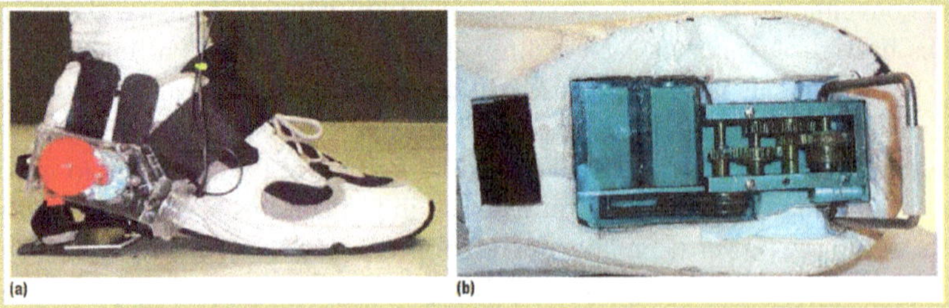

Fig. 7 Block diagram representation of Magnetic generators in shoes (a) A strap-on overshoe produced an average of 250 mW during a standard walk, powering a loud radio; (b) an assembly hosting twin motor-generators and step-up gears embedded directly into a sneaker's sole (without springs or flywheels for energy storage) produced 60 mW [28].

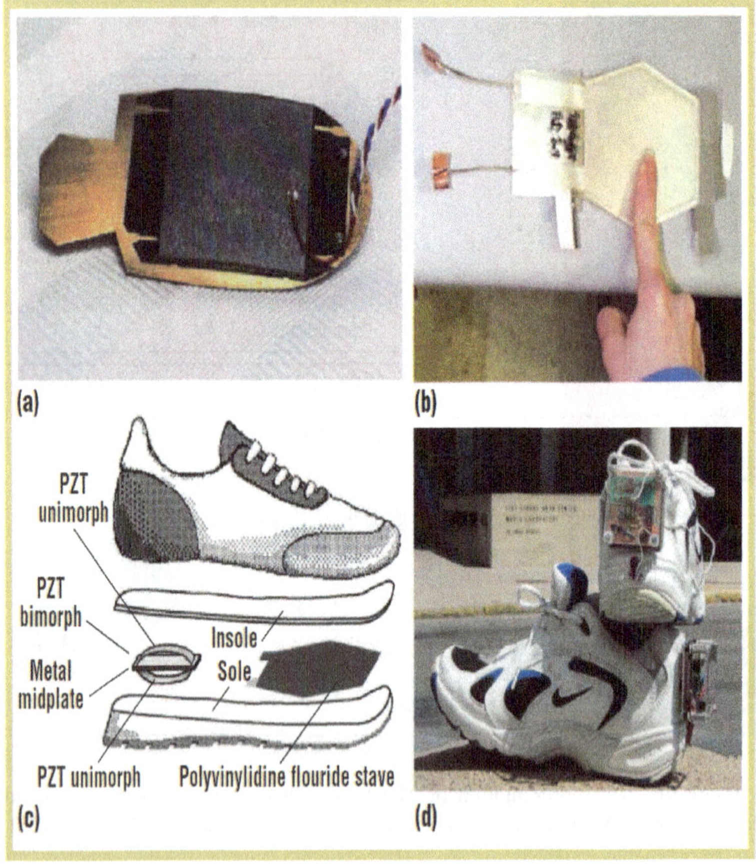

Fig. 8 Block diagram representation of Integration of (a) a flexible PZT Thunder clamshell and (b) a 16-layer polyvinyl dine fluoride bimorph stave under (c) the insole of a running shoe, resulting in (d) operational power harvesting shoes with heel-mounted electronics for power conditioning, energy storage, an ID encoder, and a 300-MHz radio transmitter [28].

The use of solar as well as wind energy in energy harvesting are some of the most preferred ways, but the problem with these two is that they are totally weather dependent, i.e., the solar energy only serves in daytime in the presence of sunlight. As far as wind energy is concerned, it depends on the wind condition in that geographical area. In wireless self-sustainable sensor node the solar panel is mostly installed outdoor to get electrical energy; the block diagram representations of a few reported systems based on solar energy harvest are shown in figure 9.

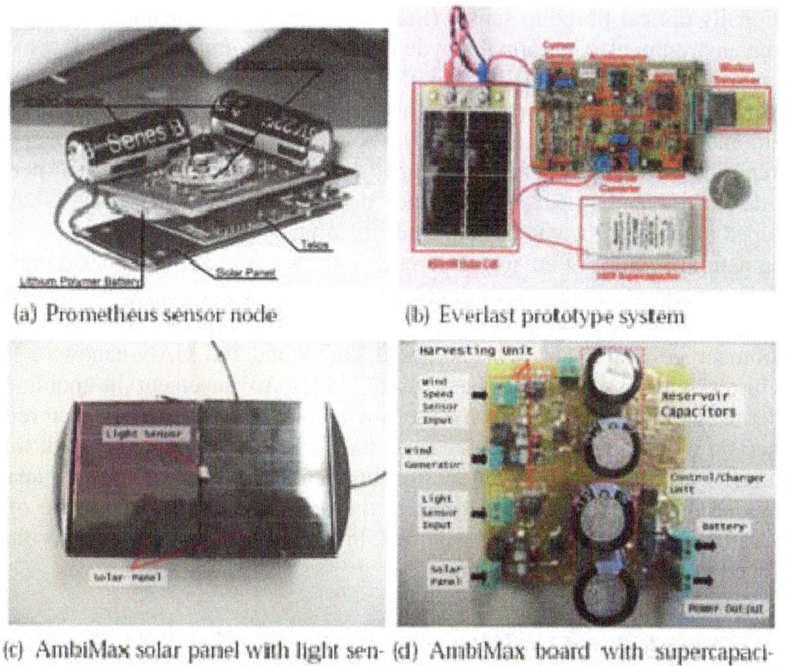

Fig. 9 Block diagram representation of solar energy harvest system [40]

Home automation is growing really fast but the important consideration which we need to make in a smart home is not only securing the home but also reducing power consumption by monitoring the home environment. By collecting information from temperature, pressure, humidity, noise, and dust using different sensors, it is possible to control the environment effectively [53]. It reduces the interference with other network and provides communication assurance. A similar study [54] uses the Zigbee- device and ADSL router.

1.2.3 Environmental Monitoring

Environment monitoring is basically the ambient parameter values collected by the sensors such as temperature, humidity, dust, gas and these parameters help us to take preventive actions.

Monitoring dangerous gas at places such as home is really essential to prevent any danger which causes risk of life. In [8], the In_2O_3 gas sensor is developed with Ga material and is found to be more effective than classical In_2O_3 sensor. It is tested with CO, CH_4, H_2 and OZONE gases and the result is satisfactory. Additionally optical fibre bio sensor (Sniffer) is good for formaldehyde detection at home environment. The formaldehyde is basically flammable aldehyde which is generated at home by our daily activities and it is harmful volatile compound. According to some medical reports at above 80 ppm level this gas causes some allergic and serious chronic disease [20]. A similar study on gas sensor devices was reported in [110], where Au nanoparticle-functionalised WO_3 nano-needles were developed and tested with ethanol detection. The wireless sensors were developed to monitor the gas level at landfill, compost and waste [58, 9].

The natural sources of environment especially water are contaminated due to a number of reasons. The sensor developed in [42] is a planer electromagnetic sensor which estimates the nitrate level in water. In a similar study of detecting environment pollutant and benzene-based compound the GAN-nanowire/TiO_2-nanocluster hybrid sensor has been used [112]. Advancement in mobile and internet technology offered us to monitor the ambient pollution level from remote distance on mobile, laptop and television; the GPRS sensor array network for air pollution is installed and mobile data acquisition server is developed to analyse and upload the information to internet in the research study [113]. Some of the software-based sensors are also developed. In [111] the social sensor algorithm is developed for earthquake monitoring.

1.2.4 Greenhouse Monitoring

The sensor based monitoring of greenhouse has attracted a lot of attention in the last decade in terms of research and development. The sensor-based greenhouse monitoring is the most efficient way to control optimum parameters but the cost of system installation still not very low and common for all.

Most of the greenhouse systems are limited to monitoring only but the system developed and implemented in [26] facilitates control of greenhouse environment as well. This system includes hardware, software and database for data analysis. The system measures the parameters such as temperature, humidity, vapour pressure and others that are linked to the growth and health of crops at micro level. The sensor data is sent to home gateway USB coordinator through wireless RF communication device. The data is logged to database server and graphical user interface has been developed for users. The most interesting part of the system is the automated control to maintain and control the essential parameter values inside the greenhouse. For automated control, the RS-485 actuator control module is connected to the actuator panel that governs the climate control equipment such as cooling, watering, etc. The lock diagram representation of the wireless sensor network for greenhouse climate control is shown in figure 10.

The web-based wireless sensors and networks have been realised in [49] for agriculture land, to detect the temperature, humidity and pH level of soil for crop growth. The system is based on the ZigBee protocol for wireless RF communication. The sensor data is fed to microprocessor MCU MSP430F149 and this data is sent to cluster head by wireless communication device CC100. The cluster head is designed with Philips microprocessor ARMLPC2131 and transceiver MC13193 (2.4 GHz). In a similar study [27] the system is implemented by CMOS sensors (Sensirion Company).

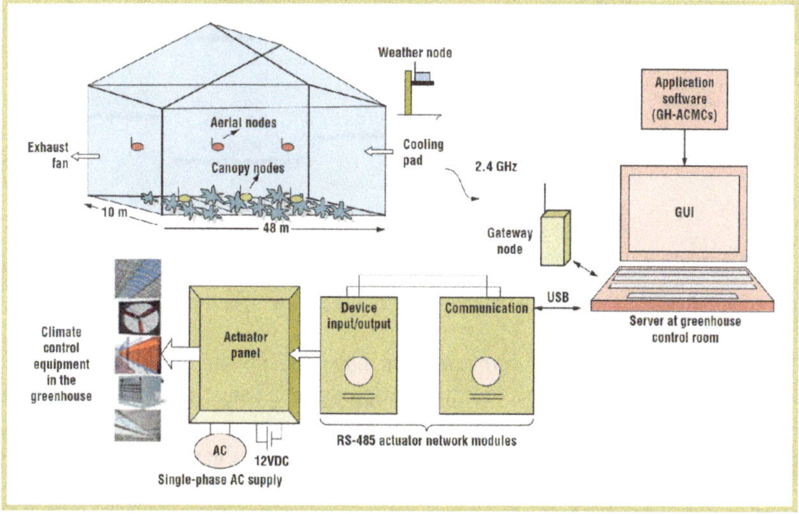

Fig. 10 Block diagram representation of wireless sensor network for greenhouse climate control [26]

The CO_2 is the major greenhouse gas which affects significantly the agricultural environment In [33] the CO_2 sensor 6004 by Telaire Company, temperature and humidity sensor SHT11 are installed and a fuzzy control system is designed to measure the environmental parameters. In a similar agricultural research, the system developed can monitor and generate SMS alerts with respect to changes in greenhouse environment for vegetables and the wireless data is linked to GSM [1]. This system uses hypothetical approach to analyze the environmental parameters and monitor the growth of plants. The block diagram representation of the relationship of the effect of humidity and temperature on vegetables is shown in figure 11.

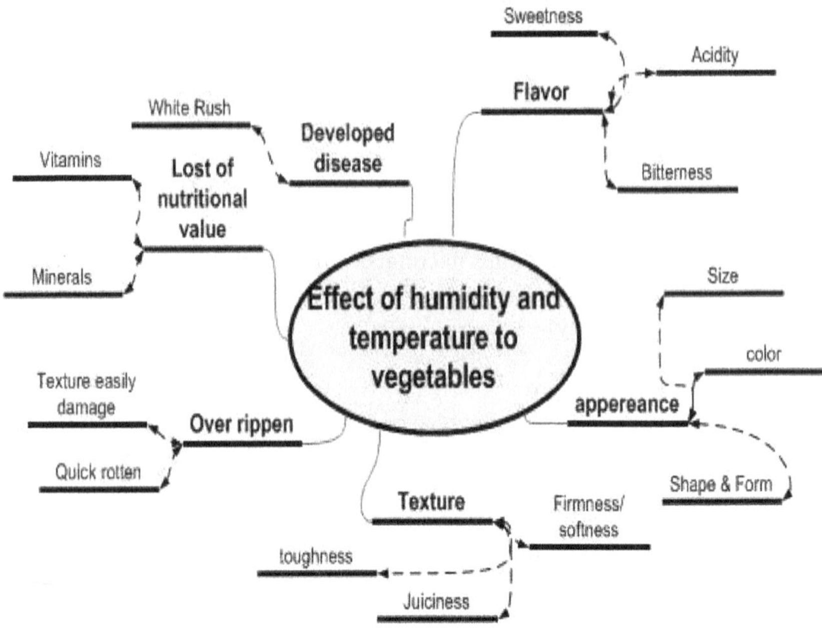

Fig. 11 Block diagram showing the effect of humidity and temperature on vegetables [1]

Generally in greenhouse monitoring we have the problems of deployment of nodes in proper positions and selecting the appropriate range according to wireless sensor networks. The whole system is divided into a hierarchy with sensor nodes at the bottom level, cluster heads at the middle level and the clustering node at the top level. For every specified number of sensors there is a cluster head and all cluster heads finally send the data to the clustering node. The sensor data is fed to microprocessor MCU MSP430F149 and this data is sent to cluster head by wireless communication device CC100.

1.2.5 Food Quality Monitoring

Poisoning due to bacteria, fungal or toxins in food can affect the taste and smell of food but early detection of bacterial infection can only be observed by sensing unit. The sensor technology has achieved excellent development in this field and there are different accurate and high performance biosensors for food testing.

In [64] interdigital sensor is modelled for seafood testing. The interdigital sensor detects the change in the sensitivity of sensor with respect to contamination as well as poisoning compound. A pictorial representation of the interdigital sensor is shown in figure 12.

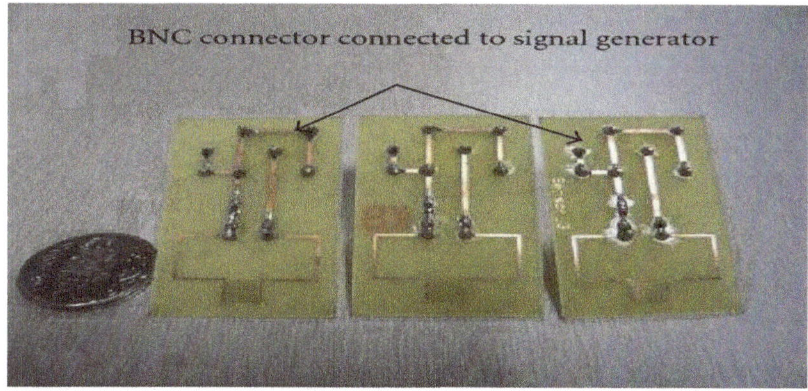

Fig. 12 PCBs of interdigital sensor [64]

In a similar study a new interdigital sensor is developed to test poisoning compound in food. These sensors are coated in four different ways to analyse the performance over the food sample which contains endotoxins or Lipopolysaccharide (responsible for germ and bacteria). These four different coating are: (a) polymetric fiber of carboxyl polymer by electro spinning, (b) spin coated carboxyl polymer, (c) dip-coated silica coating functionalized with aminopropyltrietoxysilane and (d) Dip-coated silica with thionine. The block diagram representation of thin film silicon based interdigital sensor shown in figure 13.

Fig. 13 Thin film silicon based interdigital sensor [2]

Y.Chai et al. [6] has designed wireless magnetoelastic (ME) biosensor for detecting the bacteria that produces the compound of salmonella typhimurium on fresh products such as tomato and spinach. The ME biosensor uses the ME resonator as the sensor platform and E2 phase to identify the bacterial element in the sample [6]. The block diagram representation of ME biosensor for sample testing is shown in figure 14. In another study [21] sensors are designed by three different processes to test bacterial or germ causes poisoning.

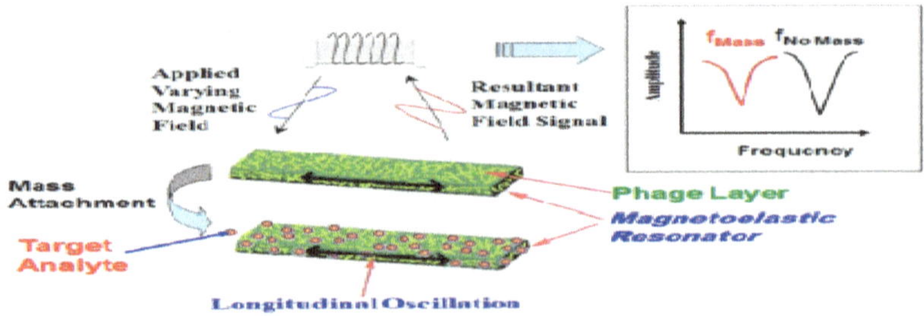

Fig. 14 The block diagram representation of ME biosensor for sample testing [6]

A system has been reported in [43] to monitor the moisture in the substrate cultivation of lettuce (plant). This moisture analysis helps to the decide irrigation by drip. The moisture sensors are placed in the depth, ranging from 0 to 15 cm in the substrate (soil) of lettuce. The EC-5 moisture sensor is used. The sensor data at different depths are collected and further correlation has been developed between parameters at different depths.

1.2.6 Monitoring of Wildlife

In wildlife monitoring, the behaviour and location of animals make it difficult due to distance, geographical area and atmospheric conditions. But by placing the sensors over the body of animals and linking that to wireless signals to satellite and cloud database can provide the location and real-time data. The wireless sensors networks and wireless multimedia sensor networks (WMSN) are providing good outcomes for monitoring animal wellness. The GPS collar and satellite remote sensing offer great results such as animal landscape interaction and social behaviour. With advancement in communication and sensing technology it is possible to monitor every possible species from remote distance with environmental and physiological parameter values, images, audio and videos [92-94]. The wireless monitoring system developed in [95] for wildlife uses ZigBee standard and various monitoring and tracking sensors such as PIR etc. The pictorial representation of animal sensing is shown in figures 15 and 16. respectively Selective audio encryption based on WMSN could provide encouraging results [98].

In the recent time some wildlife animals are at the edge of extinction. The reasons are human interference into forest area, hunters, animal move beyond the forest boundary. Installing the wireless sensor network to monitor the activity of human and animal in the restricted forest area is very essential and in any suspicious condition alarm as well as alert could be initiated. A design challenge is to reduce the number of false alarms. The border senor could be a worthy solution for the reduction of false alerts [97].

Fig. 15 (a) Image of an animal with sensing electronics and GPS device mounted on the collar, (b) Picture of camera placement in wildlife area [95]

Researchers produce different model to analyse the behaviour of animal towards the wellness such as detection of some sort of pain. The animal health monitoring system designed in [3], uses some force on animal (sheep) by stepper motor controlled pin to apply pain and measures the animal reaction by using the flexi force sensor. The whole research is to observe the behaviour of sheep at different applied force. System uses the microcontroller and signal conditioning circuit. The block diagram representation of animal health monitoring sensing system is shown in figure 17.

Fig. 16 Deployment of camera-sensor; (a) wildlife passage; (b) details of the camera-sensor emplacement; (c) picture taken by the master camera-sensor [95]

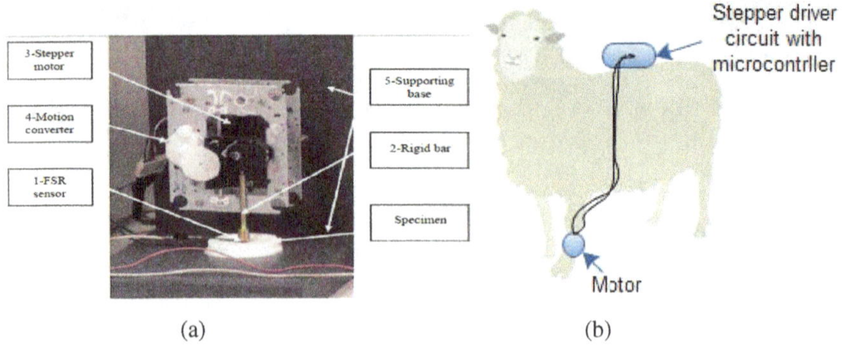

Fig. 17 The representation of (a) active probe parts with a specimen in its place, (b) reposed model for pain measurement system [3]

Wireless sensors based determination of estrus level and calving time of dairy cattle has been reported in [41]. A wireless sensor module contains acceleration sensor, RF communication device based on Zigbee standard and battery placed on the front leg of cow to get three-axial data that is sent to the coordinator attached to the local gateway computer. The motion of the cow during estrus period changes completely and the pattern can be analysed. During calving time the senor data helps to recognize the state of cow [41]. The block diagram representation of BAN for the dairy cattle cow is shown in figures 18 and 19. A similar study as reported [96] derives the behavioural pattern by 3-axis accelerometer sensor on mice.

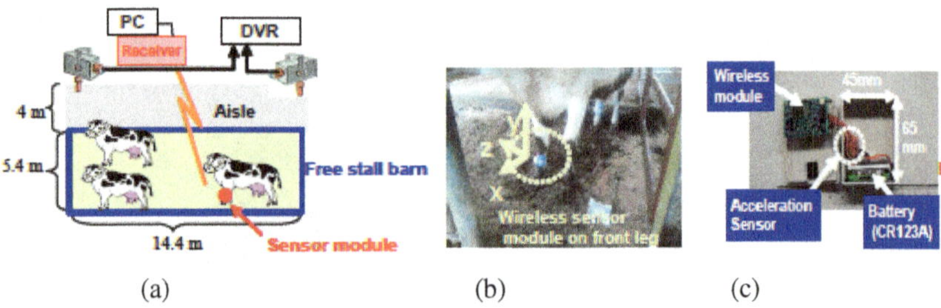

Fig. 18 The representation of (a) layout of senor unit placement over the dairy cattle, (b) real image of cow with the accelerometer sensor module and (c) Image of sensor module [41]

In [24], the progesterone analysis in milk of cow is implemented to identify body heat (before and after getting pregnant), estrous detection, postpartum ovarian status and various hormonal treatments. The advantage of this test is that it can be conducted at cow-side by the analysis kit, it is user friendly and can be completed in very optimal time. The kit shows the results in colour changes of respective field with name of particular test marked on it. There are some other sensing kits and equipment available to diagnose the behaviour of animal and other physiological parameters which can help to improve and maintain their productivity and wellness such as detecting subclinical ketosis in dairy cows by the handheld meter [103], perinatal calcium homeostasis in dairy cows to get impact of combination of 25-hydroxyvitamin D3 [104].

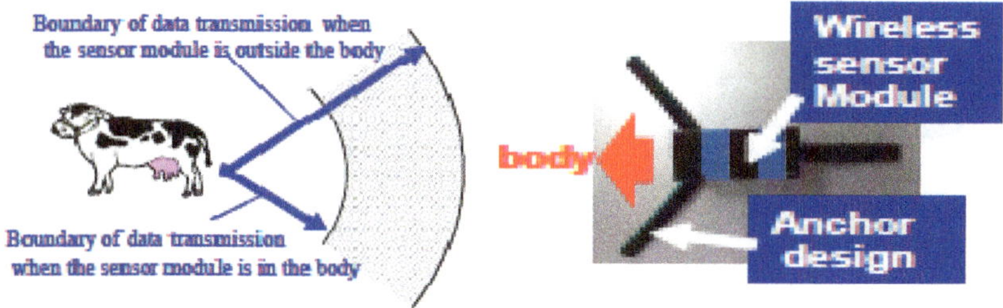

Fig. 19 The representation of (a) Area where receiver should be placed in the method based on wireless network session, (b) real image of body sensor module [41]

The animals are a part of nature so any abnormality and infection which affect them will ultimately have consequences on human being directly or indirectly [105-107].

1.2.7   Home and Healthcare

The home and healthcare sensors offer a great ambient-assisted living (AAL) environment for all. The support becomes very useful for the individual who stays alone as well as the elderly person. The healthcare sensors may be wearable as well as non-wearable sensors; Most of the times for healthcare monitoring the sensors are wearable which comes under body area network.

Placing sensors over the whole body of the person to measure body-temperature, pulse rate, heart rate, skin conductance and brain functioning, may be the best way to obtain accurate value of those physiological parameters. These parameters help us to recognize the emotion, the recovery of patient and support to individual to stay alone without any family member or close one [66]. In [51], the authors have designed protocol layer definition for all the applicable condition and priorities such as fire, injury during sports activity, soldier protection during war and etc. The size of packet which includes bits for source address, destination address, priority, gender, age, type of environments (entertainment, sports, home and study, etc.), priority level and actual payload information.

In a similar approach [10], ECG sensors are installed on the body to measure heart rates of different patients living alone at home. The block diagram representation of wireless sensor networks for monitoring physiological signals of multiple patients is shown in figure 20. The date packets from wireless patient portable unit(WPPU) are sent via wireless communication module to wireless access point unit (WAPU). The WAPU is based on Texas Instruments' (TI) wireless communication protocol. The WAPU sends the packet through the router to internet-based web server. This data is available for hospital and caregiver [71]. In a similar study in [60] reported for one patient, which also enables feature of emergency alert and support. The system designed to measure the human pulses in [48] is based on comparative analysis between the pulses of two different persons to estimate the abnormality in heart rate.

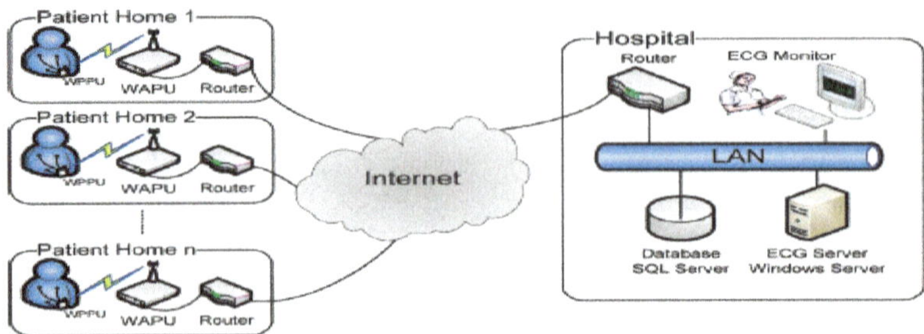

Fig. 20 The block diagram representation of wireless sensor networks for monitoring physiological signals of multiple patients [10]

Physiological parameters such as body temperature, heartbeat are kind of essential to monitor the health status of a person. Additionally, detection of fall may also be useful to monitor the health of the individual who is recovering from illness and staying alone at home environment. Sudden change in the pattern of temperature and heartbeat values indicate the problem in health and this helps the caretaker of that person to take some preventive action before anything abnormal happen.

Sometime people fall down due to some indefinite health problem or accident at home. Usually there is no one to help during fall. The monitoring system based on impact sensor is very useful to detect fall of elderly person[55]. The system in [59] measures the physiological parameters and other additional parameter such as fall detection by accelerometer and position recognition by PIR sensor. The data from sensors show the pattern of their activity and routine. The sensor in [57] is designed to detect different types of body position: lying, standing, sitting and falling. The developed MEMS sensor unit is conceptually based on 3D accelerometer sensor. On the basis of thigh inclination with respect to particular axis the body position is determined. If thigh inclination is about 80 degree with respect to Z axis, it is considered as sitting on the chair.

There are very few systems designed to diagnose the disease like Parkinson. Parkinson's disease affects directly the brain and its ultimate results come in terms of overall body movement. The research reported in [47] developed Mercury Live system, which contains wearable wireless sensors and web-based platform to directly interact with patient and provide telemedicine and proper information to caregiver. The battery life is one of the critical issue, which is limiting the performance of system [47].

The femtocell is small, low power cellular base station which is very suitable for small network at home as well as in workplace. The block diagram representations of WSN based on femtocells are shown in figures 21, 22 and 23 respectively.

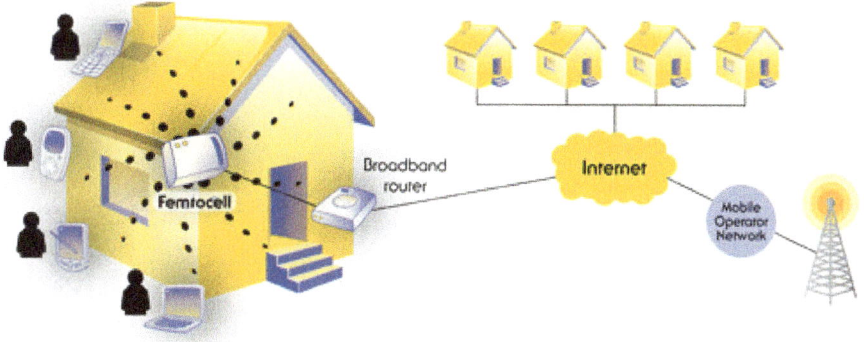

Fig. 21 The block diagram representation of wireless sensor network based on multilevel femtocells for home monitoring [19]

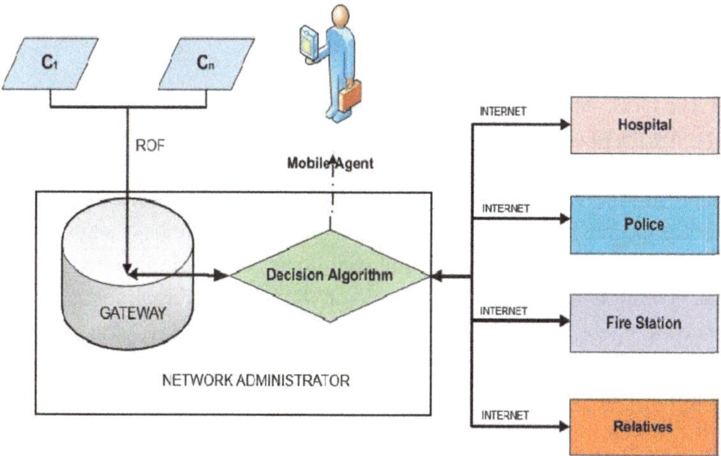

Fig. 22 The block diagram representation of wireless sensor network based on multilevel femtocells for home monitoring at data distribution level [19]

In [19], the system has been implemented with three layers. The BSN (body sensor network) is the first layer consists of sensors like fall sensor, position sensor, ECG and so on. The second layer is ASN (ambient sensor network) consists of ambient sensors like temperature, humidity sensors. The third layer is ESN (emergency sensor network) for security and emergency alert sensor like PIR, toxic gas detection sensor installed into femtocell network. Generally, the sensor data are sent to RF transceiver (USB connection) connected to internet by broadband router into home gateway. The femtocell device has been introduced before the broadband router into the home premises. The sensor data through sensor RF communication device is sent to home gateway equipped with femtocells.

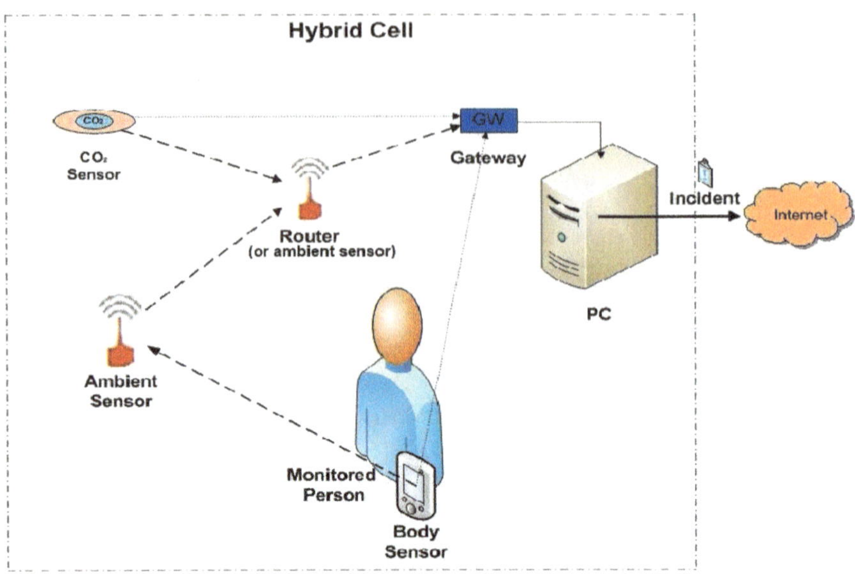

Fig. 23 The block diagram representation of wireless sensor network based on multilevel femtocells for home monitoring at home gateway [19]

Monitoring the wellness of an individual is directly linked to the living pattern. Monitoring the wellness with ambient sensor is one of the best ways, installing the temperature, PIR, force, accelerometer along with monitoring the electrical appliances which are used by the person for day-to-day lives. The usages of appliances suggest the activity of the person over the day. The sensor data are collected in web database server. Using data analysis algorithm and mathematical formulation, it is possible to generate the pattern of wellness of the person [36]. Using the wellness model and studying the activities of the person it is possible t forecast the future behaviour of the person [36, 37, 46, 50, 56]. The block diagram representation of wireless sensing units developed to monitor household appliances is shown in figure 24.

The KNX-based remote monitoring system has been implemented with temperature and proximity sensors in [65]. CAALYX has been developed and implemented to include heterogeneous sensors. It is an enhanced complete ambient assisted living experiment [68].

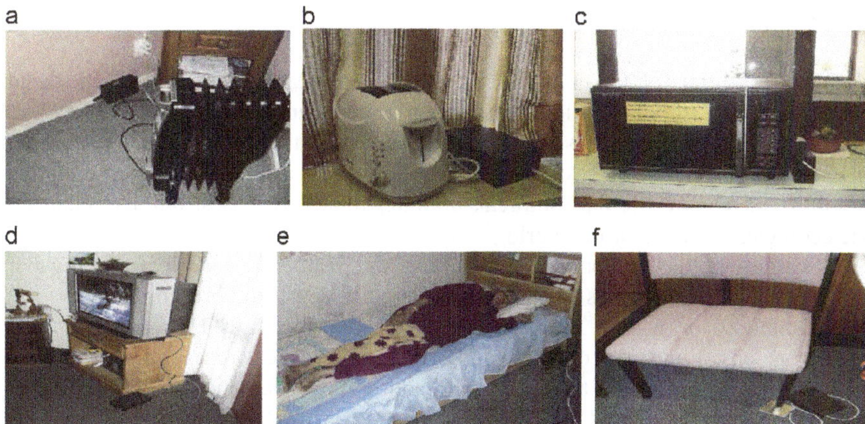

Fig. 24 The block diagram representation of ireless sensing units developed to monitor household appliances. (a) room heater, (b) toaster, (c) microwave, (d) TV, (e) bed and (f) chair [36].

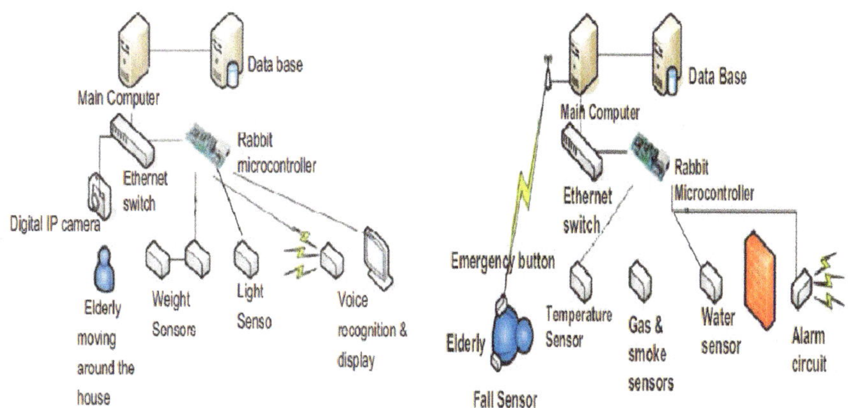

Fig. 25 The block diagram representation of a ubiquitous smart home for elderly [67]

Introducing voice recognition and image processing into home monitoring system takes it to the next level. These types of systems contain wearable sensors (physiological), non-wearable sensors (ambient sensors), voice recognition system with display, microcontroller, personal computer and database and digital camera. The system information uploaded into web-based server and individual person's record is only accessed by user ID and password (proper authentication). Apart from this, there is a GPS system which sends warning as well as emergency message to medical person and caretaker depending on the condition [67, 73]. The block diagram representation of a ubiquitous smart home for elderly is shown in figure 25. Introducing the camera into home network may be good for telemedicine and monitoring but it is not always acceptable due to privacy violation of individuals [61].

1.2.8 Water Monitoring

Water is essential for life and plays a vital role in earth's ecosystem. In recent times, with the rapid development of agriculture and industry, the excessive discharge of waste water has led to a massive degradation of the quality of water in different parts of the world. This has not only caused an increase in the mortality of aquatic life and outbreak of water-related diseases, but also changed the ecological balance of the world.

Governing bodies responsible for ensuring the health and security of freshwater systems need new methods and tools to accurately detect the presence of nutrients and pollutants in water. The current monitoring techniques used are mainly laboratory based and rely on the results of some chemical reactions. For a long-term real-time application the chemical reaction based detection is undesirable as the performance degrades over time. Moreover, with the new regulations, the available systems may not be suitable in the future to determine the amount of nutrients and pollutants in water relative to allowable levels. Environmental pollution is a huge concern of our society so the emphasis on monitoring the environment has become very important in recent times. Water, one of the most important natural resources, is increasingly polluted due to farming, industrial discharge and other reasons. It is very important to monitor water at different points, especially at the different meeting points of rivers and lakes of the country so that a thorough understanding of the relationship between soil, ground and surface water is possible. The continuous monitoring of the constituents of water will enable us to determine the sensitivity of the hydrologic system to the change in environmental parameters [1-3]. Water monitoring systems need to address a wide range of water parameters using smart and robust sensors to frequently report data input and consequent analysis. This will allow for the monitoring of water systems by taking account of the relationship between water quality and environmental parameters including sources of pollution.

The complete integrated framework of water monitoring system consists of three parts: The first part involves integrating available sensors with novel task-tailored smart high-performance sensors. There is a need to develop new sensors as the available sensors are not suitable for continuous monitoring of water under harsh condition. The second part is the combination of energy harvesting techniques such as solar, wind and hydro to obtain energy to supply the sensors for normal operation and communication. The third part is the integration of the sensors to wireless sensor nodes for centralised collection, storage, sharing and analysis of data accessible from a website. The website would be set up to update automatically with federated access by stakeholders including disaggregated information with varying levels of security from secured to open access for public availability.

The traditional method for water quality monitoring is based on laboratory measurement. The method uses field monitoring by people, which makes it not only inefficient but an improbable mode of data gathering for real-time parameters. There are a range of methods for measuring water parameters. Many of these are limited [77]. A monitoring system based on smart sensors and wireless sensor networks will provide a complimentary alternative that captures real-time information through remote monitoring. Wireless sensor networks have more flexibility in terms of design, implementation and deployment. A typical WSN water quality monitoring system would be composed of multiple wireless sensor nodes, several access points and one or more remote monitoring centers.

The water parameters considered are: temperature, humidity, pH, conductance, rain-fall, light-intensity, turbidity, nitrate, phosphate, bio-toxin, BOD, COD, TOC, nitrogen, phosphorus, potassium, etc. [2, 21, 42, 44, 74-88].

Sensors need to be located at different points distributed over a wide region of monitoring area. The fusion of sensor data is done at the wireless sensor node. The nodes are connected wirelessly with the central coordinator located at a distant point. The collected data from different nodes are stored in a computer. The data are analyzed and then the useful data are uploaded to a website for remote access.

1.3 Conclusions

In this chapter different sensing technologies for monitoring wide ranges of environment have been reviewed. It is seen that a variety of sensors are used to achieve the monitoring functionality. Still there is a need of high-performance, low-cost sensors for continuous monitoring of environment. The sensors used for monitoring external environment need continuous supply of energy. There is a need to develop new energy harvesting techniques as solar energy does not provide any guarantee of continuous supply of energy. Moreover, the waste from electronics should be kept to a minimum in design stages. Too many sensors may lead to pollution in the environment. Developing sensors which will be bio-degradable and non-hazardous will be the way in which future sensor technology should be designed and developed.

References

1. Abdul Aziz, I., Ismail, M.J., Samiha Haron, N., Mehat, M.: Remote monitoring using sensor in greenhouse agriculture. In: International Symposium on Information Technology, ITSim 2008, Kuala Lumpur, Malaysia, August 26-28, vol. 4, pp. 1–8 (2008) ISBN: 1424423279
2. Abdul Rahman, M.S., Mukhopadhyay, S.C., Yu, P.-L., Goicoechea, J., Matias, I.R., Gooneratne, C.P., et al.: Detection of bacterial endotoxin in food: New planar interdigital sensors based approach. Journal of Food Engineering 114(3), 346–360 (2013) ISSN: 0260-8774
3. Al-Bahadly, I., Mukhopadhyay, S., Alkhumaisi, K.: Pain Sensing System for Animals. Sensors & Transducers (1726-5479), 127–150 (2010) ISSN: 1726-5479
4. Bouchouicha, D., Dupont, F., Latrach, M., Ventura, L.: Ambient RF energy harvesting. In: International Conference on Renewable Energies and Power Quality, Granada (Spain), March 23-25, pp. 1–4 (2010)
5. Butala, P.M., Zhang, Y., Little, T.D., Wagenaar, R.C.: Wireless system for monitoring and real-time classification of functional activity. In: 2012 Fourth International Conference on Communication Systems and Networks (COMSNETS), Bangalore, India, January 3-7, pp. 1–5 (2012) ISBN: 1467302961

6. Chai, Y., Horikawa, S., Simonian, A., Dyer, D., Chin, B.A.: Wireless Magnetoelastic Biosensors for the Detection of Salmonella on Fresh Produce. In: 2013 Seventh International Conference on Sensing Technology (ICST), Wellington NZ, December 3-5, pp. 174–177 (2013) ISBN:1467352209

7. Chalasani, S., Conrad, J.M.: A survey of energy harvesting sources for embedded systems. In: IEEE Southeastcon, Huntsville, AL, April 3-6, pp. 442–447 (2008) ISBN: 1424418836

8. Cho, B., Korotcenkov, G., Boris, I., Golovanov, V., Karkotsky, G., Lychkovsky, Y.N.: In2O3:Ga-based Ceramics: Advantages and Shortcomings for Application in One-electrode Gas Sensors. In: 2013 Seventh International Conference on Sensing Technology (ICST), Wellington NZ, December 3-5, pp. 451–456 (2013) ISBN:1467352209

9. Collins, F., Orpen, D., Fay, C., Foley, C., Smeaton, A.F., Diamond, D.: Web-based monitoring of year-length deployments of autonomous gas sensing platforms on landfill sites. In: IEEE Sensors, Limerick, October 28-31, pp. 1620–1623 (2011) ISBN: 1424492904

10. Dilmaghani, R.S., Bobarshad, H., Ghavami, M., Choobkar, S., Wolfe, C.: Wireless sensor networks for monitoring physiological signals of multiple patients. IEEE Transactions on Biomedical Circuits and Systems 5(4), 347–356 (2011) ISSN: 1932-4545

11. Fuchs, A., Zangl, H., Moser, M.J., Bretterklieber, T.: Capacitive sensing in process instrumentation. Metrology and Measurement Systems 16(4), 557–568 (2009) ISSN: 0860-8229

12. Gooneratne, C.P., Kurnicki, A., Yamada, S., Mukhopadhyay, S.C., Kosel, J.: Analysis of the Distribution of Magnetic Fluid inside Tumors by a Giant Magnetoresistance Probe. PloS One 8(11), e81227 (2013) ISSN:1932-6203

13. He, C., Dong, G., Li, Q., Wei, H., Zhang, J., Lu, J.: Design of Automatic Force Application System and Outlier Detection for Force Sensor. In: 2013 Seventh International Conference on Sensing Technology (ICST), Wellington NZ, December 3-5, pp. 766–770 (2013) ISBN:1467352209

14. Hu, J., Shen, L., Yang, Y., Lv, R.: Design and implementation of wireless sensor and actor network for precision agriculture. In: 2010 IEEE International Conference on Wireless Communications, Networking and Information Security (WCNIS), Beijing, China, June 25-27, pp. 571–575 (2010) ISBN: 1424458501

15. Kandala, C.V., Sundaram, J.: Nondestructive measurement of moisture content using a parallel-plate capacitance sensor for grain and nuts. IEEE Sensors Journal 10(7), 1282–1287 (2010) ISSN:1530-437X

16. Kelly, S.D.T., Suryadevara, N., Mukhopadhyay, S.C.: Towards the implementation of IoT for environmental condition monitoring in homes. IEEE Sensors Journal 13(10), 3846–3853 (2013) ISSN:1530-437X

17. Kulagina, N.V., Lassman, M.E., Ligler, F.S., Taitt, C.R.: Antimicrobial peptides for detection of bacteria in biosensor assays. Analytical Chemistry 77(19), 6504–6508 (2005) ISSN: 0003-2700

18. Lee, Y., Park, S., Park, J., Koh, W.-G.: Micropatterned assembly of silica nanoparticles for a protein microarray with enhanced detection sensitivity. Biomedical microdevices 12(3), 457–464 (2010) ISSN: 1387-2176

19. Maciuca, A., Popescu, D., Strutu, M., Stamatescu, G.: Wireless sensor network based on multilevel femtocells for home monitoring. In: 2013 IEEE 7th International Conference on Intelligent Data Acquisition and Advanced Computing Systems (IDAACS), Berlin, September 12-14, pp. 499–503 (2013) ISBN: 1479914266

20. Miyajima, K., Arakawa, T., Kudo, H., Mitsubayashi, K., Yamashita, T., Ye, M., et al.: " Gas-Phase Biosensor with High Sensitive & Selective for Formaldehyde Vapour". In: 2013 Seventh International Conference on Sensing Technology (ICST), Wellington NZ, December 3-5, pp. 447–450 (2013) ISBN: 1467352209

21. Mohd Syaifudin, A., Mukhopadhyay, S.C., Yu, P.-L., Haji-Sheikh, M.J., Chuang, C.-H., Vanderford, J.D., et al.: Measurements and performance evaluation of novel interdigital sensors for different chemicals related to food poisoning. IEEE Sensors Journal 11(11), 2957–2965 (2011) ISSN:1530-437X

22. Mukhopadhyay, S.: Prediction of Thermal Condition of Cage-Rotor Induction Motors under Non-Standard Supply Systems. International Journal on Smart Sensing and Intelligent Systems 2(3), 381–395 (2009)

23. Niyato, D., Hossain, E., Rashid, M.M., Bhargava, V.K.: Wireless sensor networks with energy harvesting technologies: a game-theoretic approach to optimal energy management. IEEE Wireless Communications 14(4), 90–96 (2007) ISSN: 1536-1284

24. O'Connor, M.L.: Milk progesterone analysis for determining reproductive status, p. 98-5. The Pennsywania State University Publication DAS (1998)

25. Ortiz, P.M.R., Jia, Y., Vargas, N., Cabrera, C.: Label-free Capacitance DNA Sensing. In: 2013 Seventh International Conference on Sensing Technology (ICST), Wellington NZ, December 3-5, pp. 169–173 (2013) ISBN:1467352209

26. Pahuja, R., Verma, H., Uddin, M.: A Wireless Sensor network for greenhouse climate control. IEEE Pervasive Computing 12(2), 49–58 (2013) ISSN: 1536-1268

27. Pang, N.: ZigBee Mesh network for greenhouse monitoring. In: 2011 International Conference on Mechatronic Science, Electric Engineering and Computer (MEC), Jilin, August 19-22, pp. 266–269 (2011) ISBN: 1612847196

28. Paradiso, J.A., Starner, T.: Energy scavenging for mobile and wireless electronics. IEEE Pervasive Computing 4(1), 18–27 (2005) ISSN: 1536-1268

29. Ranhotigamage, C., Mukhopadhyay, S.C.: Field trials and performance monitoring of distributed solar panels using a low-cost wireless sensors network for domestic applications. IEEE Sensors Journal 11(10), 2583–2590 (2011) ISSN: 1530-437X

30. Rathod, K., Parikh, N., Parikh, A., Shah, V.: Wireless automation using Zigbee protocols. In: 2012 Ninth International Conference on Wireless and Optical Communications Networks (WOCN), Indore, India, September 20-22, pp. 1–5 (2012) ISBN: 1467319880

31. Sahota, H., Kumar, R., Kamal, A., Huang, J.: An energy-efficient wireless sensor network for precision agriculture. In: 2010 IEEE Symposium on Computers and Communications (ISCC), Riccione, Italy, June 22-25, pp. 347–350 (2010) ISBN: 1424477549

32. Shi, W.: SERS from ZnO Nanorod Arrays and its Application for detecting N719. In: 2013 Seventh International Conference on Sensing Technology (ICST), Wellington NZ, December 3-5, pp. 444–446 (2013) ISBN:1467352209

33. Shuying, M., Yuquan, M., Lidong, C., Shiguang, L.: Design of a new measurement and control system of CO2 for greenhouse based on fuzzy control. In: 2010 International Conference on Computer and Communication Technologies in Agriculture Engineering (CCTAE), Chengdu, June 12-13, pp. 128–131 (2010) ISBN: 1424469449

34. Siuli Roy, A., Bandyopadhyay, S.: Agro-sense: precision agriculture using sensor-based wireless mesh networks. In: First ITU-T Kaleidoscope Academic Conference on Innovations in NGN: Future Network and Services, K-INGN 2008, Geneva, May 12-13, pp. 383–388 (2008) ISBN: 9261124410

35. Sulaiman, S., Manut, A., Nur Firdaus, A.: Design, fabrication and testing of fringing electric field soil moisture sensor for wireless precision agriculture applications. In: International Conference on Information and Multimedia Technology, ICIMT 2009, Jeju Island, December 16-18, pp. 513–516 (2009) ISBN: 076953922X

36. Suryadevara, N., Mukhopadhyay, S., Wang, R., Rayudu, R.: Forecasting the behavior of an elderly using wireless sensors data in a smart home. Engineering Applications of Artificial Intelligence 26(10), 2641–2652 (2013) ISSN: 0952-1976

37. Suryadevara, N.K., Mukhopadhyay, S.C.: Wireless sensor network based home monitoring system for wellness determination of elderly. IEEE Sensors Journal 12(6), 1965–1972 (2012) ISSN: 1530-437X

38. Tang, J.-Y., Wang, M.-H., Chen, M.-K., Jang, L.-S.: Glucose Detection Using an Electro-Optical Fluidic Device Based on Pulse Width Modulation. In: 2013 Seventh International Conference on Sensing Technology (ICST), Wellington NZ, December 3-5, pp. 325–329 (2013) ISBN:1467352209

39. Tirelli, P., Borghese, N., Pedersini, F., Galassi, G., Oberti, R.: Automatic monitoring of pest insects traps by Zigbee-based wireless networking of image sensors. In: 2011 IEEE Instrumentation and Measurement Technology Conference (I2MTC), May 10-12, pp. 1–5 (2011) ISSN: 1091-5281

40. Wan, Z., Tan, Y., Yuen, C.: Review on energy harvesting and energy management for sustainable wireless sensor networks. In: 2011 IEEE 13th International Conference on Communication Technology (ICCT), Jinan, September 25-28, pp. 362–367 (2011) ISBN: 978-1-61284-306-3

41. Watanabe, T., Sakurai, A., Kitazaki, K.: Dairy cattle monitoring using wireless acceleration-sensor networks. In: 2008 IEEE Sensors, Lecce, October 26-29, pp. 526–529 (2008) ISSN: 1930-0395

42. Yunus, M.A.M., Mukhopadhyay, S.C., Ibrahim, S.: Planar electromagnetic sensor based estimation of nitrate contamination in water sources using independent component analysis. IEEE Sensors Journal 12(6), 2024–2034 (2012) ISSN: 1530-437X

43. Zhigang, L., Jizhang, W., Yunfeng, X., Pingping, L.: Study on the Number and Placement for Moisture Monitoring Sensor in Lettuce with Substrate Cultivation. In: 2013 Fourth International Conference on Digital Manufacturing and Automation (ICDMA), Qingdao, June 29-30, pp. 1319–1322 (2013)

44. Zia, A.I., Rahman, M.S.A., Mukhopadhyay, S.C., Yu, P.-L., Al-Bahadly, I., Gooneratne, C.P., et al.: Technique for rapid detection of phthalates in water and beverages. Journal of Food Engineering 116(2), 515–523 (2013) ISSN: 0260-8774

45. Bordencea, D., Valean, H., Folea, S., Dobircau, A.: Agent based system for home automation, monitoring and security. In: 2011 34th International Conference on Telecommunications and Signal Processing (TSP), Budapest, August 18-20, pp. 165–169 (2011) ISBN: 978-1-4577-1410-8

46. Casas, R., Marín, R.B., Robinet, A., Delgado, A.R., Yarza, A.R., McGinn, J., Picking, R., Grout, V.: User modelling in ambient intelligence for elderly and disabled people. In: Miesenberger, K., Klaus, J., Zagler, W.L., Karshmer, A.I. (eds.) ICCHP 2008. LNCS, vol. 5105, pp. 114–122. Springer, Heidelberg (2008)

47. Chen, B.-R., Patel, S., Buckley, T., Rednic, R., McClure, D.J., Shih, L., et al.: A web-based system for home monitoring of patients with Parkinson's disease using wearable sensors. IEEE Transactions on Biomedical Engineering 58, 831–836 (2011) ISSN: 0018-9294

48. Chen, C.-M.: Web-based remote human pulse monitoring system with intelligent data analysis for home health care. Expert Systems with Applications 38(3), 2011–2019 (2011) ISSN: 0957-4174

49. Chi, T., Chen, M., Gao, Q.: Implementation and study of a greenhouse environment surveillance system based on wireless sensor network. In: International Conference on Embedded Software and Systems Symposia, ICESS Symposia 2008, Sichuan, July 29-31, pp. 287–291 (2008) ISBN: 978-0-7695-3288-2

50. Costa, Â., Castillo, J.C., Novais, P., Fernández-Caballero, A., Simoes, R.: Sensor-driven agenda for intelligent home care of the elderly. Expert Systems with Applications 39(15), 12192–12204 (2012) ISSN: 0957-4174

51. Domingo, M.C.: Throughput efficiency in body sensor networks: A clean-slate approach. Expert Systems with Applications 39(10), 9743–9754 (2012) ISSN: 0957-4174

52. Ferrigno, L., Paciello, V., Pietrosanto, A.: A Bluetooth-based proposal of instrument wireless interface. IEEE Transactions on Instrumentation and Measurement 54(1), 163–170 (2005) ISSN: 0018-9456

53. Folea, S., Bordencea, D., Hotea, C., Valean, H.: Smart home automation system using Wi-Fi low power devices. In: 2012 IEEE International Conference on Automation Quality and Testing Robotics (AQTR), Cluj-Napoca, May 24-27, pp. 569–574 (2012) ISBN: 1467307017

54. Gill, K., Yang, S.-H., Yao, F., Lu, X.: A zigbee-based home automation system. IEEE Transactions on Consumer Electronics 55, 422–430 (2009) ISSN: 0098-3063

55. Gupta, G.S., Mukhopadhyay, S., Devlin, B., Demidenko, S.: Design of a low-cost physiological parameter measurement and monitoring device. In: Instrumentation and Measurement Technology Conference Proceedings, IMTC 2007, Warsaw, Poland, May 1-3, pp. 1–6. IEEE (2007) ISBN: 1424405882

56. Hamza, N., Touati, F., Khriji, L.: Wireless biomedical system design based on ZigBee technology for autonomous healthcare. In: Proc. Int. Conf. Commun., Comput., Power (ICCCP 2009), pp. 15–18 (2009)

57. Jia, F., Sun, Y., Yu, J., Wang, X.: A home monitoring system for elderly people based on MEMS sensors and wireless networks. In: 2013 IEEE Sensors, pp. 1–4 (2013) ISBN: 1930-0395

58. Jimenez, M., Jimenez, A., Lozada, P., Jimenez, S., Jimenez, C.: Using a Wireless Sensors Network in the Sustainable Management of African Palm Oil Solid Waste. In: 2013 Tenth International Conference on Information Technology: New Generations (ITNG), pp. 133–137 (2013) ISBN: 0769549675

59. Junnila, S., Kailanto, H., Merilahti, J., Vainio, A.-M., Vehkaoja, A., Zakrzewski, M., et al.: Wireless, multipurpose in-home health monitoring platform: Two case trials. IEEE Transactions on Information Technology in Biomedicine 14(2), 447–455 (2010) ISSN: 1089-7771

60. Lai, C.-C., Lee, R.-G., Hsiao, C.-C., Liu, H.-S., Chen, C.-C.: A H-QoS-demand personalized home physiological monitoring system over a wireless multi-hop relay network for mobile home healthcare applications. Journal of Network and Computer Applications 32(6), 1229–1241 (2009) ISSN: 1084-8045

61. Lee, H., Kim, Y.-T., Jung, J.-W., Park, K., Kim, D., Bang, B., et al.: A 24-hour health monitoring system in a smart house. Gerontechnology 7(1), 22–35 (2008) ISSN: 1569-111X
62. Lewis, F.L.: Wireless sensor networks. In: Smart Environments: Technologies, Protocols, and Applications, pp. 11–46 (2004)
63. Malhi, K., Mukhopadhyay, S.C., Schnepper, J., Haefke, M., Ewald, H.: A Zigbee-based wearable physiological parameters monitoring system. IEEE Sensors Journal 12(3), 423–430 (2012) ISSN: 1530-437X
64. Mohd Syaifudin, A., Jayasundera, K., Mukhopadhyay, S.: A low cost novel sensing system for detection of dangerous marine biotoxins in seafood. Sensors and Actuators B: Chemical 137(1), 67–75 (2009) ISSN: 0925-4005
65. Nazabal, J., Falcone, F., Fernández-Valdivielso, C., Mukhopadhyay, S., Matias, I.: Accessing KNX Devices using USB/KNX Interfaces for Remote Monitoring and Storing Sensor Data. International Journal of Smart Home 7(2), 105–110 (2013) ISSN: 1975-4094
66. Quazi, M., Mukhopadhyay, S., Suryadevara, N., Huang, Y.: Towards the smart sensors based human emotion recognition. In: 2012 IEEE International Instrumentation and Measurement Technology Conference (I2MTC), pp. 2365–2370 (2012) ISBN: 1457717735
67. Raad, M., Yang, L.T.: A ubiquitous smart home for elderly. Information Systems Frontiers 11(5), 529–536 (2009) ISSN: 1387-3326
68. Rodrigues, A., Resende, C., Carvalho, L., Saleiro, P., Abrantes, F.: Performance analysis of an adaptable home healthcare solution. In: 2011 13th IEEE International Conference on e-Health Networking Applications and Services (Healthcom), pp. 134–141 (2011) ISBN: 1612846955
69. Al-Ali, A.-R., Zualkernan, I.A., Lasfer, A., Chreide, A., Abu Ouda, H.: GRPS-based distributed home-monitoring using internet-based geographical information system. IEEE Transactions on Consumer Electronics 57(4), 1688–1694 (2011) ISSN: 0098-3063
70. Barbato, A., Capone, A., Rodolfi, M., Tagliaferri, D.: Forecasting the usage of household appliances through power meter sensors for demand management in the smart grid. In: 2011 IEEE International Conference on Smart Grid Communications (SmartGridComm), pp. 404–409 (2011) ISBN: 1457717042
71. Rotariu, C., Manta, V.: Wireless system for remote monitoring of oxygen saturation and heart rate. In: 2012 Federated Conference on Computer Science and Information Systems (FedCSIS), pp. 193–196 (2012) ISBN: 1467307084
72. Williams, E.D., Matthews, H.S.: Scoping the potential of monitoring and control technologies to reduce energy use in homes. In: Proceedings of the 2007 IEEE International Symposium on Electronics & the Environment, pp. 239–244 (2007) ISBN: 142440861X
73. Xu, L., Zheng, X., Guo, W., Chen, G.: A Cloud-based monitoring framework for Smart Home. In: 2012 IEEE 4th International Conference on Cloud Computing Technology and Science (CloudCom), pp. 805–810 (2012) ISBN: 1467345113
74. Kotsilieris, T., Karetsos, G.T.: A mobile agent enabled wireless sensor network for river water monitoring. In: Proceedings of the Fourth International Conference on Wireless and Mobile Communication, pp. 346–351 (2008), doi:10.1109/ICWMC.2008.71

75. Anvari, A., Reyes, J.D., Esmaeilzadeh, E., Jarvandi, A., Langley, N., Navia, K.R.: Designing an Automated Water Quality Monitoring System for West and Rhode Rivers. In: Proceedings of the 2009 IEEE Systems and Information Engineering Design Sysmposium, Charlottesville, USA, April 24, pp. 131–136 (2009)
76. Alippi, C., Camplani, R., Galpetri, C., Roveri, M.: A Robust, Adaptive, Soalr-Powered WSN Framework for Aquatic Environmental Monitoring. IEEE Sensors Journal 11(1), 45–55 (2011)
77. Korostynska, A.O., Mason, A., Al-Shamma'a, A.I.: Monitoring Pollutants in Wastewater: Traditional Lab Based versus Modern Real-Time Approaches. In: Mukhopadhyay, S.C., Mason, A. (eds.) Real-Time Water Quality Monitoring. SSMI, vol. 4, pp. 1–24. Springer, Heidelberg (2013)
78. Md Yunus, M.A., Mukhopadhyay, S.C.: Novel Planar Electromagnetic Sensors for Detection of Nitrates and Contamination in Natural Water Sources. IEEE Sensors Journal 11(6), 1440–1447 (2011)
79. Md Yunus, M.A., Mukhopadhyay, S.C.: Development of Planar Electromagnetic Sensors for Measurement and Monitoring of Environmental Parameters. Meas. Sci. Technol. 22, 025107(9pp) (2011), doi:10.1088/0957-0233/22/2/025107; Md Yunus, M.A., Mukhopadhyay, S.C., Punchihewa, A.: Application of Independent Component Analysis for Estimating Nitrate Contamination in Natural Water Sources Using Planar Electromagnetic Sensor. In: Proceedings of the 2011 International Conference on Sensing Technology, ICST 2011, Palmerston North, New Zealand, November 28-December 1, 2011, pp. 564–569 (2011) ISBN 978-1-4577-0167-2
80. Md Yunus, M.A., Mukhopadhyay, S.C.: A New Method for Monitoring Ammonium Nitrate Contamination in Natural Water Sources Based on Independent Component Analysis. In: Proceedings of the 2011 IEEE Sensors Conference, October 29-31, pp. 1066–1069 (2011) IEEE catalog Number CFP11SEN-CDR, ISBN Number 978-1-4244-9288-6
81. Mukhopadhyay, S.C., Jiang, J.-A. (eds.): Wireless Sensor Networks & Ecol. Monit. SSMI, vol. 3. Springer, Heidelberg (2013)
82. Mukhopadhyay, S.C., Mason, A. (eds.): Real-Time Water Quality Monitoring. SSMI, vol. 4. Springer, Heidelberg (2013)
83. Md Yunus, M.A., Mukhopadhyay, S.C., Rahman, M.S.A., Zahidin, N.S., Ibrahim, S.: The selection of novel planar electromagnetic sensors for the application of nitrate contamination detection. In: Mukhopadhyay, S.C., Mason, A. (eds.) Real-Time Water Quality Monitoring. SSMI, vol. 4, pp. 171–196. Springer, Heidelberg (2013)
84. Md Yunus, M.A., Mukhopadhyay, S.C., Punchihewa, A., Ibrahim, S.: The Effect of Temperature Factor on the Detection of Nitrate based on Planar Electromagnetic Sensor and Independent Component Analysis. In: Mukhopadhyay, S.C. (ed.) Smart Sensing Technology for Agric. & Environ. Monit. LNEE, vol. 146, pp. 103–118. Springer, Heidelberg (2012)
85. Yunus, M.A., Mukhopadhyay, S.C.: Design and application of low cost system based on planar electromagnetic sensors and impedance spectroscopy for monitoring of nitrate contamination in natural water sources. In: Kanoun, O. (ed.). Lecture Notes on Impedance Spectroscopy, Measurement, Modeling and Applications, vol. 2, pp. 83–102. CRC Press, Taylor and Francis Group, ISBN 978-0-415-69838-2
86. Mendez, G.M., Yunus, M.A.M., Mukhopadhyay, S.C.: A WiFi based Smart Wireless Sensor Network for Monitoring an Agricultural Environment. In: Proceedings of IEEE I2MTC 2012 Conference, Graz, Austria, May 13-16, pp. 2640–2645 (2012) IEEE Catalog number CFP12MT-CDR, ISBN 978-1-4577-1771-0

87. Haefke, M., Mukhopadhyay, S.C., Ewald, H.: A Zigbee Based Smart Sensing Platform for Monitoring Environmental Parameters. In: Proceedings of IEEE I2MTC 2011 Conference, IEEE Catalog number CFP11MT-CDR, ISBN 978-1-4244-7934-4, Hangzhou, China, May 10-12, pp. 1549–1556 (2011) IEEE Catalog number CFP11MT-CDR, ISBN 978-1-4244-7934-4

88. Gungor, V.C., Hancke, G.P.: Industrial wireless sensor networks: Challenges, design principles, and technical approaches. IEEE Transactions on Industrial Electronics 56(10), 4258–4265 (2009) ISSN 0278-0046

89. Alippi, C., Anastasi, G., Di Francesco, M., Roveri, M.: Energy management in wireless sensor networks with energy-hungry sensors. IEEE Instrumentation & Measurement Magazine 12(2), 16–23 (2009) ISSN 1094-6969

90. Tan, Y.K., Panda, S.K.: Review of energy harvesting technologies for sustainable wireless sensor network. In: Sustainable Wireless Sensor Networks, pp. 15–43. INTECH Publisher (2010)

91. Handcock, R.N., Swain, D.L., Bishop-Hurley, G.J., Patison, K.P., Wark, T., Valencia, P., Corke, P., O'Neill, C.J.: Monitoring animal behaviour and environmental interactions using wireless sensor networks, GPS collars and satellite remote sensing. Sensors 9(5), 3586–3603 (2009)

92. Juang, P., Oki, H., Wang, Y., Martonosi, M., Peh, L.S., Rubenstein, D.: Energy-efficient computing for wildlife tracking: Design tradeoffs and early experiences with ZebraNet, pp. 96–107. ACM (2002)

93. Urbano, F., Cagnacci, F., Calenge, C., Dettki, H., Cameron, A., Neteler, M.: Wildlife tracking data management: a new vision. Philosophical Transactions of the Royal Society B: Biological Sciences 365(1550), 2177–2185 (2010)

94. Garcia-Sanchez, A.-J., Garcia-Sanchez, F., Losilla, F., Kulakowski, P., Garcia-Haro, J., Rodríguez, A., López-Bao, J.-V., Palomares, F.: Wireless sensor network deployment for monitoring wildlife passages. Sensors 10(8), 7236–7262 (2010)

95. Venkatraman, S., Jin, X., Costa, R.M., Carmena, J.M.: Investigating neural correlates of behavior in freely behaving rodents using inertial sensors. Journal of Neurophysiology 104(1), 569 (2010)

96. Sun, Z., Wang, P., Vuran, M.C., Al-Rodhaan, M.A., Al-Dhelaan, A.M., Akyildiz, I.F.: BorderSense: Border patrol through advanced wireless sensor networks. Ad Hoc Networks 9(3), 468–477 (2011) ISSN: 1570-8705

97. Wang, H., Hempel, M., Peng, D., Wang, W., Sharif, H., Chen, H.-H.: Index-based selective audio encryption for wireless multimedia sensor networks. IEEE Transactions on Multimedia 12(3), 215–223 (2010) ISSN: 1520-9210

98. Zhang, F., Zhang, Y., Silver, J., Shakhsheer, Y., Nagaraju, M., Klinefelter, A., et al.: A batteryless 19μW MICS/ISM-band energy harvesting body area sensor node SoC. In: 2012 IEEE International Solid-State Circuits Conference Digest of Technical Papers (ISSCC), San Francisco, CA, February 19-23, pp. 298–300 (2012) ISSN: 0193-6530

99. Sharma, V., Mukherji, U., Joseph, V., Gupta, S.: Optimal energy management policies for energy harvesting sensor nodes. IEEE Transactions on Wireless Communications 9, 1326–1336 (2010)

100. Li, P., Wen, Y., Liu, P., Li, X., Jia, C.: A magnetoelectric energy harvester and management circuit for wireless sensor network. Sensors and Actuators A: Physical 157, 100–106 (2010)

101. Liu, R.-S., Sinha, P., Koksal, C.E.: Joint energy management and resource allocation in rechargeable sensor networks. In: 2010 Proceedings of the IEEE INFOCOM, San Diego, CA, March 14-19, pp. 1–9 (2010) ISSN: 0743-166X
102. Voyvoda, H., Erdogan, H.: Use of a hand-held meter for detecting subclinical ketosis in dairy cows. Research in Veterinary Science 89(3), 344–351 (2010) ISSN: 0034-5288
103. Wilkens, M., Oberheide, I., Schröder, B., Azem, E., Steinberg, W., Breves, G.: Influence of the combination of 25-hydroxyvitamin D3 and a diet negative in cation-anion difference on peripartal calcium homeostasis of dairy cows. Journal of Dairy Science 95(1), 151–164 (2012)
104. Erskine, R.J., Corl, C.M., Gandy, J.C., Sordillo, L.M.: Effect of infection with bovine leukosis virus on lymphocyte proliferation and apoptosis in dairy cattle. American Journal of Veterinary Research 72(8), 1059–1064 (2011)
105. Tolkamp, B.J., Haskell, M.J., Langford, F.M., Roberts, D.J., Morgan, C.A.: Are cows more likely to lie down the longer they stand? Applied Animal Behaviour Science 124(1), 1–10 (2010)
106. Legrand, A., Schütz, K., Tucker, C.: Using water to cool cattle: Behavioral and physiological changes associated with voluntary use of cow showers. Journal of Dairy Science 94(7), 3376–3386 (2011) ISSN: 0022-0302
107. Shen, F., Wang, J., Xu, Z., Wu, Y., Chen, Q., Li, X., Jie, X., Li, L., Yao, M., Guo, X.: Rapid flu diagnosis using silicon nanowire sensor. Nano Letters 12(7), 3722–3730 (2012)
108. Guo, Z., Nam, S., Park, S., Yoon, J.: A highly selective ratiometric near-infrared fluorescent cyanine sensor for cysteine with remarkable shift and its application in bioimaging. Chemical Science 3(9), 2760–2765 (2012)
109. Vallejos, S., Stoycheva, T., Umek, P., Navio, C., Snyders, R., Bittencourt, C., Llobet, E., Blackman, C., Moniz, S., Correig, X.: Au nanoparticle-functionalised WO3 nanoneedles and their application in high sensitivity gas sensor devices. Chemical Communications 47(1), 565–567 (2011)
110. Sakaki, T., Okazaki, M., Matsuo, Y.: Earthquake shakes Twitter users: real-time event detection by social sensors, pp. 851–860. ACM (2010)
111. Aluri, G.S., Motayed, A., Davydov, A.V., Oleshko, V.P., Bertness, K.A., Sanford, N.A., Rao, M.V.: Highly selective GaN-nanowire/TiO2-nanocluster hybrid sensors for detection of benzene and related environment pollutants. Nanotechnology 22(29), 295503 (2011)
112. Al-Ali, A., Zualkernan, I., Aloul, F.: A mobile GPRS-sensors array for air pollution monitoring. IEEE Sensors Journal 10(10), 1666–1671 (2010) ISSN 1530-437X

Chapter 2
Micro Motes: A Highly Penetrating Probe for Inaccessible Environments

Elena Talnishnikh, J. van Pol, and H.J. Wörtche

Abstract. We present a novel approach for mapping and monitoring otherwise non accessible environments by means of injecting millimeter sized sensor motes. Prerequisite for successful operation is the development of self-sustained and fully functional sensor motes designed for deployment in situations of radio isolation and the absence of the Global Positioning System. Based on the results of a pilot project we discuss the principle feasibility of the approach, indications for the achievable penetration of underground oil reservoirs and methodologies to extract system information based on mote sensor data.

If our approach proves to be successful, we are convinced that many industries and the society in general will benefit from the availability of this novel sensor technology, eventually leading to more effective and smarter usage of available resources, contributing to economical maintenance of large-scale infrastructures and the assessment and the control of calamities.

2.1 Introduction

In this paper, we discuss a novel approach to the exploration of difficult-to-access geological structures, located underground and extending over large distances. The developments were initiated by the request of the heavy oil industry seeking technologies to map so called *wormholes* and explore their structure. A request which up to now could not be matched by state of the art exploration technologies [1].

Wormholes are generated during Cold Heavy Oil Production with Sand (CHOPS) [2], that is the deliberate extraction of oil together with sand, water and gas, which are present in the formation. As significant volumes of sand and fines are being removed from the reservoir during production, wormholes or high-porosity, high-permeability channels that start to grow from the well and can propagate hundreds of meters into the formation, as

Elena Talnishnikh · J. van Pol · H.J. Wörtche
INCAS³
Dr. Nassaulaan 9/9401 HJ Assen, The Netherlands

© Springer International Publishing Switzerland 2015
H. Leung and S.C. Mukhopadhyay (eds.), *Intelligent Environmental Sensing,*
Smart Sensors, Measurement and Instrumentation 13, DOI: 10.1007/978-3-319-12892-4_2

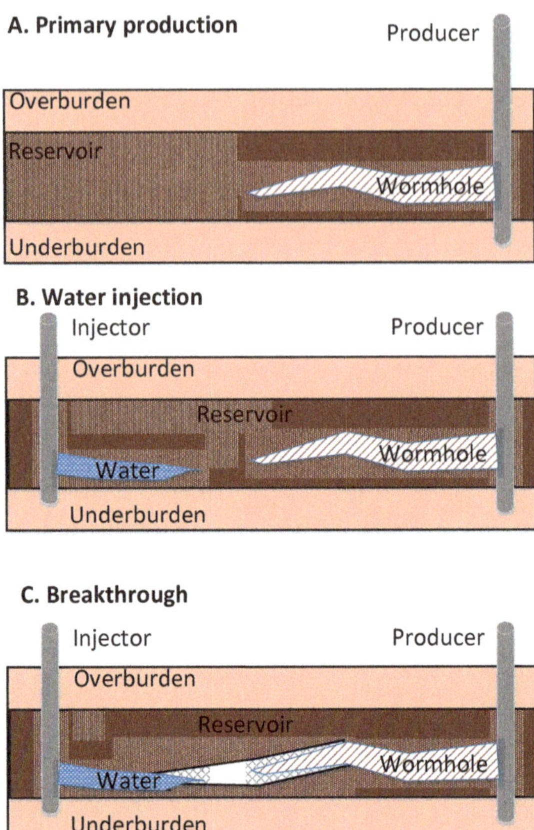

Fig. 1 Schematic representation (A) of a wormhole developing during the production. Due to sand production a weakened zone around a well-bore perforation is generated and a wormhole starts propagating into the reservoir. Shape, size, structure, branching factor and number of wormholes are not known. Water injection is a regularly practice in order to maintain reservoir pressure at first place, dilute and guide residual oil towards the production well as shown in (B). Normally, the injection fluid is isolated or separated from the production. Sometimes it may lead to unwanted reservoir conditions such as breakthrough when injected fluid finds a path to the production well and creates a shortcut between two wells as shown in (C) or even more wells.

schematically depicted in Fig. 1(A). In consequence the formation matrix becomes weaker, the pressure decreases, and potentially a portion of the formation can collapse and leave a void in the reservoir. Water injection through special injection wells is used to maintain reservoir pressure and to displace the residual oil to an adjacent production well, Fig. 1(B). Often water injection becomes inefficient because wormholes affect fluid transport within a reservoir in unpredictable manner. In some cases it can cause production and surface processing problems when the injected fluid gains direct access to the production well [3], Fig. 1(C).

Field tracer tests implying the injection and the propagation of tracer fluids (dyes) [4] and laboratory tests [5, 6] yielded evidence of wormholes in CHOPS reservoirs. However, efforts to provide detailed structural information such as dimensions, shapes or topologies failed. Standard remote methods, like detection of structural information by means of seismic measurements, are not feasible because of the presumably small diameters of wormholes compared to the applied seismic wave lengths [7]. Remote installation and application of wireless sensor networks as proposed and discussed in Ref. [8] are excluded because oil sands are basically non-penetrable for electromagnetic radiation, the attenuation e.g. for 100 MHz radio waves is 15.5 dB/m [9], which effectively suppresses any wireless communication within a sensor network.

On a more abstract level, this particular problem can be generalized as acquisition of information from a class of systems, natural or man-made, that are inaccessible for direct or remote measurements, non-penetrable for electromagnetic radiation and are shielded from satellite based positioning systems. To satisfy this information request the measurement of parameters determining the internal structure and the condition of a system dependent on location and transfer time are necessary. Commonly a flow is associated with the system, e.g. CHOPS reservoirs in operation feature a four-phase flow consisting of gas, water, sand and oil. Apart of CHOPS reservoirs, this class of systems includes underground water reservoirs, geothermal reservoirs, underground and surface flow line networks, formations with mining-induced fractures or major industrial mixing tanks.

2.2   Conceptual Approach

Focus of our work is the development of a methodology to study inaccessible environments based on a minimum of a priori information. Our approach is twofold and on the one hand based on dedicated sensor mote hardware development which is the main subject of the present document. But our approach also requires methodologies to develop models allowing to analyze the sensed data and to convert the data into the requested system information. In this respect and under given conditions especially the required localization of sensor motes while traveling in the systems poses a major challenge and requires significant efforts.

A typical feature of geological systems like wormholes is the combination of vast size with complex and small-dimensioned internal structure. The dimension and the mass of motes have to be chosen such that a possibly complete penetration of the system is achieved, i.e. in case of a wormhole ten thousands of cubic meters have to be penetrated by millimeter sized sensor motes. Sensor motes are injected directly into the system, where they go with the flow while sensing and memorizing the sensed data until they are eventually retrieved from the system. To achieve this goal, massive numbers of motes have to be injected. Motes, therefore, do not only have to be robust to withstand geological and process related conditions but also have to be cost effective, which can only be achieved in industrial mass production. Robustness and cost efficiency require to restrict mote capabilities to a minimum, which goes along with the limited electronics payload of millimeter sized motes in combination with the weight limitations opposed by the neutral buoyancy requirement to keep motes floating. Excluding propulsion mechanisms or any movable part in the design of the sensor motes decreases the risk of motes being destroyed in the system as well as it facilitates cheap and easy mass production.

The benefit of using an existing flow as a carrier of the motes is the dynamics required to access and to penetrate the system under investigation. As a result motes are expected to measure and store sensed data depending on the travel time through the system, which is, in its turn, directly correlated to the flow behavior. System information is generated through the offline analysis of the memorized data read out from the retrieved motes in conjunction with deduced retrieval probabilities of the motes and correlated travel times. The analysis will yield statistical distributions of system parameters, which will be confronted with distributions generated by simulations based on models describing the structure of the system, the fluid dynamics and the mote interaction with the system. The comparison of *a priori* information with distributions built on the obtained experimental data will deliver verification and refinement of the structural models underlying the simulations hereby revealing the required system information [10]. We foresee multiple iterations of the mote injection-retrieval process before reaching the desired goal of gaining structural information with sufficient precision and building an overall model of the system and while doing so we expect that the design and functionality of motes will be progressively refined aiming at optimal performance of motes for the specific application.

2.2.1 *Localization Problem*

For mapping a structure and providing the input required for model based simulations of the system structure and the system dynamics mote sensing data have to be correlated with reliable time and location data. While correct time stamping for a single mote can be realized based on readily available electrical circuits or oscillators. The determination of the location or at least the relative position of a single mote in contrast poses a challenge which has not been solved yet for conditions we discuss in the present context. Standard approaches such as Global Position Systems (GPS), coordinate grids based on reference beacons and markers or Inertial Measurements Units (IMU) based navigation are not applicable for the localization of motes at all or not with the required precision because of deep underground conditions or the absence of radio communication due to physical or chemical conditions imposed by the system or the fluid, e.g. high salinity water. For deep underground oil reservoirs flushed with water, localization by means of the GPS is excluded because radio absorption prohibits connecting to satellites. Beacons can be installed only at very limited locations, such as access and/or the recovery points of motes but not in the reservoir. The main limitation of IMUs is that they accumulate location errors (or drift) over periods of operation, which can be days in the present context. This makes absolute localization of motes in the system based on standard technologies very difficult - if not impossible.

At present we investigate options provided by ultrasound technologies that are commonly used to detect objects and to measure distances in liquids. Underwater acoustical applications benefit from the capability of acoustical waves to penetrate the medium and yield location of objects and surface features by reflection measurements. We intent to apply ultrasound ranging measurements for the relative localization of the motes floating through the voids of a system [11]. The individual sensor motes perform ranging measurements to neighboring motes and to the environment boundaries by emitting ultrasound pulses and receiving the reflections. The analysis of detected reflections is expected to show the relative position of a single mote within a group of motes and eventually the cumulated information on the local distribution of a collective of motes will yield information about the structure of the system.

An illustrative and simplified example is given in Fig. 2, where a few motes are shown passing through a narrow wormhole causing mainly reflections from the environment to be detected, in the case of a bigger wormhole more motes will float through the system at once and both types of reflections, caused

Fig. 2 Illustration of a collective of motes equipped with ultrasound transceivers to probe the internal structure of a wormhole. Motes are indicated as filled black circles, the omnidirectional emitted ultrasound waves are symbolized by concentric circles. Only few motes will pass through a narrow wormhole 1 and detect fast reflections from the environmental boundaries. The wider wormhole 2 allows for many more motes to float through and a spectrum of reflections caused by motes and the wormhole boundaries will be detected.

by the environment and neighboring motes will be detected. For successful implementation of the method it is required to detect the time when pulses are emitted and reflections are detected, and to discriminate between reflections caused by neighboring motes and the system boundaries. Knowing the speed of the ultrasound in the medium, the time information can be converted into distances between motes. To determine the relative positions of the motes, lateration techniques can be used as, for example, discussed in Ref. [12].

For the purpose of identification, ranging pulses will be emitted in a specific frequency band using orthogonal frequency division multiplexing (OFDM) [13] or signal coding. Each mote will store the frequency and arrival time of received pulses from other motes within their sensing radius. It should be noted that only a limited number of frequencies can be used because ultrasound transducers have a limited operating bandwidth, resulting in a trade-off in the amount of non-overlapping frequency bands versus the pulse lengths of the ultrasound signals, as discussed in [14]. It also means that in the case when many sensor motes are used, the same ID will be assigned to more than one mote and, consequently, it will require proper identification of corresponding interacting pairs of motes.

Distinguishing between true and false interaction between motes requires solving an optimization problem of finding a pair of points with specific geometrical conditions. For example, it can be done applying neighbor matching method as discussed in detail [15]. When a set of distance matrices is collected from retrieved motes, the neighbor matching algorithm searches for the same measured distance at two different frequencies assigned to two particular motes. In order to illuminate false matches among identified pairs, a third mote can be introduced into a matching condition. In this case, if all three motes have measured reflections from each other, the pair of motes is identified correctly.

This approach allows to keep the hardware of a single mote simple and easy to implement while using complex data analysis upon retrieval of motes from the reservoir. However, a trade-off has to be made between the total number of motes and the number of not unique ID's which can be used.

2.2.2 Aspects of Ultrasound Implementation in Micro Motes

Ultrasound is widely applied in medical applications, marine investigations or geological explorations. Commonly, conventional ultrasound transceivers are rather big and operate at frequencies varying from tens of kilohertz to few megahertz. Often these systems are connected to a host providing required supplies and data acquisition, e.g. surface ships, submarines or unmanned underwater vehicles.

Miniaturization of ultrasound transceivers is mainly driven by medical applications , e.g. endoscopic or intravascular studies. These devices are guided through the body from outside and connected directly to a processor unit or computer providing supplies, controls and data readout.

Typically two types of transducers are used, piezoelectric transducers (PZT) and Capacitive Ultrasonic Transducers (CMUTs). CMUTs have better acoustic impedance matching with the propagation medium and provide wider bandwidth than PZT. On the other hand PZT have better transduction efficiency.

In order to perform initial feasibility studies we have developed a setup consisting of five macroscopic prototype motes. A proof of principle can be performed on macroscopic scale in an (aqueous or gaseous) environment without compromising on the sensitivity for testing the performance on omnidirectional emission, ranging measurements, communication, multiplexing and signal coding. Compared to an application in aqueous environment the setup is scaled for operating in the air. The scaling is given in table 1.

Table 1 Scaling from water to air

	aqueous	gaseous
environmental dimensions	0.5 m	5 m
speed of sound	1500 m/s	330 m/s
frequency	2 MHz	40 kHz
sound wavelength	0.8 mm	8 mm
transducer size	1 mm	1 cm

An actual macroscopic sensor mote is depicted in Fig. 1.3. An omnidirectional 2D emission is achieved by a set of 20 ultrasound transducers (Series 7000 Transducers by SensComp) arranged in an outer ring. The transducers have been chosen because they feature a bandwidth of 25% at (f_0 =50kHz, BW(-6dB)=12kHz) required for signal encoding. The radius of the outer ring has been optimized to provide a possibly flat emission characteristic depending on the beam angle of a single transducer.

2.3   Proof of Principle

In November 2012 we performed a feasibility study at a suspended CHOPS site, located in Alberta, Canada [16]. The objective of the field trial was twofold. First objective was to show that in principle millimeter sized plastic shells, blank sensor motes, capable of carrying electronics could be injected into and retrieved from a CHOPS reservoir. Second objective was to investigate and to establish the optimum conditions for the mote design and the

Fig. 3 A scaled macroscopic prototype mote optimized to study feasibility of embedding and applying ultrasound technologies in (microscopic) sensor motes in aqueous environments.

well operation to successfully pass and retrieve the motes through a CHOPS reservoir.

2.3.1   The Test Site

The test site consisted of two wells, an injector (IW) and a producer (PW). Both wells were about 500 m deep, inclined at 47° and separated by about 345 m at reservoir depth as it is shown in Fig. 4. This site was a subject of waterflood operation until decline in the production and until breakthrough happened between two wells. The existence of wormholes at the time of the experiment had been confirmed by executing dye tests, i.e. by injecting a fluid colored with a dye into the IW and waiting for the dye to be produced in the PW. On the surface the PW and the IW were connected to the same stock tank (heavy oil tank) by means of flow lines transporting fluid. The fluid flow was regulating by a pump installed between the heavy oil tank and the IW. An other pump, a progressive cavity pump (or PCP pump), was installed in the PW transporting the fluid from the reservoir up to the surface. For the purpose of retrieving motes a screen was installed between the PW and the heavy oil tank. The site installation was organized so that the flow circulation was maintained almost constantly in a closed loop system of IW, the reservoir, PW, heavy oil tank, not taking into account possible losses of

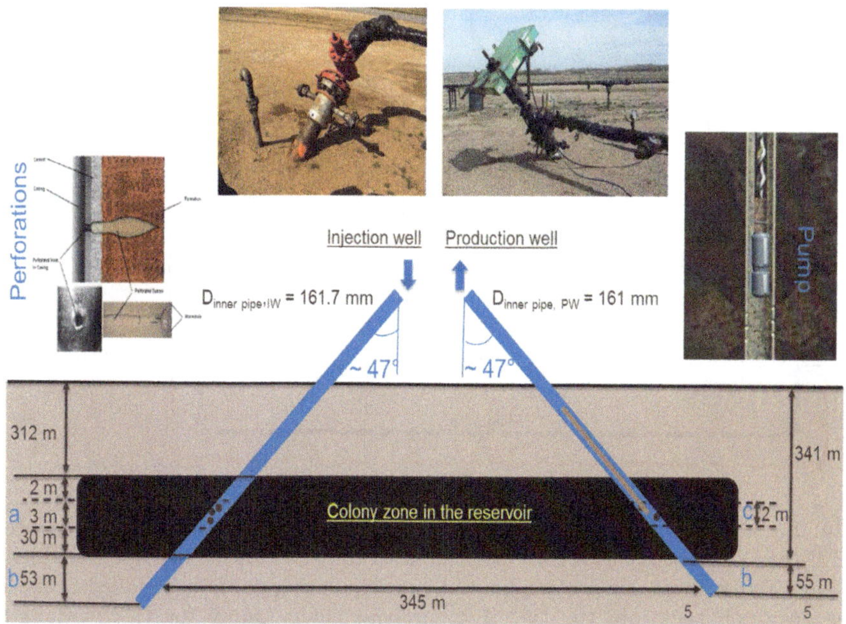

Fig. 4 Producer and injector wells and cross section of the suspended CHOPS reservoir used for the pilot studies. The location of the well perforations and the PCP pumps are indicated.

fluid in the reservoir itself. A simplified diagram of the field experiment is shown in Fig. 5.

2.3.2 Prototype Blank Motes

A first set of prototype motes were developed taking into account requirements posed by the reservoir and by equipment installed on the site. Reservoir parameters given in table 2 dictate the choice of a suitable, chemically and mechanically robust, material for the shell of motes, which provides necessary protection from the environment and at the same time does not compromise the functionality of motes. The size and the shape of the motes is defined by the requirements to pass the pumps and to pass through the perforation holes, which initial seize was estimated to be 12 mm.

Motes of varying size (5, 9, 12 mm) and shape (spherical, elongated and even cubic) were designed in order to determine the maximum size for the motes to travel through reservoir, and associated pumps and piping. Ahead to the field test, robustness of these motes was tested in a horizontal PCP pump identical to that installed in the producer at the pump manufacturer facility. The PCP pump had a typical design consisting of a helical rotor

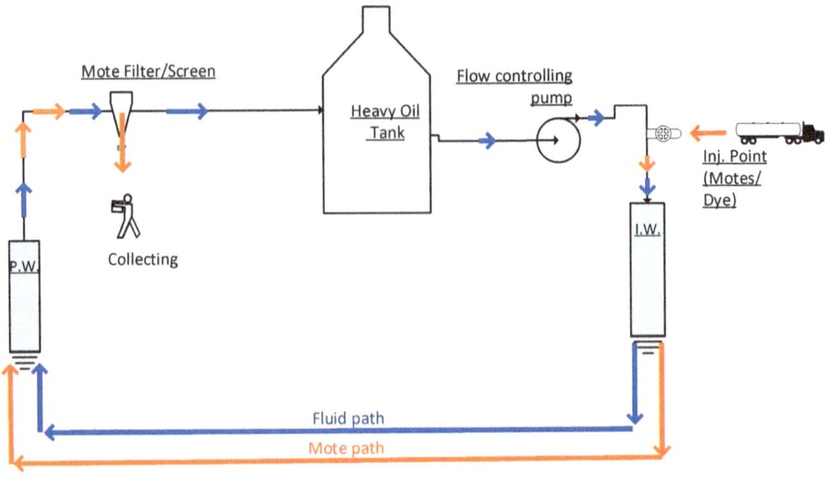

Fig. 5 Simplified diagram of the field experiment. On the surface the producer (PW) and the injector (IW) are connected to the stock tank (heavy oil tank) by means of flow lines transporting fluid. A flow regulating pump is installed between the heavy oil tank and the IW. A progressive cavity pump (or PCP pump) is installed in the PW, not shown, controlling the fluid transported from the reservoir up to the surface. In order to catch the sensor motes a screen was installed between the PW. and the heavy oil tank. According to the experimental procedure the sensor motes were injected directly into the IW, traveled through the reservoir, came up the PCP in the PW. and ended up in the screen. Blue arrows indicate the path of the fluid from the IW. into the reservoir up to the PW. and back to the IW. via the heavy oil tank. Orange arrows show the path of the sensor motes from the IW. into the screen via the reservoir and the PW. It should be noted that the fluid circulated in a closed loop while the motes only went from the IW to to the screen in PW).

Table 2 CHOPS reservoir parameters

	typical value
reservoir depth	450 m - 850 m
net pay	2 m - 15 m
porosity	25 % - 35 %
reservoir temperature	20°C - 35°C
oil density	940 kg/m^3 - 990 kg/m^3
oil viscosity	100 cp - 10000 cp
oil salinity	50 g/m^3 - 150 g/m^3

Fig. 6 The main photograph (A) shows the field site with both, the injector (IW) and the producer (PW), wells. Installed screen for the mote capture is shown in the small photograph on the top right (B). Few of collected motes are in the low right insert (C).

made of steel with hard smooth surface and a stator that is a metal tube filled with a molded elastomer with complex cavities inside. Motes were sent through the pump with water flow at pressure of 55 kPa - 70kPa. Results of the tests in PCP pump showed that motes smaller than 9 mm could pass through a PCP pump with negligible probability to be ground up and without damages to the pump itself. Consequently, the largest motes for the actual field experiment were limited to outer dimensions of 9 mm or less and only spheres and capsules were used. The specification of the motes are given in table 3.

For the actual field tests hollow motes with two different shapes (spheres and elongated), three different diameters (5 mm, 7 mm, 9 mm) and two different densities were developed. The shells were made of industrial glass fiber reinforced plastic and were capable of withstanding pressures up to 10 MPa and temperatures up to 80°C. The density of motes was chosen such that one class was neutrally buoyant within tolerances determined by slight variation of the water salinity, for the other class the density was chosen to be slightly positively buoyant, i.e. the motes were floating. In consequence this led to a variety of 12 different types of motes. A fraction of the 7 mm and 9 mm motes were equipped with custom made RFID chips. Additionally, solid 5 mm and 7 mm spherical motes were produced from standard ABS plastic, featuring neutral buoyancy.

Table 3 Design specifications of the prototype blank motes

	spheres			capsules		
characteristic length	diameter (D)			diameter (D) & total length (L)		
outer diameter	9.0 mm	7.0 mm	5.0 mm	9.0 mm & 18.2 mm	7.0 mm & 14.0 mm	5 mm & 15.0 mm
wall thickness	0.9 mm	0.8 mm	1.0 mm	1.0 mm	1.0 mm	1.0 mm
relative gravity (relative to the water of 1.04 g/cm^3)	0.9 and 1.0			0.9 and 1.0		
material	IXEF 1022	Bayblend 165		IXEF 1022	Bayblend 165	
assembly	2 parts			2 parts		
operating temperature	$-30°$C to $+80°$C			$-30°$C to $+80°$C		
max pressure on the casing	300 bar	100 bar	100 bar	300 bar	100 bar	100 bar
chemical environment	heavy oil, sand, HCl			heavy oil, sand, HCl		

2.3.3 The Field Test

During the field test the fluid circulation in the reservoir was controlled through the pump at the surface, maintaining the flow in the whole system between injector and producer well at the rate of several tens of cubic meter of water per day. Motes were flushed down the IW with a flush of high pressured water generated by a water pressure truck. In total 25000 motes were subsequently flushed into the injector, with time intervals of several hours between the different mote injections. Additionally, we run dye tests to monitor conditions of the reservoir and to establish a minimum time required for the injected water to pass through the reservoir and to establish a time baseline for mote transition through the same reservoir. For that purpose the mounted screen, Fig. 6(B), was checked regularly. After the injection phase of the experiment was accomplished, monitoring of the screen continued for a few weeks more. Eventually a total of about 8% of the initially injected motes was retrieved. The condition of the retrieved motes varied from intact to severely damaged, Fig. 6(C). The majority of RFID motes split up, however, some chips were still functional after spending a significant time in the reservoir and wells. Swabbing of the injector well at late stage showed that in total about 6% - 8% of motes stayed in the injector. It also indicates that 90% of motes went into the reservoir. The sensor motes spent significant longer times time in the system than the dye. This could be caused by irregular morphology and shapes of wormholes in the reservoir. The main main results of the field trial can be summarized:

- to the best of our knowledge we established the first experimental proof that a solid object of mm-size can pass through a heavy oil reservoir,
- motes took significantly longer to pass from the IW to the PW than the dye,
- in total about 8% of sensor motes were retrieved from the reservoir,
- wormholes feature open channels of at least 7 mm size and
- the well equipment was not damaged by sensor motes.

Overall the achieved results of the field trial were promising [17] and showed that despite hash conditions and the unknown structure of the reservoir it is feasible to apply micro motes for system exploration. A graphical overview of the experimental procedure and the obtained results is given in Fig. 7.

2.4   Conceptual Design of a First Generation of Sensor Motes

Based on results of the field trial, general requirements were establish for the development of a first generation of sensor motes. Objective was the development of a general mote design serving as technology bases for subsequent, increasingly sophisticated sensor mote generations that can easily be adapted for applications and that are ranging from a complex sensor motes capable of performing measurements individually and autonomously to motes featuring rather limited functionality and bound for collective operation.

Xploring WiseMotes^{TM} represent a first generation of fully functional motes, equipped with a limited sensing capabilities and not capable to determine the mote location. Xploring WiseMotes^{TM} are designed to measure and to store basic physical parameters as a function of time, to operate for

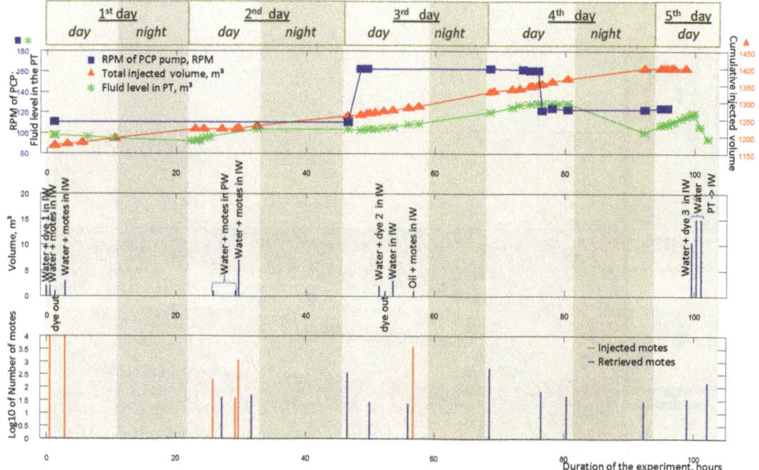

Fig. 7 Overview of the field pilot experiment. The first graph on the top indicates fluid level in the PT (green line), revolutions per minute of the PCP pump (blue line), volume of the injected fluid from the PT into the IW (red line). The second graph in the middle shows injection of motes, dye, water and oil. The last graph on the bottom indicates injected motes and their number (red bars) and retrieved motes and their number (blue bars).

several hours, to make the sensed data offline accessible via wireless read-out and to provide wireless energy charging and programming [18]. Selected functionalities are optionally included in the design, for example, self-testing, adding extension with external sensor ID, using antenna circuit as sensor. A spherical prototype under development will have 7 mm outer diameter and shell thickness of 1 mm allowing mass payload of 50 mg and resulting in total weight of 190 mg.

The primary sensing function of Xploring WiseMotes^{TM} is an environmental temperature sensor. The temperature readings are digitized and stored in an on-board memory. There are various ADC topology options for low power dissipation (for example Successive Approximation). However, it is also important to keep the chip area of the ADC circuit to a minimum to comply with the size restrictions. A simple digital *sequential machine*, the brain of the of the sensor mote, is foreseen to control the temperature sensor, the wireless transceiver and the data storage. To ensure low power dissipation, it is integrated in the design that various parts of the Xploring WiseMotes^{TM} are separately activated depending on the mode of operation. Typically, most circuits will be in the *sleep* mode until activated. Specific digital parts, however, like the sequential machine need to remain active (at low power) at all times to facilitate communication.

In order to keep the integrity of the Xploring WiseMotes^{TM} emphasis is placed on the implementation of antenna systems for different purposes, i.e. charging, offline data reading and programming. One of the major challenges

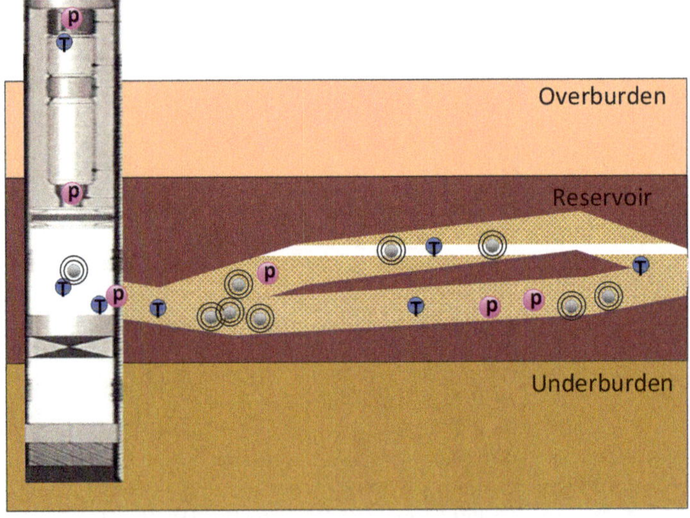

Fig. 8 Schematic representation of the deployment of motes featuring different sensor functionalities. Pink motes measure the system pressure, the environmental temperature is measured by motes in blue. Grey motes are equipped with ultra-sound technology.

in communicating with the Xploring WiseMotes^{TM} is the effectiveness of the antenna system. To begin with, the antenna configuration needs to fit into the physical limitations of the mote. The effectiveness of the candidate antenna arrangements needs to be characterized for the frequencies of interest: 125kHz (ISM) and 13.56MHz (NFC) bands. Moreover, in an optimum scenario the antenna should feature three gain matched axes to ensure directional invariant communication with the mote, for example, while charging.

The implementation of a wireless transceiver depends on a number of factors, such as the frequency choice that is in turn influenced by the mentioned above antenna system. There are also limitations on power dissipation, related to battery capacity, and chip area, related to the package size. In order to optimize chip area and power consumption, the modulation format and scheme should be chosen based on requirements for budget and complexity of realization. The modulation scheme could be passive (load modulation) or active (magnetic) coupling. A future task will be to chose the optimum modulation format and scheme and the corresponding transceiver architecture best suited for the Xploring WiseMotes^{TM} requirements.

A complex communication protocol are not required because Xploring WiseMotes^{TM} only need to communicate with the *Reader* at the end of their journey. Nevertheless, there are various modes of operation that need to be pre-defined for every mote. For example, every mote needs to be activated just before it is used. Measurement cycles need to be defined for the active operation mode. At the end of the measurement cycle, sensor data on every surviving mote needs to be read by the Reader.

2.5   Conclusive Remarks

We presented a novel approach on mapping geological structure non accessible to the state of the art technologies. Our approach is based on micro sensor motes. First feasibility studies delivered promising results for the application of millimeter sized motes for characterization of CHOPS reservoirs.

There is a realistic chance, that the technology, allowing for sand grain seized processors featuring embedded sensors will emerge in the coming years. This technology would offer exploration and mapping of even more complicate to access systems, like hydraulic fracturing layers or deep underground thermal reservoirs.

The hampered communication represents the biggest challenge, especially under deep underground conditions. A challenge which is interfering with the requirements of a reliable and controlled system exploration due to requirements to capture data from stuck motes and the need to establish the collective dynamics to get insight in the degree of system penetration. While the approach based on ultra-sound technologies has been discussed above we are also investigating options offered by multiple injections of motes into the system as shown in Fig. 8, where the functionality of the motes is going to be adapted from generation to generation. The specific changes in functionality are determined during offline analysis and embedded in each new generation.

From this perspective future motes might develop in a powerful tool to explore systems based on a minimum of precognition and so far to accessible for the state of the art technology.

Acknowledgements. INCAS[3] is co-financed by the European Union (European Fund for Regional Development), the Dutch Ministry of Economic affairs (Peaks in the Delta), the Province of Drenthe and the Municipality of Assen.

Special thanks to Prof. Dr. Ir. Peter Baltus, Director Centre for Wireless Technology, Technical University of Eindhoven and Prof. Dr. Ir. Jan Bergmans, Technical University of Eindhoven for stimulating, constructive and valuable discussions.

Thanks to PEZY Group for their expertise in plastics engineering and production of motes for the field test.

Thanks to KUDU Industries INC. Canada for their time and provided facilities for testing motes in PCP pumps. The authors would like to express their gratitude to KUDU pumps, for providing their test bench for these test, as well as to the three engineers who spend a large part of their Saturday morning to support the tests.

Thanks to PTRC for introducing to the world of heavy oil recovery.

References

1. Vogt, C. (ed.): Summary MoreWise Workshop, INCAS[3], Assen, The Netherlands (2011)
2. Dusseault, M.B.: CHOPS: Cold Heavy Oil Production with Sand in the Canadian Heavy Oil Industry. MBDC Inc., Waterloo, Canada (2002), `http://www.energy.alberta.ca/OilSands/1189.asps`
3. Chen, Z.: Heavy Oils, Part II. SIAM News 39(4) (May 2006)
4. Tremblay, B., Sedgwick, G., Vu, D.: A Review of Cold Production in Heavy Oil Reservoirs. In: EAGE-10th European Symposium on Improved Oil Recovery, Bringhton, UK, August 18-20 (1999)
5. Tremblay, B.G., Sedgwick, G., Forshner, K.: Imaging of sand production in a water-saturated horizontal sand pack by X-ray computed tomography. AACI Research Program Report 9697-6 (1996)
6. Tremblay, B., Sedgwick, G., Vu, D.: CT imaging of wormhole growth under solution - gas drive. SPE Reservoir Engineering Journal 2(1), 37–45 (1999)
7. Lines, L., Chen, S., Daley, P.F., Embleton, J., Mayo, L.: Geophysics 22(5), 459 (2003)
8. Wilson, M., Morgan, Y.: Technical Investigative White Paper, PTRC, Regina (2011)
9. Cook, J.C.: Geophys. 40, 865–885 (1975)
10. Talnishnikh, E.: Method and system for mapping a three-dimensional structure using motes, Patent Pending, INCAS[3], Assen, The Netherlands (2014)
11. Talnishnikh, E., et al.: Motes for environment mapping, Patent Pending, INCAS[3], Assen, The Netherlands (2013)
12. Schlupkothen, S., Dartmann, G., Ascheid, G.: Exploration With Massive Sensors Swarms. In: IEEE GlobalSIP Symposium on Controlled Sensing For Inference: Applications, Theory and Algorithms, Austin, Texas (December 2013)

13. Alard, M., Lassalle, R.: Principles of modulation and channel coding for digital broadcasting for mobile receivers. EBU Tech. Rev. 224, 3–25 (1987)
14. Demi, L., Verweij, M.D., van Dongen, K.W.A.: Parallel transmit beamforming using orthogonal frequency division multiplexing applied to harmonic imaging–a feasibility study. IEEE Transactions on Ultrasonics, Ferroelectrics and Frequency Control 59(11), 2439–2447 (2012)
15. Duisterwinkel, E., Demi, L., Dubbelman, G., Talnishnikh, E., Wörtche, H.J., Bergmans, J.W.: Environment mapping and localization with an uncontrolled swarm of ultrasound sensor motes. J. Acoust. Soc. Am. 134, 4185 (2013)
16. Smith, M.: When tiny is mighty. New Technology Magazine (December 2013)
17. Talnishnikh, E. (ed.): MoreWise Field Trial, internal report, INCAS[3], Assen, The Netherlands (2013)
18. Talnishnikh, E., et al.: Sensor system, mote and a mote-system for sensing an environmental parameter. Patent Pending, INCAS[3], Assen, The Netherlands (2014)

Chapter 3
A Multi-sensor Smart System for Vulcanic Ash Monitoring

B. Andò, S. Baglio, and V. Marletta

Abstract. The ash fall-out phenomenonfollowing the explosive activity of volcanoes represents a considerablefactor of risk for people and a serious hazard for air traffic.

Researchers at Department of Electronic and Information Engineering of the University of Catania are facing the venture of developinga low-cost smart multi-sensor system providing a real time and continuous information on the ongoing ash fall-out phenomenonwith particular regards toash granulometryclassification and flow-rate estimation. Moreover, the monitoring system must be selective in respect to volcanic ash against others sediments.

Ash granulometry detection exploits a piezoelectric transducer to convert ash impacts into electrical signals. Receiver Operating Characteristic (ROC) analysis has been adopted as a theoretical support to properly implement the classification approach.

The proposed idea for ash flow-rate estimation is to measure the time interval required to collect a fixed amount of ash inside an instrumented beaker. Finally, to implement the selectivity task the intrinsic paramagnetic property of ash particles is exploited.

The low-cost adopted sensing methodologies as respect to traditional instrumentation, the real time and continuous monitoring, the selectivity as respect to volcanic ash and the multi-parameter estimationare considered main claims and novelties of the methodology proposed in the following. Moreover, the low-cost architecture allows to implement a distributed monitoring network which can guarantee high spatial resolution information on ash fall-out phenomenon.

B. Andò · S. Baglio · V. Marletta
Dipartimento di Ingegneria Elettrica,
Elettronica e Informatica (DIEEI)
University of Catania, Italy
e-mail: bruno.ando@dieei.unict.it

© Springer International Publishing Switzerland 2015
H. Leung and S.C. Mukhopadhyay (eds.), *Intelligent Environmental Sensing,*
Smart Sensors, Measurement and Instrumentation 13, DOI: 10.1007/978-3-319-12892-4_3

The latter represents a very attractive goal as it is a mandatory information to be usedby models to predict the time-space evolution of ash transportation.

3.1 Introduction

Volcanic ash clouds and associated ash fall-out phenomenarepresent a relevant risk factor for people living close to an active volcano and sometimes hundreds of kilometers away. Ash clouds are known to cause extensive damages to roads, sanitation systems [1], agriculture [2], health [3] and the daily activities of people living on the slopes of the volcano. Moreover, ash is a real hazard to air traffic [4], especially when the airport is located near the active volcano. As an example, this is the case of the international Fontanarossa airport in Catania, in the south of Italy, close to the mount Etna, the largest active volcano in Europe.

The worldwide increase in volcanic explosion phenomena with consequent atmospheric ash dispersion, and in particular after the eruption of the Icelandic Eyafjallajokull volcano in 2010, has heightened the attention to this problem and to the need of regulation for air transport safety. In response to this subject, the International Civil Aviation Organization (ICAO) has issued guidance on managing flight operations into, or near, areas interested by actual or forecast volcanic clouds [5].

Very often during volcanic eruptions, due to ash plumes spewed by the volcano and the resulting volcanic ash fall-out, many airportsare declared inappropriate for take-offs and landings, thus creating great inconvenience to passengers along with financial loss for airlines and airport operators.

In the past, during such crises, the decision to open or close the airport by aviation authoritiesresults from subjective assessments, and therefore contained high levels of uncertainty and risk. Scientific and technical commissions havethen been established with the aim of regulating flight operations at airports in the presence of volcanic ash [6].

Monitoring of ash fall-out is then of great interest to meet logistical needs, as well as to achieve a better understanding of such phenomenon and its ruling mechanisms.

As an example, the National Institute of Geophysics and Volcanology in Catania is equipped with systems for monitoring ash clouds and forecasting their space-time evolution [7]. The latter provides decision support for aeronautical authorities in order to significantly reduce the factors of unpredictability, and therefore the impact of this phenomenon on airport infrastructure and flight operations.

Traditional approaches for the monitoring of volcanic ash employ high-cost instrumentation, typically based on satellites [8]or X-Band dual-polarization radars [9]. Moreover, reliable instruments for ash granulometry classification, generally based on high-cost infrared cameras [10] and providing information on ash suspended in atmosphere, have been developed and deployed.

Anyway, solutions above mentioned allow for achieving discrete measurements at a restricted number of monitoring stations, thus guaranteeing a low spatial resolution of the information collected. Moreover, such systems are difficult to be installed and maintained and are often used to perform spot and discrete measurements.

Low cost networks of multi-sensor nodes would play a fundamental role in performing an effective monitoring of ash fall-out phenomena.

To cope with this need researchers at the DIEEI in Catania have developed a low cost multi-sensor node oriented to the implementation of a distributed early warning approach for the monitoring of volcanic ash. The main idea behind the early warning approach is the possibility to provide rough but well spatially-distributed information on incoming phenomena. Furthermore, in cases of specific need, accurate measurements can be performed by installing dedicated high-cost instrumentation in critical sites identified by the early warning system.

The developed node is able to measure main quantities of interest for the considered phenomenon such as the ash flow-rate and the ash granulometry, by exploiting a self-powered and low-cost μ-controller based architecture [11-14]. The node is intended to be integrated into a sensor network which will provide a distributed information useful to predict the time-space evolution of ash fall-out by models taking into account both meteorological quantities (e.g. wind speed and direction) and characteristics of the detected ash (flow-rate and granulometry) [7]. Such forecast is useful to implement an optimal planning of actions required to both restore the airport functionalities and manage air traffic during the ongoing phenomenon.

The activity described is conducted within the SECESTA project, the Italian acronym for "A sensor network for the monitoring of volcanic ash fall-out for the safety of air transport". The project is funded under the POR FESR Sicily 2007–2013 program and it exploits the synergic cooperation between research institutes and small–medium sized enterprises with the aim of developing a Wireless Sensor Network for the monitoring of ash fall-outalong the area spreading from main craters of the Etna volcano to the Fontanarossa airport (the shadowed area in Figure 1).

This approach in the field of volcanic ash monitoring is innovative, in particular when taking into account its capability to provide experts with a time continuous awareness of the ash fall-out phenomenon with a high degree of spatial resolution. The latter, along with themulti-parametric approach for the characterization of ash fall-out and the low-cost features of the monitoring nodes are the main advantages of the developed solution.

The sensing approachesadoptedfor ash granulometry classification,ash flow-rate estimation and the discrimination of volcanic ash from other types of sediments [11-14] will be discussed in the following sections together with experimental evidences.

Fig. 1 The area interested by ash fall-out phenomena spreading from main craters of the Etna volcano to the Fontanarossa airport (© OpenStreetMap contributors)

The main idea for the granulometry classification is to use a piezoelectric transducer to convert ash impacts into electrical signals whose characteristics are strictly related to ash granulometry.Ash flow-rate estimation is based on the measurement of time intervals required to collect a fixed amount of ash inside an instrumented beaker. The ash selectivity task exploits the intrinsic paramagnetic property of ash particles.

In the next Section an overview of the multi-sensor node is given. The sensing strategy developed for ash granulometry estimation is deeply discussed in Section 3 along with theoretical considerations and experimental results. Moreover, the methodology for particle classification based on Receiver Operating Characteristic (ROC) analysis is presented.A detailed description and assessments of sensing methodologies developed for ash flow-rate monitoringand ash discrimination is given in Sections 4 and 5, respectively. Concluding remarks are given in Section 6.

3.2   The Multi-sensor Node

The multi-sensor node with the on-board electronics isschematized in Figure 2, while a true pictureof the laboratory prototype aimed to the investigation of the proposed sensing methodologies is shown in Figure 3.

Basically, the system consists of a funnel-shaped structureto convey falling volcanic ash to aninstrumentedtank. A piezoelectric sensor is placed within the funnel-shaped structure with the purpose of converting the impacts of ash grains into electrical signals [11, 12]. As it will be highlighted in Section 3, the piezoelectric transducer is installed with a slope of 45° with respect to the direction of fall (vertical axis) to reduce multiple impacts of the same ash particle.

To cope with the task of ash flow-rate estimation, a dedicated array of coupled infrared (IR) diode-phototransistors allows for the monitoring of ash levels in the tank [13, 14].

To perform tank empty operations, asuitable shutter system has been placed at the bottom of the tank.

A low-power digital magnetometer is used to provide the system with the required selectivity as respect to exogenous sediments [14, 15]. The presence of volcanic ash particles in the tank will perturb the magnetic field produced by a permanent magnet, thus producing a variation of the magnetic sensor output.

Wind and rain sensors are also used to provide information on meteorological quantities. The latter are used in models in order to predict the space-time evolution of the observed phenomenon.

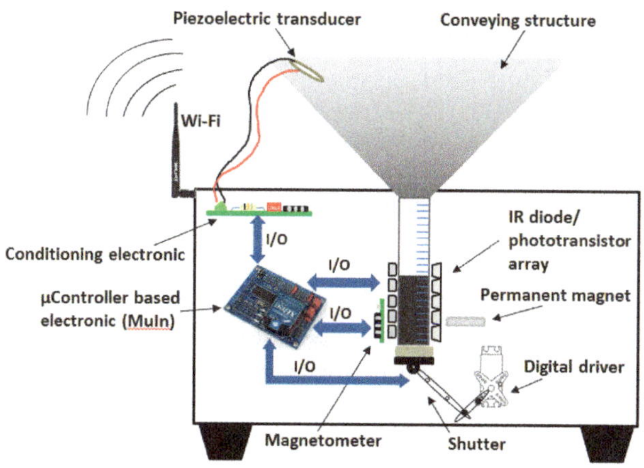

Fig. 2 Schematization of the multi-sensor node with the on-board electronic

Signals provided by sensors are acquired and processed by a μ-controller architecture employing a Droids Multi Interface Board (MuIn) equipped with a PIC18F2520 Microchip running at 40MHz. Data are then transferred by a wireless transmission protocol (IEEE 802.15.4) to a dedicated PC station for the sake of debugging during the development phase of the system. Nodes are poweredby the supply network, if available, or by battery based solutions. Alternative energy sources (photovoltaic and/or aeolic) will be employed in order to extend the battery life time.

In the final node a further shutter system will be placed on top of the ash collector with the purpose of shielding the system against hostile agents (water, dust, sand, soil or bird droppings) that could damage the system. Opening of the shutter can be remotely controlled by operators, or automatically triggered when explosive phenomena are observed.

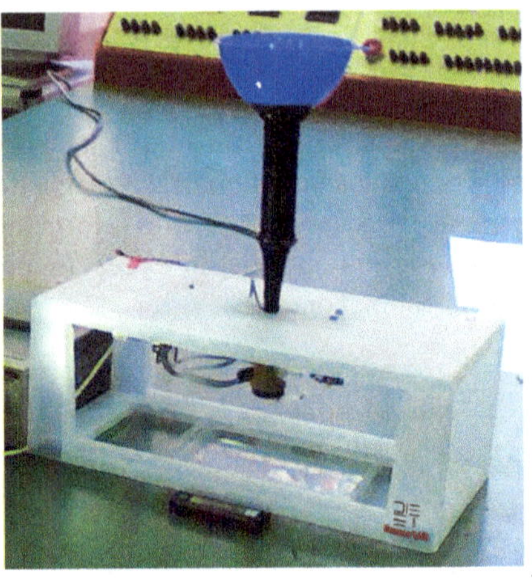

Fig. 3 A picture of the multi-sensor node prototype for ash fall-out monitoring, developed at DIEEI laboratories in Catania [12]

With the aim of rough event estimation pursued by an early warning system in mind, it should be highlighted that the low-cost features are mainly related to the sensing technology adopted, as it is easily demonstrated from a rough comparison with other monitoring systems cited in the SOTA review [8-10], such as dedicated satellites, instruments based on X-band radar and IR cameras.

Moreover, conversely to traditional high cost instrumentation, the system does not require complicate installation efforts.

The system has been designed in order to cope with needs of outdoor operation in terms of both mechanics and sensing methodology, to assure robustness against hostile environmental conditions. Anyway it must be considered that the whole multi-sensor system is intended to be encapsulated within a waterproof box, beingthe conveying structurethe only interface between the multi-sensor node and the outdoor environment.

3.3   The Methodology for Ash Granulometry Classification

Basically, the idea underpinningthe methodology for ash granulometry classification is to exploit the relationship between particle granulometry and the physical force provided by their impacts on a piezoelectric transducer. The latter is in charge to convert ash impacts into electrical signals, which will provide information about ash granulometry. As a premise, it is important to note that the ash density (which fixes the relationship between granulometry and the mass of each particle) is an intrinsic characteristic of the volcano, and it can be considered as a known quantity.

The use of such a sensing strategy, with respect to other sensing methodologies (e.g. capacitive or inductive), provides an information related to each particle bumping into the piezoelectric device. It avoids an indirect estimation of the average particle size, e.g. obtained by measuring the volume of collected ash.

Apart from alreadymentioned advantages, the proposed sensing solution for ash granulometry classification is trulylow-cost and it offers high reliability due to the sensing architecture adopted.

Concerning the importance of the monitoring of the ash granulometry, it must be remarked that the awareness of ash dimensions and its distribution throughout the area of interest are fundamental for forecasting the time-space evolution of ash clouds and fall-out.

3.3.1 Modelling and Design of the Ash Granulometrydetection System

This section presents an analytical model for the collisionforce produced by ash particles impacting on the piezoelectric, together with a model describing the impulse response of the piezoelectric transducer.

A schematization of the interaction forces between the volcanic ash particle and the piezoelectric transducer is shown in Figure 4. Each ash grain impacts on the piezoelectric surface with a force F thus generating an electrical signal. The interaction force F can be resolved into two forces: a tangential component F_t, whose effect on the piezoelectric sensor can be assumed to be negligible, and a normal (to the slope) component F_p. Moreover, assuming that both ash particles and the piezoelectric transducer are linearly elastic solids and modelling the ash particle as a sphere of radius R_A, the perpendicular force F_p can be determined by the Hertz law [16, 17]:

$$F_p = 1.2644\, \rho_A^{3/5} \left(\frac{1}{k_P + k_A} \right)^{2/5} R_A^2\, v_A^{6/5}$$

(1)

where $\rho_A = 2$ kg/m^3, k_A and v_A represent the density, the stiffness and the terminal settling velocity [18] of the ash particle, respectively, and k_p represents the stiffness of the piezoelectric transducer.

The impact of a ash particle produces an impulsive stress on the piezoelectric transducer which will start to oscillate at its natural frequency f_n, thus producing an oscillating voltage with the same frequency. The amplitude of this voltage signal decreases with an exponential rate as shown in Figure 5. The predicted trend for the impulse response is given by:

$$V_{out} = A \cdot \frac{e^{-\xi \omega_n t}}{\sqrt{1 - \xi^2}} \sin(2\pi f_n t)$$

(2)

where ξ is the system damping, f_n is the natural frequency and A depends on the impulsive stress generated by the F_p force by the relationship [19]:

$$A = g_{33} F_p\, d / S$$

(3)

where $g_{33}=22e^{-3}$ V*m/N is the piezoelectric coefficient in the stress direction, $d=0.23e^{-3}$ m and $S=4.155e^{-4}$ m^2 are the thickness and the surface area of the transducer, respectively.

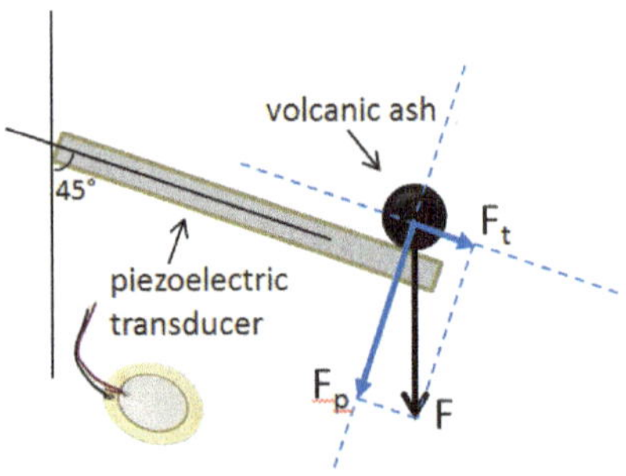

Fig. 4 Schematisation of the interaction forces between the volcanic ash particle and the piezoelectric transducer [12]

Accordingly, Eq. (2) states that the voltage generated by the piezoelectric sensor depends on the ash granulometry through the term A, which is related to the impact force and hence to the particle radius. In fact, the impact (and consequently the A parameter) will also be related to the velocity and the density of the falling particles. Nevertheless, ash density is usually a characteristic of the volcanic area while, close to the ground, ash particles reach a constant terminal velocity [18]. The latter property is strictly related to the ash dimensions, and hence will not randomly affect the classification procedure.

Figure 5shows the response generated by the piezoelectric sensor in case of three ash granulometries, (indicated as small, medium and large), with an average radius of around 0.4 mm, 1 mm and 2 mm, respectively. As clearly evidenced, the amplitude of the response depends on the ash granulometry.

The Nelder–Mead optimisation algorithm [20] with the following function J:

$$J = \sum_{size=1}^{3} \left(\frac{\sqrt{\Sigma_i^N \left(V_{out} - V_{out,pred}\right)^2}}{N} \right) \begin{matrix} size: \\ 1 = small \\ 2 = medium \\ 3 = large \end{matrix} \tag{4}$$

defined as the sum of the root mean square of the residuals between the observed responses V_{out} and predicted responses $V_{out,pred}$ for the three particle size classes considered, has been adopted in order to fit the model (2) to the observed responses.The following parameter values have been estimated:

$\xi = 0.0098$, $f_n = 3.49$ kHz and $A_s = 2.30$ V, $A_m = 3.20$ V, $A_b = 11.5$ V for small, medium and large particles, respectively.

Figures 5a, 5c and 5e show the fitting between the predicted and experimental impulse responses for the three ash sizes, while details are presented in Figures 5b, 5d and 5f.

Equations (1)–(3) have been used for the design of the sensing system, including the conditioning electronics.

It must be born in mind that volcanic ash granulometry depends on the monitoring site, wind conditions and is strictly related to the specific volcano.

With the general view of using the proposed approach at different sites and for different volcanoes, model (1)–(3) becomes a fundamental tool to predict the system's behaviour under different operating conditions and to develop a suitable design for the sensing system.

Some consideration about the proposed sensing strategy could be useful. From a statistical point of view, the probability of concurrent impacts on the piezoelectric is very low, in any case consecutive impacts of particles on the piezoelectric will cause consecutive impulsive solicitations and then consecutive but observable signature in the piezoelectric response. The analysis of these signatures can be exploited for the sake of granulometry classification.

The probability of sediments laying on the piezoelectric sensor is very low due to the intrinsic topology of the tilted piezoelectric sensing tool. In any case, the device's responsivity (in terms of voltage impulse amplitude as a consequence of the ash impact force) can be considered independent on the sensor mass (which can change for the effect of sediments on the sensor surface) as stated by model (2)-(3).The effect of sediments on the natural frequency does not affect the classification methodology. Moreover, masking effects of the sensing surface due to deposited sediments can be considered negligible due to the high ratio between the sensor section and the particle dimension.

3.3.2 Synthesis and Characterization of the Sensing System

The developed granulometry sensor prototype adopts the low-cost piezoelectric transducer 7BB-35-3L0 by Murata, which has a round piezoelectric diaphragm with a diameter of 35 mm. Dedicated conditioning electronics consisting of a charge amplifier followed by an inverting amplifier has been designed and developed. The charge amplifier furnishes a voltage signal proportional to the charge generated by the piezoelectric transducer. A schematic of the electronics is shown in Figure 6a, where C_{eq}=14 nF and R_{eq}=67 kΩ are the equivalent components of the piezoelectric transducer. Values of the feedback resistance R_f=1.2 MΩ and capacitance C_f=22nF of the charge amplifier have been chosen to fix a low cut-off frequency, f_t^{chA}=6 Hz, which is compatible with the specific application, as shown in Figure 6b.

The cut-off frequency, f_t^{chA}, was defined according to experts in order to make the piezoelectric system also sensitive to seismic activity, with the purpose of implementing a trigger system for the automatic opening of the top shutter. It must be observed that volcanic or seismic activities take place at lower frequencies (from Hz to some hundreds Hz) with respect to the device's natural frequency, f_n, thus making the effect of ash fall-out distinguishable from ground vibrations.

The electronic gain of the amplifier has been conveniently fixedto meet the specifications of the μ-controller device.

A power consumption of the sensing system (including the conditioning electronics for the piezoelectric transducer and the μC board) of about 10 mA @ 3.3 V has been measured. A higher power budget is required by the wireless transmitter module. The radio module adopted is a XBee-PRO embedded RF module by Digi @2.4GHz, whose power consumption in transmission mode is 215 mA @ 3.3V.

In order to validate the sensing methodology and to assess prototype performances when executing the ash granulometry classification task, real experiments have been performed in laboratory by repeatedsingle particle impacts. To perform the experiment volcanic ashes from Etna volcano with three different typical size (small, medium and large) ranging from 0.4 mm to 2.0 mm, shown in Figure 7, have been used.

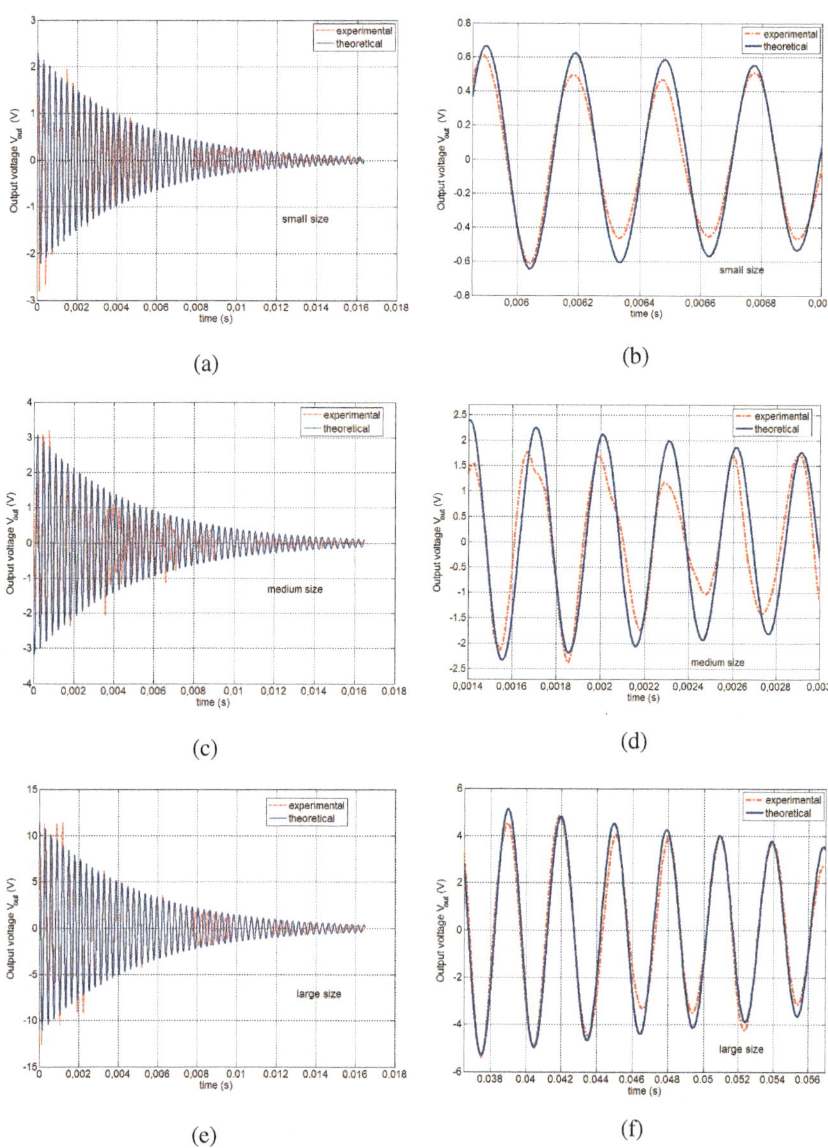

Fig. 5 (a), (c), (e) Experimental impulse responses of the piezoelectric sensor and fitting obtained through model (2) for the three ash sizes (small, medium, large). A natural frequency of 3.49 kHz has been estimated for the piezoelectric sensor. (b), (d), (f) Details of the device impulse response for the three ash classes [12]

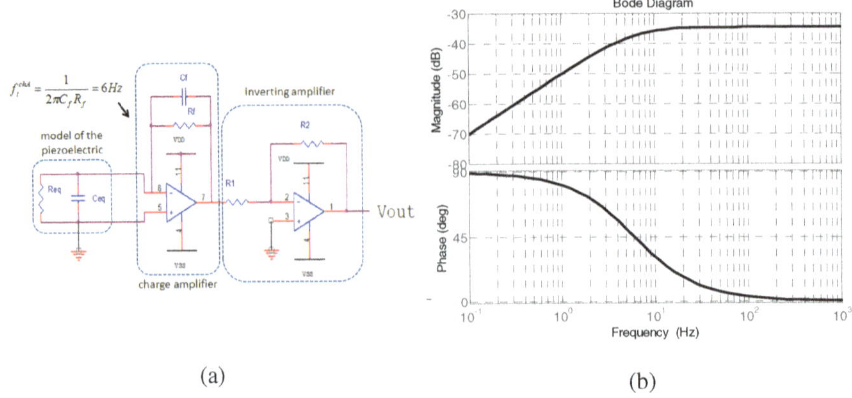

(a)   (b)

Fig. 6 (a) Schematic of the conditioning circuit for the piezoelectric transducer. Values of components are: C_{eq}=14 nF, R_{eq}=67 kΩ, C_f =22 nF, R_f=1.2 MΩ; R_1 and R_2 are trimmers of 10 kΩ and 100 kΩ, respectively, used to appropriatelytune the amplifier gain. (b) Frequency response of the charge amplifier [12]

Ash particles have been dropped on the sensor from a height of at least 4 meters to ensure their terminal velocity is reached. Wind conditions have not been taken into account also due to the consideration that wind will not compromise the functionality of the sensing node while it will define the ash fall-out distance.

Figure 8a shows a typical output signal from the readout electronics. A suitable algorithm has been developed to automatically window the raw signal in order to extract information related to the impact of ash particles on the piezoelectric. The algorithm uses a threshold mechanism to detect the useful part of the signal to be considered. Examples of signals obtained for the three granulometries investigated are shown in Figures 8b, 8c and 8d.

As can be observed, the signal peak value increases with ash particle size. The maximum signal amplitude has been chosen as the granulometry classification parameter due to the simplicity of extracting such a quantity from the observed signals, which is also strategicto embed the automatic classification task into the sensor node.

Figure 9 shows the distributions of voltage peak amplitudes, A, for the three granulometries investigated. As it can be observed, the ash classification task is not trivial. Actually, distributions are partially overlapped and consequently a small sized ash particles could be classified as medium or vice versa, as well as a medium grain potentially can be recognized as large and vice versa.

Anyway, as shown in Figure 9, on average the considered output quantity, A, are well separated. Moreover, as shown in Figure 10 and according to model (1)–(3), A increases with the size of the ash particles, following the square law reported in the figure itself.

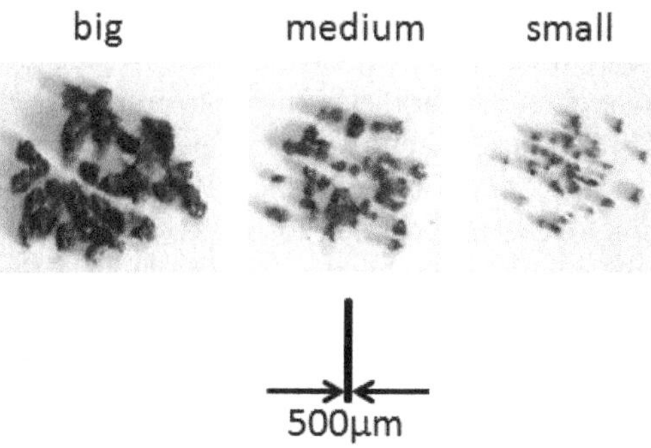

Fig. 7 Volcanic ash with the three different granulometries used to perform experimental investigations [12]

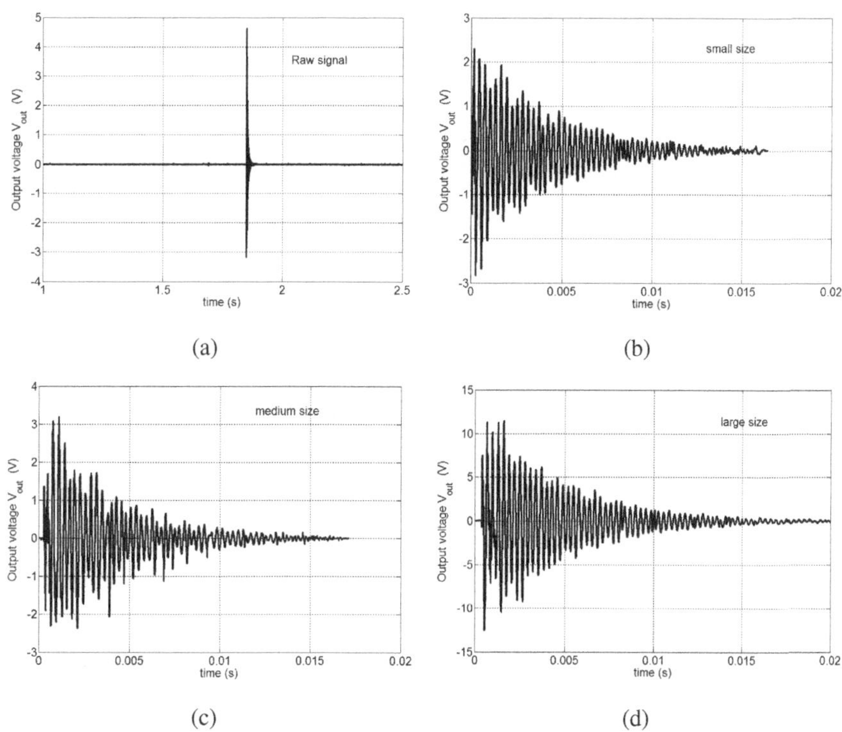

Fig. 8 (a) Raw output voltage signal from the conditioning electronics for the piezoelectric; (b), (c) and (d) The output voltage signal after windowing for the case of small, medium and large sized ash particles, respectively [12]

The situation depicted in Figure 9 is typical of a two-class prediction problem, namely a binary classification. The case under investigation can be hence approached as two separate binary classification problems which require two thresholds, *Th1* and *Th2*. A useful approach to properly fix threshold values can be based on the Receiver Operating Characteristics (ROC) analysis [21,22].

In the next section a brief introduction to the ROC curve theory will be given together with its application to the ash granulometry classification task.

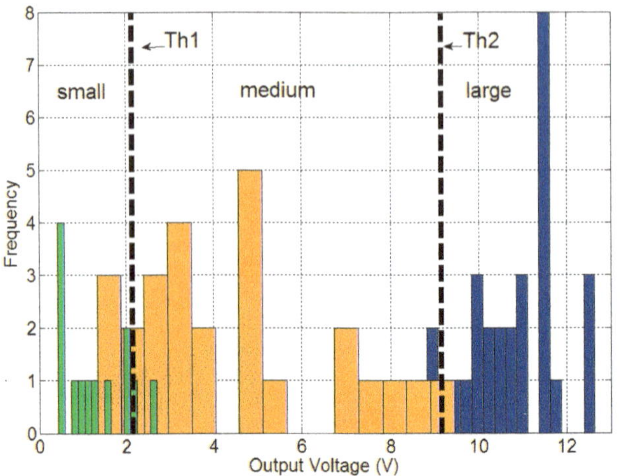

Fig. 9 Distribution of output voltage peaks [12]

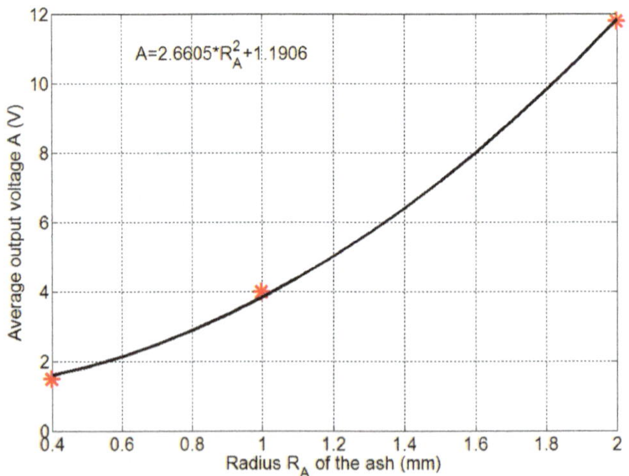

Fig. 10 output voltage vs. the size (R_A) of ash particles. Symbols represent real observations, while the continuous curve is the model fitting [12]

3.3.3   ROC Analysis as a Methodology for Ash Granulometry Classification

Receiver Operating Characteristics (ROC) analysis is a widely used technique for visualizing, organizing and selecting classifiers based on their performance [23]. Typical applications are in the fields of signal detection theory, in medicine to evaluate clinical tests and the behaviour of diagnostic systems and in the machine learning research field [23, 25].In the following a brief introduction to the ROC analysis is given with reference to the simple case of a two-class prediction problem. In such a case, the classifier is required to map each instance (i.e. samples to be classified) to one of the two classes (positive and negative classes). A situation like the one shown in Figure 11a with an overlap between two classes is typically observed.The best case rarely exists, i.e. two classes are perfectly separated with a well-defined cut-off value.

For every possible cut-off point or criterion adopted to discriminate between two populations, there will be four possible outcomes:

- some instances correctly classified as positive (TP = True Positive fraction)
- some instances wrongly classified as negative (FN = False Negative fraction)
- some instances correctly classified as negative (TN = True Negative fraction)
- some instances wrongly classified as positive (FP = False Positive fraction)

Two indexes of interest can be defined:
-the True Positive Rate (TPR) or Sensitivity as the ratio between the number of true positives and the total positive instances:

$$TPR = \frac{TP}{TP + FN} = Sensitivity \tag{5}$$

-the False Positive Rate (FPR), or 1-Specificity, as the ratio between the number of false positives and the total negative instances:

$$FPR = \frac{FP}{TN + FP} = 1 - Specificity \tag{6}$$

A ROC curve is a 2D plot of the TPR (sensitivity) *vs.* FPR (1-specificity) for a binary classifier system, as its discrimination threshold varies. It is a graphical representation which illustrates the performance of the classifier. An example of a ROC curve is shown in Figure 11b. ROC graphs are used to depict the trade-off between the hit rates (benefits) and false alarm rates (costs) of classifiers [24, 25].

To choose a point on the ROC curve with a higher *TPR* means to move the threshold to the left (towards lower values) with the consequence that on one hand more positive instances will be classified as positive (*TP* increases and *FN* decreases) but on the other hand more negative instances will be wrongly classified as positive (*FP* increases and *TN* decreases).In the case of perfect discrimination (no overlap in the two distributions) the ROC curve passes through the upper left corner (100% sensitivity, 100% specificity). Therefore the closer the ROC curve is to the upper left corner, the higher the overall accuracy of the classifier.

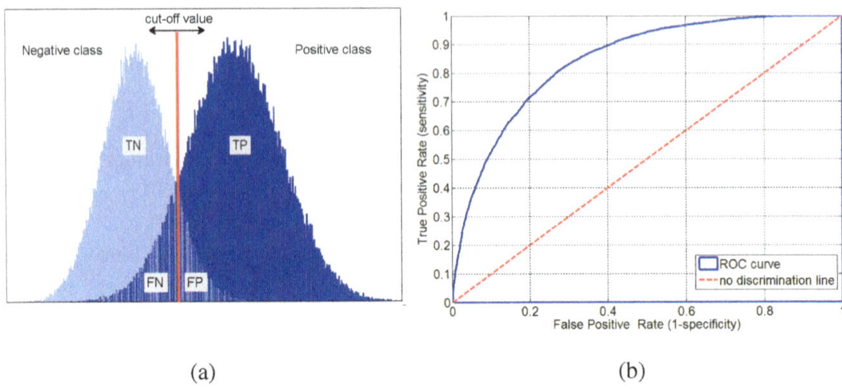

Fig. 11 (a) Example of binary classification; (b) The ROC curve related to the example in (a) [12]

The worst case is that in which the two distributions are completely overlapped, thus making the classification task impossible.The latter case is represented in Figure 11b by the no discrimination line.

In the following the ROC analysis is applied to the ash granulometry classification task.With reference to Figure 9, when the two distributions of small and medium particles are taken into account and a threshold value *Th1* is fixed, the four fractions previously introduced will be defined as:

TP = fraction of *A* quantity due to medium ash particles correctly classified as medium

FN = fraction of *A* quantity due to medium ash particles classified as small

TN = fraction of *A* quantity due to small ash particles correctly classified as small

FP = fraction of *A* quantity due to small ash particles classified as medium.

Similarly, when the two distributions of medium and large particles are taken in account and a threshold value *Th2* is fixed, the four fractions will be defined as:

TP = fraction of *A* quantity due to large ash particles correctly classified as large

FN = fraction of *A* quantity due to large ash particles classified as medium

TN = fraction of *A* quantity due to medium ash particles correctly classified as medium
FP = fraction of *A* quantity due to medium ash particles classified as large.

Applying the above theory of ROC curves to the ash granulometry problem, taking in account the distribution of voltage peaks shown in Figure 9, curves in Figure 12 have been obtained.

Considering that each point in the ROC curve is related to a threshold value, the next step consists of fixing suitable classification thresholds. As an example, if the classification problem for the small and medium sized particles is taken in account and accordingly to the above considerations, to move the threshold to the left (high *TPR*) means increasing the number of small particles classified as medium, and to decrease the number of medium particles classified as small. Of course, it is easy to deduce the consequence of choosing a point on the ROC curve with a low *TPR*.

As emerges from Figure 12b and consistently with information in Figure 9, a suitable threshold value, *Th2*, discriminating large sized grains from medium sized grains with very good performances (close to the best case of separated distributions) can be easily defined.Similarly, a threshold value, *Th1*, to discriminate small ash particle from medium particles could be also defined but with lower performance than the previous case.

The results obtained for the case of the study presented, and in general the proposed methodology, can provide strategic input to experts in order to properly fix classification criteria. The latter could be definitively tailored by the position of the monitoring station in the route between craters and airports, as well as by the interest of properly detecting the amount of small ash particles rather than medium or large grains.

Although the operative context (position of the monitoring station and weather conditions) should be taken into account, some preliminary criteria to properly choose classification thresholds can be drawn.

On the basis of the above considerations, in case the classification problem between small and medium ash particles is addressed, moving the threshold to the right affects the right classification of the medium size ash particles while it improves the correct classification of the small size ash particles. Just the opposite happens in case of moving the threshold to the left.Since small particles can be transported over long distances, effect of the last choice would be the missing of warning alarms with a consequently high risk for the safety of air transport. For the specific application addressed, a correct classification of small size particles is mandatory also at the expense of false alarms due to medium particles classified as small ash.

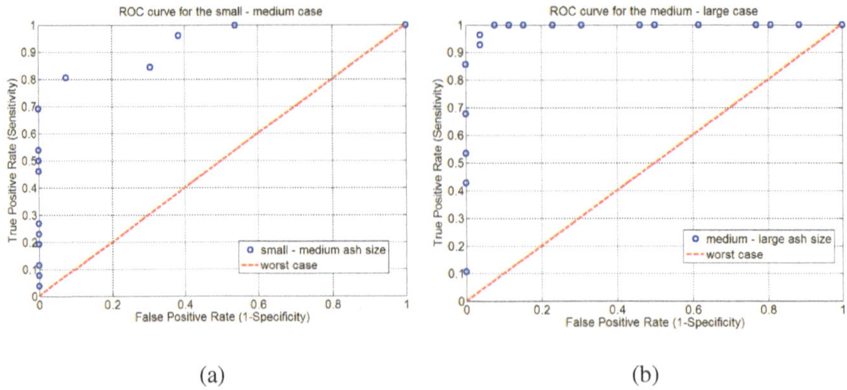

Fig. 12 curves for the volcanic ash granulometries classification. (a) Curve for the discrimination between small and medium sized grains; (b) curve for the discrimination between medium and large sized grains [12]

The same considerations can be applied to the classification problem between medium and large ash particles. Also in this case the right classification of medium particles can be crucial for air transport safety with respect to the correct classification of large ash.

3.4 Flow Rate Measurement

The ash flow-rate is estimated by measuring the time, Δt, required by the falling ash to fill a known volume, ΔV, of a tank. Known the time Δt and the corresponding volume ΔV of collected ash (here expressed in ml) the ash flow-rate ϕ can be estimated by:

$$\phi = \frac{\Delta V}{\Delta t}[ml/s] \tag{7}$$

As schematized in Figure 2,the volume ΔV of collected ash is measured by an array of coupled infrared (IR) diodes-phototransistors.A suitable architecture has been designed fulfilling specifications related to granulometries and ash flow-rate and taking into account a number of constraints, such as the distance between consecutive IR couples. Details on the system design are given in the following Section.

3.4.1 Design of the Sensing Architecture

A suitable sizing of the convoying structure and the tank aimed to collect the volcanic ash was a mandatory first step in the design of the structure. To this purpose specifications about typical granulometries and ash mass flows observed in three locations at different distance from the craters of the volcano, reported in

Table 1, have been taken into account.A schematization of lateral and top views of the collector and the tank is shown in Figure 13 where $d1$ and $d2$ are the diameters of the convoying structure and the tank, respectively, $A1$ is the top section area of the convoying structure and $A2$ is the area of the tank section. By expert suggestions, the tank diameter $d2$was fixed ten times the average granulometry of the ash in order to avoid occlusion problems.

Through this section the case of study of ash with an average granulometry of 1 mm (which fix $d2=10$ mm), is considered.

Sizing of the convoying structure has been performed by taking into account specifications on the ash mass flow P in Table 1 and the maximum updating time,t_{max}, for the flow-rate estimation.

The design flow for the sizing of the top section area of the convoying structure is presented in the following.

With reference to Figure 13, the flow-rate through the top section $A1$ is given by:

$$\Phi = P \cdot A1 \quad [kg/h] \tag{8}$$

In the time t_{max} the volume of collected ash is:

$$V_{t\,max} = \Phi \cdot t_{max} \quad [kg] \tag{9}$$

being V_{tmax} the amount of ash required to fill the volume between contiguous couples of IR. The following expression of V_{tmax} can be written:

$$V_{t\,max} = A2 \cdot h \cdot \rho \quad [kg] \tag{10}$$

where h is the distance between contiguous couples of IR detectors and ρ is the density of the volcanic ash.

Table 1 Specifications on typical granulometriesand ash mass flow observed in three locations at different distance from the craters

Location name	RifugioSapienza	Nicolosi	Catania
Distance from craters [km]	$5-10$	$15-20$	$25-30$
Expected ash granulometry [mm]	$0.130-4$	$0.060-0.5$	$0.030-0.125$
Expected ash mass flow-rate[kg/m²*h]	0.450	0.120	0.090

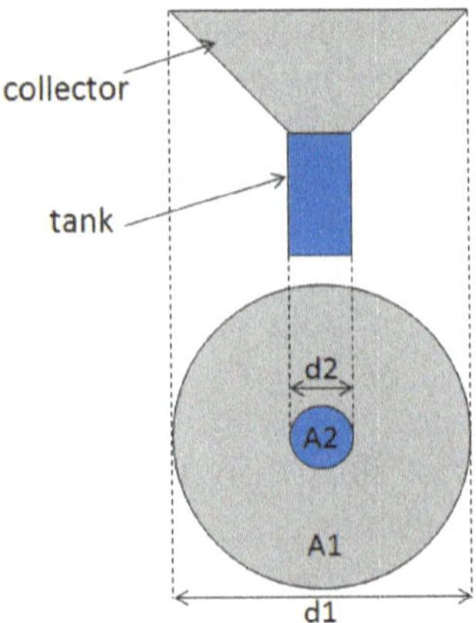

Fig. 13 Schematization of lateral and top views of the collector and the tank

As already stated above, the ash density (which fixes the relationship between granulometry and the mass of each particle) is an intrinsic characteristic of the volcano. For the case of study here considered, accordingly with experts, a density $\rho=1000$ kg/m^3 was used.By combining equations (8), (9) and (10) the following relationship supporting the design of the convoying structure and assuring an update of the ash flow-rate measurement in t_{max}can be obtained:

$$A1 = \frac{A2 \cdot h \cdot \rho}{P \cdot t_{\max}} \quad [m^2] \tag{11}$$

For the case considered, a collector with a section $A1 = 0.0104$ m^2 and a cylindrical tank with a height of 10 cm and a diameter $d2 = 10$ mm was used.

An array of aluminum gallium arsenide (AlGaAs) infrared emitting diodes SEP8736 by Honeywell and the NPN silicon phototransistors SDP8436 by Honeywell have been installed with a step $h = 5$ mmto avoid cross influence. The devices are designed to mechanically match one each other and are characterized by a narrow emitting and acceptance angle, making them well suited to applications in which adjacent channel crosstalk must be avoided. A dedicated electronics has been realized to drive the IR devices and to properly process outputs signals [13, 14].

Figure 14 shows estimations of the top collector area $A1$(left axis) or the collector diameter, $d1$ (right axis), as a function of the ash mass flow Pin the range [0.012 – 0.2] kg/m^{2}h, in case of $h = 5$ mm, and for different values of t_{max} in the range [10 s – 600 s]. Results shown in Figure14 are useful to design the convoying structure. The top collector area $A1$ is defined by fixing distance between contiguous IR couples, the expected mass flow-rate P and the t_{max}time.

As an example, for the case of study considered (P=0.120 kg/m^{2}h) the choice of a t_{max}= 20 s leads to a conveying structure with a top section diameter of $d1$ = 870 mm.

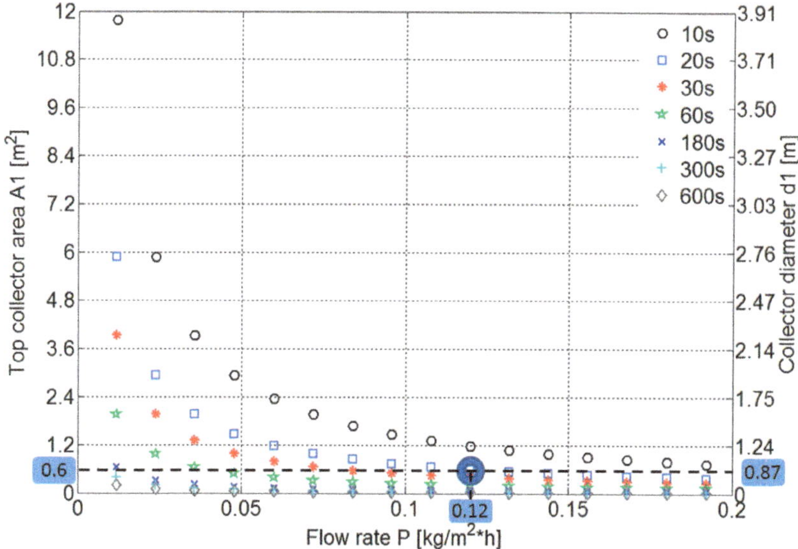

Fig. 14 Trend of the top collector area $A1$ (left axis) or diameter $d1$(right axis) *vs* the ash mass flow P, in case of a distance between contiguous couples of IR detectors of 5 mm, for different values of t_{max} in the range [10 s – 600 s]. A particular in the range [0.07 - 0.16] kg/m^{2}h of the ash mass flow is also shown

3.4.2 *Characterization of the Flow-Rate Sensor*

In order to characterize the sensor behavior in terms of amount of ash ΔV_i (i = 1 to 5) required to activate each couple of IR detectors, experimental surveys have been performed by injecting a continuous flow of ash into the tank. Experimental investigations have been performed using volcanic ash from Etna volcano.Experiments have been repeated 10 times for each level and for the three considered classes of granulometry. A dedicated injection system equipped with a graduated tank with a resolution of 0.1 ml was used. The estimation of ash volume values, ΔV_i, was performed by computing the amount of ash injected into the tank (as the difference between the initial and the final amount in the graduated tank). The time required by the ash to fill the volume between contiguous couples of IR detectors have been estimated by a firmware routine implemented into the μC and exploiting the state of the recorded IR detectors.

Data dispersions confined within 0.11 ml have been estimated as the standard deviation of experimental observations [14].

Further experiments have been performed with the aim to assess performances of the ash flow-rate sensor developed. In order to avoid long time observations a t_{max} of 20 s was fixed.

Experiments have been repeated 30 times by recording times, $t1$, $t2$ and $t3$, required by the ash injected into the tank to fill volumes between the first and the second IR couple, the second and the third IR couple and between the third and the fourth IR couple, respectively. Estimations of flow-rate ϕ have been performed by applying model (7) to different quantities of collected ash. As an example, Table 2 shows results obtained in case of five different flow-rate estimations for a nominal value of 0.030 ml/s assured by the experimental setup.

Some considerations about results in Table 2 are mandatory. First of all, as it can be easily evinced, the third and the fifth estimations are characterized by a lower dispersion from the nominal value than other cases. The flow-rate estimation based on the full scale collected volume of volcanic ash is the best case. It arises that, ash flow-rate estimations performed by taking into account the times recorded to fill a single volume, as in the cases of the first, the second and the fourth estimations, are characterized by a high dispersion and hence by a poor accuracy. Estimations obtained by considering longer waiting times, as in the cases of the third and fifth flow-rate estimations, are characterized by higher accuracy due to the lower dispersions of measurements. On the other hand these estimations require longer waiting times than the previous case.

Uncertainty of the flow-rate ϕ measurements can be evaluated by the following expression:

$$u_c^2(\phi) = \left(\frac{\partial \phi}{\partial V} \right)^2 u^2(\Delta V) + \left(\frac{\partial \phi}{\partial \Delta t} \right)^2 u^2(\Delta t) \tag{12}$$

where $u^2(\Delta V)$ and $u^2(\Delta t)$ represent uncertainties of volume and time interval measurements, respectively. An uncertainty of 0.075e-3 ml/s has been evaluated for the best case considered.

Table 2 Comparison of five methods for the estimation of the flow-rate in case of a nominal value of 0.030 ml/s

	Mean [ml/s]	Variance [ml/s]
V1/t1	0.03089	1.69e-9
V2/t2	0.03091	4.28e-9
(V1+V2)/(t1+t2)	0.03090	1.58e-9
V3/t3	0.03089	3.53e-9
(V1+V2+V3)/(t1+t2+t3)	0.03090	1.21e-9

3.5 Volcanic Ash Discrimination

Basaltic volcanic ash particles (as in the case of Etna volcano)havea paramagnetic behavior. Such intrinsic characteristichas been exploited to confers selectivity features to the monitoring system [12]. To this purpose, a small low-power digital magnetometer MAG3110 by Freescale Semiconductor with a sensitivity of 0.1 μT has been used. The devicecontains a magnetic transducerand a ASIC managing a digital I2C communications. Aschematization of the adopted sensing methodology is shown in Figure 15. A permanent magnet has been used for biasing the output of the magnetometer.The presence of volcanic ash causes a variation of the magnetometer output signal. Two examples,in case of 0.6 ml and 0.8 ml of volcanic ash, are shown in Figure16. An experimental setup has been developed to investigate the best relative positions of the magnetometer and the biasing permanent magnet as respect to the tank, assuring the best performance of the device especially in terms of responsivity. To this purpose, taking into account the setup schematization in Figure 15, experiments have been performed by observing the sensor response as a function of distances $\delta1$ and $\delta2$ between the magnetometer and the tank and between the tank and the permanent magnet, respectively.Moreover, the device behavior has been investigated for different quantity of collected ash. As an example, the observed variations of the magnetic field ΔH for different values of distances $\delta1$ and $\delta2$, due to the presence of 0.6 ml and 0.8 ml of volcanic ash in the tank, respectively, are reported in Table 3. As it can be observed, the device responsivity increases for decreasing values of $\delta2$ and $\delta1$. For distances $\delta1$lower than 4 cm the magnetometer resulted blinded. On the basis of obtained results, distances $\delta1$ and $\delta2$, for the developed prototype, have been fixed to 4 cm and 0 cm, respectively.

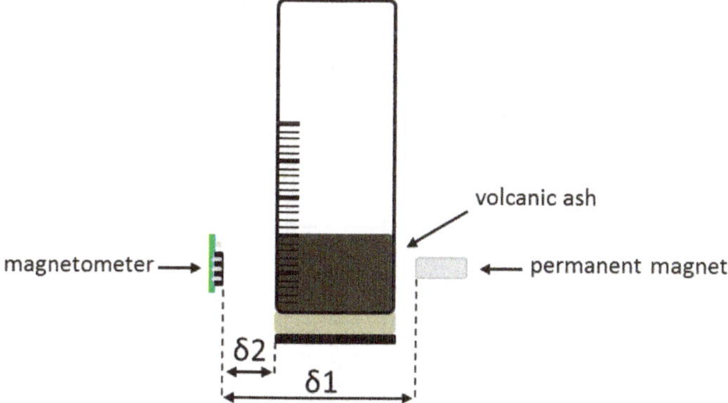

Fig. 15 Schematization of the methodology adopted to confer ash selectivity to the measurement system

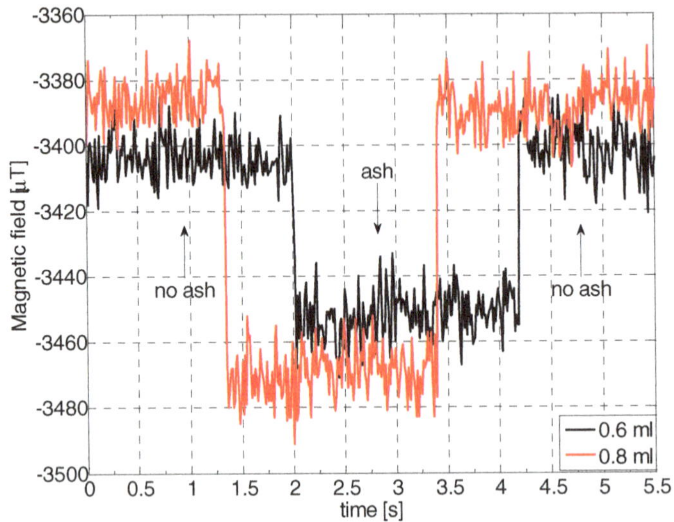

Fig. 16 Signals provided by the magnetometer in case of 0.6 ml and 0.8 ml of volcanic ash. The presence of the volcanic ash produces a visible increasing of the signal level

Table 3 Observed variations of the magnetic field, ΔH, due to the presence of 0.6 ml and 0.8 ml of volcanic ash in the tank, for different values of distances $\delta 1$ and $\delta 2$

0.6 ml					
ΔH [µH]		$\delta 1$ [cm]			
		4	6	9	10
$\delta 2$ [cm]	0	80	80	80	80
	1	20	20	15	15
	2	15	10	5	5
	3	0	0	0	0
	4	0	0	0	0
	5	0	0	0	0
	6	0	0	0	0
0.8 ml					
ΔH [µH]		$\delta 1$ [cm]			
		4	6	9	10
$\delta 2$ [cm]	0	95	50	35	55
	1	20	10	10	10
	2	10	0	0	0
	3	0	0	0	0
	4	0	0	0	0
	5	0	0	0	0
	6	0	0	0	0

Experiments with sand and water have been also performed with the aim to confirm the system selectivity against exogenous sediments. Experiments consisted in exposing, consecutively, the same quantity of ash, sand and water between the magnetometer and the biasing permanent magnet. Conversely to the effect of volcanic ash which produces a visible variation of the magnetometer output, no variations are observed for other sediments, thus confirming the selectivity of the system as respect to exogenous sediments.

3.6   Conclusions

In this chapter a low-cost smart multisensor system for the monitoring of ash fall-out phenomenon developed at the DIEEI of the University of Catania, under the SECESTA project, is presented. In particular, the methodologies adopted for ash granulometry detection, ash flow-rate estimation and its discrimination from other sediments have been discussed and supported by experimental results. The main idea for ash granulometry classification is to use a piezoelectric transducer to convert ash impacts into electrical signals which convey information about the ash granulometry. Experimental results showing the capability of the proposed approach are presented. ROC curve analysis has been used as a theoretical support to suitably fix thresholds for the ash granulometry classification approach.

The ash flow-rate is estimated by an array of coupled infrared (IR) diodes - phototransistors used to measure the time required to fill a known volume of a tank. A novel approach, exploiting a magnetic sensor,is used to implement the system selectivity to volcanic ash. Obtained experimental results confirm the suitability of the proposed sensing methodologies both in terms of flow-rate estimation and selectivity against exogenous sediments.

Main claims and novelties of the methodology proposed consist in the low-cost sensing strategies, as respect to traditional instrumentation, the multi-sensor approach for the monitoring of ash fall out phenomena and the selectivity as respect to volcanic ash.

Above features of the developed multi-sensor node will enable the implementation of a early warning sensor network for a real time monitoring of ash fall-out phenomenon with a high spatial resolution.Such amount of information is strategic for ash dispersion forecasting.

Acknowledgment. This work has been developed under the SECESTA project of the POR FESR Sicilia 2007-2013, (CUP: G53F11000040004). SECESTA is the Italian acronym for "A sensor network for the monitoring of volcanic ash fall-out for the safety of air transport".

Authors wish to thanks, for the precious collaboration, Dr. M. Coltelli of the INGV, Catania-Italy.

References

[1] Mayer, W.H.: The mitigation of ash fall damage to public facilities: Lessons learned from the 1980 eruption of Mt. St. Helens, Washington Federal Emergency Management Agency, Region X (1984)

[2] Neild, J., O'Flaherty, P., Hedley, P., Underwood, R., Johnston, D., Christenson, B., Brown, P.: Agriculture recovery from a volcanic eruption, MAF Technical Paper 99/2 (1998)

[3] Horwell, C., Baxter, P.: The health hazards of volcanic ash - A guide for the public, `http://www.ivhhn.org/pamphlets.html`

[4] Flight Operations Briefing Notes - Operating Environment - Volcanic Ash Awareness, Rev.01 (September 2006), `http://www.airbus.com`

[5] ICAO, Management of flight operations with known or forecast volcanic cloud contamination, Guidance Material, ICAO Document, vers. 3.1 (December 2010)

[6] ENAC, APT 15 - Operazioni volo su aeroporti in presenza di nube di cenere vulcanica and annex Doc. APT-ETNA ed. 1, ENAC Circular (July 23, 2003) (in Italian)

[7] Scollo, S., Prestifilippo, M., Spata, G., D'Agostino, M., Coltelli, M.: Monitoring and forecasting Etna volcanicplumes. Nat. Hazards Earth Syst. Sci. 9, 1573–1585 (2009), doi:10.5194/nhess-9-1573-2009

[8] Marchese, F., Corrado, R., Genzano, N., Mazzeo, G., Paciello, R., Pergola, N., Tramutoli, V.: Assessment of the robust satellite technique (RST) for volcanic ash plume identification and tracking. In: Second Workshop on the Use of Remote Sensing Techniques for Monitoring Volcanoes and Seismogenic Areas (USEReST 2008), pp. 1–5 (2008)

[9] Marzano, F.S., Picciotti, E., Vulpiani, G., Montopoli, M.: Synthetic signatures of volcanic ash cloud particles from X-band dual-polarization radar. IEEE Trans. Geosci. Remote Sens. 50(1), 193–211 (2012)

[10] Prata, A.J., Bernardo, C.: Retrieval of volcanic ash particle size, mass and optical depth from a ground-based thermal infrared camera. J. Volcanol. Geotherm. Res. 186, 91–107 (2009)

[11] Andò, B., Baglio, S., Marletta, V.: A smart multisensor system for the ash fall-out monitoring. In: Proceedings of the 26th European Conference on Solid-State Transducers (Eurosensors 2012), Krakow, Poland, September 9-12 (2012)

[12] Andò, B., Baglio, S., Marletta, V., Medico, S.: A Smart Multisensor System for Ash Fall-Out Monitoring. Sensors and Actuators A: Physical 212, 13–22, doi:10.1016/j.sna.2013.03.027.

[13] Andò, B., Baglio, S., Marletta, V.: Selective Measurement of Volcanic Ash Flowrate. In: Proceedings of the IEEE International Conference on Instrumentation and Measurement, I2MTC 2013, Minneapolis, MN, USA, May 6-9, pp. 1367–1371 (2013), doi:10.1109/I2MTC.2013.6555637

[14] Andò, B., Baglio, S., Marletta, V.: Selective Measurement of Volcanic Ash Flowrate. IEEE Transactions on Instrumentation and Measurement (in press)

[15] Ando, B., Baglio, S., Pitrone, N., Trigona, C., Bulsara, A.R., In, V., Coltelli, M., Scollo, S.: A novel measurement strategy for volcanic ash fallout estimation based on RTD Fluxgate magnetometers. In: IEEE Instrumentation and Measurement Technology Conference Proceedings (I2MTC 2008), pp. 1904–1907 (2008)

[16] Stronge, W.J.: Impacts mechanics. Cambridge University Press (2000)

[17] Gao, L., Yan, Y., Lu, G.: On-line measurement of particle size distribution using piezoelectric sensors. In: IEEE International Instrumentation and Measurement Technology Conference (I2MTC), pp. 1154–1158 (May 2012)

[18] Ando, B., Coltelli, M., Prestifilippo, M., Scollo, S.: A lab-scale experiment to measure terminal velocity of volcanic ash. IEEE Trans. Instrum. Meas. 60(4), 1340–1347 (2011)

[19] Measurement Specialties Inc., Piezo film sensors technical manual, Measurement Specialties Inc. (1999), http://www.meas-spec.com

[20] Lagarias, J.C., Reeds, J.A., Wright, M.H., Wright, P.E.: Convergence properties of the Nelder–Mead simplex method in low dimensions. SIAM J. Optim. 9(1), 112–147 (1998)

[21] Fawcett, T.: ROC graphs: Notes and practical considerations for data mining researchers, HP Technical Report HPL-2003-4, HP Laboratories (2003)

[22] Slaby, A.: ROC Analysis with Matlab. In: 29th International Conference on Information Technology Interfaces (ITI 2007), pp. 191–196 (2007)

[23] Fawcett, T.: An introduction to ROC analysis. Pattern Recognition Letters 27, 861–874 (2006)

[24] Hand, D., Till, R.J.: A simple generalisation of the area under the ROC curve for multiple class classification problems. Mach. Learn. 45, 171–186 (2001)

[25] Qin, Z.-C.: ROC analysis for predictions made by probabilistic classifiers. In: Proceedings of the Fourth International Conference on Machine Learning and Cybernetics, Guangzhou, August 18-21 (2005)

Chapter 4
Portable High Frequency Surface Wave Radar OSMAR-S

Hao Zhou and Biyang Wen

Abstract. The Ocean State Monitoring and Analyzing Radar (OSMAR-S) is a portable high frequency surface wave radar (HFSWR) adopting the compact crossed-loop/monopole antenna. It is designed and used for monitoring of sea surface currents, waves and winds. This chapter mainly describes the principles and processing steps of sea state extraction with such a small-aperture radar. Mapping of radial current velocities and wind directions is achieved by a direction finding process, whereas the wave height is extracted by an empirical method after the beamforming. OSMAR-S is has a function of automatic frequency selection (AFS) and interference suppression, which makes it an intelligent instrument in the complex electromagnetic environment. Some comparison experiments are demonstrated to show the good performances of OSMAR-S in sea state monitoring.

4.1 Introduction

The basic form of motion of the surface seawater can be regarded as sea wave fluctuations superimposed on the overall surface flows. The ocean current is a continuous, directed movement of seawater generated by the forces acting upon this mean flow, such as breaking waves, wind, Coriolis effect, cabbeling, temperature and salinity differences and tides caused by the gravitational pull of the Moon and the Sun [1]. The surface current is generally wind-driven. The ocean current is a very important factor for the global and regional climates. The knowledge on sea currents and waves is always needed in a wide range of human's applications from weather forecast, oceanic engineering such as the design and operational safety of harbors, ships, and offshore structures, to coastal management including coastal stability and pollution.

Hao Zhou · Biyang Wen
Wuhan University
Wuhan 430072, China

© Springer International Publishing Switzerland 2015
H. Leung and S.C. Mukhopadhyay (eds.), *Intelligent Environmental Sensing,*
Smart Sensors, Measurement and Instrumentation 13, DOI: 10.1007/978-3-319-12892-4_4

Because the sea surface currents and waves influence so many processes and operations at sea, many techniques have been invented for measuring sea states in different temporal and spatial scales. Moored or ship-towed buoys can provide direct and *in-situ* current and wave measurements with the highest accuracy and the smallest scales [2], X-band nautical radar can measure sea states within a few kilometers with some larger scales [3], while satellite-borne altimeter can provide measurements over an extremely large area with relatively low resolutions in both time and space [4]. High frequency (HF) surface wave radar (HFSWR) just fills the large gap of these means, which can provide low-cost sea state measurement by remote sensing with a time resolution of a few tens of minutes and a spatial resolution of several kilometers [5]. The vertically polarized HF radio wave (between 3 and 30 MHz) can diffract along the sea surface due to the good electrical conductivity of the seawater, so the HFSWR achieves a capability of over-the-horizon detection. For example, with average transmit power of 10 Watts, the HFSWR can detect sea currents up to 100 km at 13 MHz, 200 km at 7.5 MHz, and 350 km at 5 MHz. Due to the great advantages in sea state monitoring, HFSWR has been continuously receiving attention from lots of researchers [6-10] and been built and tested worldwide in various situations. Now it is widely accepted and also expected to play a more important role in the modern sea surveillance network, where different instruments based on different mechanisms work together to provide measurements that can meet the needs of various applications with a higher accuracy and confidential level.

When HF radio waves are transmitted towards the sea surface, the sea echo spectrum will show Doppler frequency shifts due to the motion of the seawater. The sea waves can be modeled as the summation of a series of sinusoid waves with different wavelengths, directions, amplitudes and initial phases. All the wave components will contribute to the echo energy, but only those whose wavelengths equal one half of the radio wavelength will produce a significantly strong spectrum. The echo signals reflected along the entire surface of the successive half-radio-wavelength sea waves are precisely in-phase cumulated to produce a resonance. This phenomenon is similar as the Bragg scattering of X-ray from a crystal lattice and was firstly explained by Crombie in 1955 [11]. The Doppler and the Bragg effects are the two most important physics mechanisms that enable the high frequency radar to be an efficient tool for sea state measurements. The wavelengths of the sea waves are typically between 60 and 150 meters, which just fall into the resonance region of the HF scattering and thus result in very strong echoes. By analysis of the Doppler spectra of the sea echoes, the sea surface dynamic parameters such as the sea surface current, wave height and period, and wind direction and speed can be extracted.

Radar is a device that can be used for perception of the range, velocity and bearing of a certain target. In the case of an oceanographic HFSWR, the target is the sea surface water, which is continuously distributed and quite different from the hard target like a ship. Since there are lots of waves undertaking different sea states in the illuminated sea area, the radar coverage region should be divided into small cells to identify the local sea state parameters individually. The size of the cell is defined as the radar's range resolution by the directional resolution. The range resolution is determined by the frequency bandwidth of the radar waveform, while the directional resolution is determined by the type and aperture of the receive antenna array. Restricted by the existing antenna theory and techniques, a whip antenna should be at least one-fourth of the working wavelength to gain a high efficiency. Accordingly, the whip antenna at the HF band usually has a relatively large size. When multiple whip antennas are combined into a phased-controlled array for receiving, it is not surprising that the array aperture will be more than one hundred meters [12, 13], and sometimes it even extends to above one kilometer [14]. To reduce the demand for antenna field, loop antennas have been used and turned out to be a very efficient approach for antenna array miniaturization. The crossed-loop/monopole antenna has achieved great success in sea state measurement [15, 16]. The three antenna elements, say two crossed loops and one monopole, are compactly equipped on one vertical pole so that the array can be setup in almost all situations. Quite different from the phase-controlled array, this array is amplitude-controlled, which depends on the antenna patterns. Although this type of antenna array has a relatively pool ability of beam-forming (BF) and thus cannot achieve a sufficient spatial resolution in the wave height estimate, it really performs well on surface current mapping. The myth of its success is the use of the super-resolution direction finding (DF) algorithm. There are also other efforts to reduce the array size, such as incorporating loop antennas into the phased array [17, 18] and reducing the array aperture with fewer antennas [19, 20]. Because reduced array scale means fewer costs but lower spatial resolution, the selection of a suitable antenna array is almost always the most important task in the HFSWR application.

The Ocean State Monitoring and Analyzing Radar, type S, (OSMAR-S) is a portable HFSWR developed by Wuhan University, China in 2006. It adopts compact crossed-loop/monopole antenna following the CODAR system, as shown in Figure 1. This chapter mainly describes how it works to achieve the sea surface current and wave height measurements. Firstly the principles of the sea state measurements with an HFSWR are briefly described in Section 2. Then the surface current mapping process is described in Section 3 and the method to extract the wave height in Section 4. To account for the radio frequency interference (RFI) which frequently occurs, the operational automatic frequency selection (AFS) system and signal processing method is also described in Section 5. Finally some field experimental results are demonstrated in Section 6 to show the good performance of the radar.

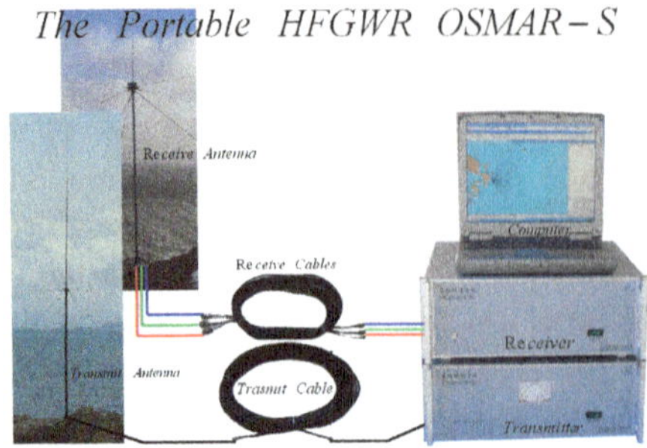

Fig. 1 An overview of the OSMAR-S system (left) and the vehicle-borne radar system (right)

4.2 Principle of Sea State Sensing

Due to Barrick's famous work on the first- and second-order radar cross-section (RCS) equations [21, 22], the quantitative explanation of the Doppler spectra and thus the inversions of the sea current and wave height become feasible. It has formed the solid basis of nearly all the operational oceanographic HFSWR sea state measurements.

4.2.1 Barrick's First-Order RCS Equation

According to the Bragg scattering effect, the sea wave of half the radar wavelength will produce a resonance peak on the Doppler spectral. The phase velocity of the sea wave is readily given by

$$V_p = \sqrt{\frac{gL}{2\pi}\tanh(\frac{2\pi h}{L})} \quad , \tag{1}$$

where $g = 9.8\text{ms}^{-2}$ is the gravitational acceleration, h is the depth of the water, and L is the wavelength of the sea wave. For simplicity hereinafter only the deep water condition, say $h > L/2$, is considered, thus (1) reduces to

$$V_p = \sqrt{\frac{gL}{2\pi}} . \tag{2}$$

Accordingly the Doppler shift of the Bragg peak is

$$f_B = \frac{2V_p}{\lambda} = \frac{2}{\lambda}\sqrt{\frac{gL}{2\pi}} = \sqrt{\frac{gf_0}{\pi c}} , \tag{3}$$

with λ being the radar wavelength, f_0 being the radar frequency and $c = 3 \times 10^8$ m/s being the speed of light.

It is noticeable that the sea waves are always propagating towards all directions. So, there are two Bragg peaks on both sides of the Doppler spectrum, i.e., the positive peak generated from the sea wave approaching the radar and the negative peak from that receding from the radar. When there is a certain sea surface current, both Bragg peaks will be shifted in the same direction corresponding to the radial velocity of the seawater, V_r. The overall Doppler shifts of the resonance peaks can be denoted as

$$f_{d,\pm} = \pm f_B + f_c,\tag{4}$$

where $f_c = 2V_r / \lambda$ is the Doppler shift due to the current velocity. Therefore, the radial current velocity can be estimated by calculating the shift of the Bragg peaks relative to the standard Bragg lines. If the radar has an ability of directional resolution, a radial current map can be finally plotted.

Barrick accomplished a detailed analysis of the radio wave scattering from the sea surface by the perturbation method. His first-order RCS equation further quantitatively relates the Bragg peak power to the sea wave height spectrum, which reads

$$\sigma^{(1)}(\omega) = 2^6 \pi k_0^4 \sum_{m=\pm 1} S(-2m\vec{k}_0)\delta(\omega - m\omega_B - \omega_c),\tag{5}$$

where $\omega = 2\pi f$ is the circular frequency, ω_B and ω_c are the circular frequency counterparts of f_B and f_c respectively, $k_0 = 2\pi / \lambda$ is the radar wavenumber, $S(\cdot)$ is the directional wave height spectrum and $\delta(\cdot)$ is the Dirac impulse function.

Although the definite amplitude information is not required in current mapping, it really contains much information on the wind direction. Since the Bragg wave belongs to the short wave, it will be easily blown to the direction in accordance with the wind. For a fully-developed sea surface dominated by wind-driven sea waves, the wave height spectrum $S(\vec{k})$ can be modeled as the product of the non-directional wave height spectrum $F(k)$ and the directional spreading factor $g(\theta)$ [23], i.e.,

$$S(\vec{k}) = F(k)g(\theta)\tag{6}$$

where $\vec{k}$ is the sea wave vector which can also be written in the polar form (k, θ) with k being the wavenumber. The positive- and negative-side Bragg peak pair gives two samples of the wave spectrum, and the term of non-directional wave height spectrum can be eliminated by evaluating the positive-to-negative Bragg peak ratio. By assuming a certain model for the directional spreading function, the wind direction can be estimated up to a left-right ambiguity due to the symmetry of the function. For example, a cardiod angular distribution of wave energy [23] is given by

$$g(\theta) = A\cos^s(\frac{\theta - \theta_w}{2}) \tag{7}$$

where θ_w is the direction of maximum wave energy which is also assumed to keep the same as the wind direction, s is an even-integer spreading parameter, and A is a constant required for the normalization $\int_{-\pi}^{\pi} g(\theta)d\theta = 1$ for different values of s. Corresponding to the approaching and receding Bragg waves in the direction of the radar beam, θ, the Bragg ratio [24] is thus given by

$$R_B = \frac{\sigma^{(1)}(\omega_B + \omega_c)}{\sigma^{(1)}(-\omega_B + \omega_c)} = \frac{\cos^s[(\theta - \theta_w + \pi)/2]}{\cos^s[(\theta - \theta_w)/2]} = \tan^s[(\theta - \theta_w)/2]. \tag{8}$$

Therefore the wind direction can be estimated by

$$\theta_w = \theta \pm 2\arctan^s(R_B). \tag{9}$$

The ambiguity may be finally eliminated by minimizing the overall error of the estimated wind field from multiple beams.

4.2.2   Barrick's Second-Order RCS Equation

Barrick further derived an equation for the second-order Bragg scattering from the sea surface. There are two main coupling effects: one is the electromagnetic (EM) double scattering by two waves which propagate in perpendicular directions; and the other is the hydrodynamic scattering where two first-order sea waves interact to produce a second-order wave satisfying the Bragg condition. In both cases, backscatters will occur only if the two sea waves, with wave vectors $\vec{k}$ and $\vec{k}'$, respectively, satisfy the relation

$$\vec{k} + \vec{k}' = -2\vec{k}_0, \tag{10}$$

where $\vec{k}_0$ is the incident radar wave vector in the direction of the narrow radar beam. The second-order RCS equation is finally given by

$$\sigma^{(2)}(\omega) = 2^4\pi k_0^4 \sum_{m,m'=\pm 1} \int_0^{2\pi} \int_{-\infty}^{\infty} |\Gamma|^2 S(m\vec{k})S(m'\vec{k}')\delta(\omega - m\sqrt{gk} - m'\sqrt{gk'})kdkd\theta, \tag{11}$$

where Γ is the coupling coefficient.

Lots of research work has been done on the inversion of this equation [25-28]. Barrick also proposed an efficient method to calculate the significant wave height via the ratio of the overall power in the second-order spectral region to that in the first-order region [29], i.e.

$$r_0 = \frac{\int_{\Omega_2} \sigma^{(2)}(\omega)/W(\omega)\mathrm{d}\omega}{\int_{\Omega_1} \sigma^{(1)}(\omega)\mathrm{d}\omega} \tag{12}$$

$$H_{S,0} = \alpha_0 \frac{4\sqrt{2}}{k_0}(r_0)^{0.5} \tag{13}$$

where W is a dimensionless weighting function to eliminate the coupling coefficient of the double scatter, Ω_1 and Ω_2 respectively indicate the first- and second-order spectral regions, and α_0 is a correction factor introduced to correct for biases in the estimates because of simplifications in the theory and account for a weak dependence of W on the radar look angle that was not fully removed during the derivation process. The mean wave period is estimated by

$$T_p = 2\pi \frac{\int_{\Omega_2} \sigma^{(2)}(\omega)/W(\omega)\mathrm{d}\omega}{\int_{\Omega_2} |\omega - \omega_B| \sigma^{(2)}(\omega)/W(\omega)\mathrm{d}\omega}. \tag{14}$$

This method needs no *a-priori* knowledge of the sea state and the calculation is not complicated, so till now it has been a prevailing method used in oceanographic radars.

To further simplify the calculation, other formulations based on the ratio of the un-weighted second-order to first-order spectral power are also used [30-32]. For example, the wave height can be estimated empirically by

$$r = \frac{\int_{\Omega_2} \sigma^{(2)}(\omega)\mathrm{d}\omega}{\int_{\Omega_1} \sigma^{(1)}(\omega)\mathrm{d}\omega} \tag{15}$$

$$H_S = \alpha \cdot r^{0.5}. \tag{16}$$

where α is the regression coefficient to be determined by fitting between the spectral ratio and the field buoy wave data. Although some losses of accuracy may result in certain directions, e.g. the crosswind directions, these empirical methods do perform well in a wide variety of situations.

4.3 Current Mapping in OSMAR-S

Mapping of the total vector current field relies on radial current maps from two or more properly configured radar sites. The sea region of interest is separated to many small cells, and for every given sea cell, the radars can give radial current velocity projections in different directions so that the final vector velocity can be estimated by the geometric relations.

4.3.1 Radial Current Mapping

For radar capable of forming narrow receive beams, the radial current mapping process is relatively easy. The main task is just to locate the Bragg peak and read out its Doppler shift relative to the standard Bragg line, and thus the radial current velocity, after the beamforming (BF) on each range cells. However, for radar with a broad receive beam like OSMAR-S, the process is much more complex. Direction finding (DF) is executed instead for all the first-order, or Bragg, spectral points respectively. In most situations, the sea surface current field has a simple structure so that there are only one or two directions with the same radial velocities, making it feasible to use the extremely compact crossed-loop/monopole antenna for current extraction. Generally, there are four steps in the radial current mapping of OSMAR-S.

Step 1 - Selecting valid spectral points

The upper limit of the radial current velocity, $v_{\max}$, is often preset according to the local historical oceanographic data. The Doppler spectra of the monopole within this range around the standard Bragg line are compared with a signal power threshold to screen out the valid spectral points. The signal power threshold is calculated by a preset signal-to-noise power ratio (SNR) threshold (in dB) plus the noise level, which is estimated by the average power of the spectra beyond twice the Bragg frequency. This selection process is shown in Figure 2. Both spectral sides are included to account for the randomness of scattering.

Step 2 - Forming the snapshots

For each range cell on the antenna element indexed by $i(i = 1,2,3)$, the short-time Fourier transform (STFT) [33], $F_i(t, f)$, of the time series are iscalculated with a moving Hamming window. Then the profiles of the STFT at a given frequency f constitute the 3-by-N_f snapshot matrix $X(f) = \left[X_1^T(f), X_2^T(f), X_3^T(f) \right]^T$, with $X_i(f) = [F_i(1, f), \cdots, F_i(N_f, f)]^T$ being the STFT profile vector, N_f being the number of temporal samples the STFT and the superscript ^{T} being the transpose operator.

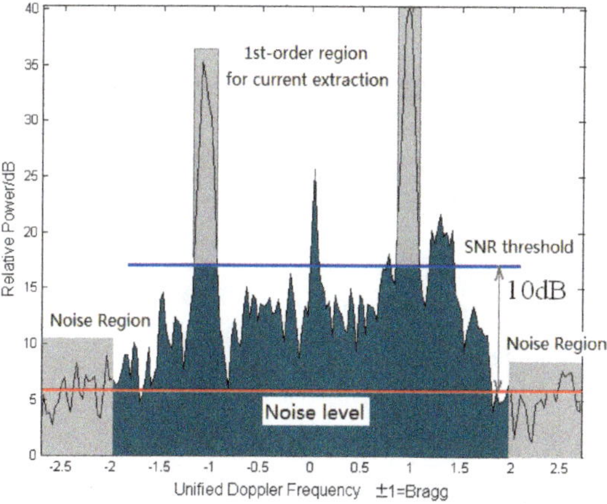

Fig. 2 Example of location of the first-order spectral region for current extraction

Step 3 - Calculating the directions of arrival (DOAs)

The MUltiple SIgnal Classification (MUSIC) algorithm [34] is used here. The correlation matrix of the array snapshots is given by $\boldsymbol{R}_{XX} = \dfrac{1}{N_f} \boldsymbol{X}(f)\boldsymbol{X}^H(f)$ with the superscript H being the conjugate transpose operator. Then the eigenvalue decomposition (ED) is performed on it, which leads to $\boldsymbol{R}_{XX} = \boldsymbol{U}\boldsymbol{S}\boldsymbol{U}^H$, with $\boldsymbol{S} = \mathrm{diag}[\lambda_1, \lambda_2, \lambda_3]\,(\lambda_1 \geq \lambda_2 \geq \lambda_3)$ containing the eigenvalues and $\boldsymbol{U} = [\boldsymbol{U}_1, \boldsymbol{U}_2, \boldsymbol{U}_3]$ containing the eigenvectors. The number of incident sources, K, is restricted to be either one or two and determined by testing the condition $\dfrac{\lambda_1}{\lambda_2} > \dfrac{\lambda_2}{\lambda_3}$. If it is true, it is decided $K = 1$, otherwise $K = 2$. The MUSIC spatial spectrum is given by

$$P_m(\theta) = \frac{1}{\displaystyle\sum_{i=K+1}^{3}\left|\boldsymbol{U}_i^H \boldsymbol{a}(\theta)\right|^2}, \tag{17}$$

where $\boldsymbol{a}(\theta)$ is the antenna pattern matrix given by

$$\boldsymbol{a}(\theta) = [\cos(\theta + \frac{\pi}{4}), \sin(\theta + \frac{\pi}{4}), 1]. \tag{18}$$

with the zero direction pointing to the bisector of the two loop mainlobe directions, as shown in Figure 3. The DOAs are thus estimated by seeking the spectral maxima. Consequently, a complete set of velocity/direction pairs is obtained.

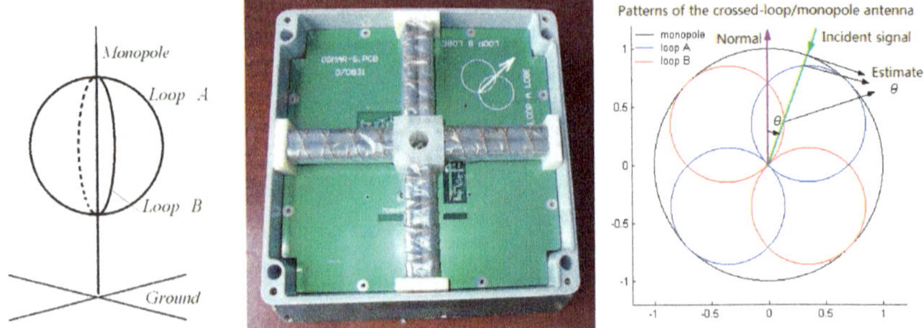

Fig. 3 Sketch map of the crossed-loop/monopole antenna (left), electronic board of the crossed-loop (center) and the directional patterns (right)

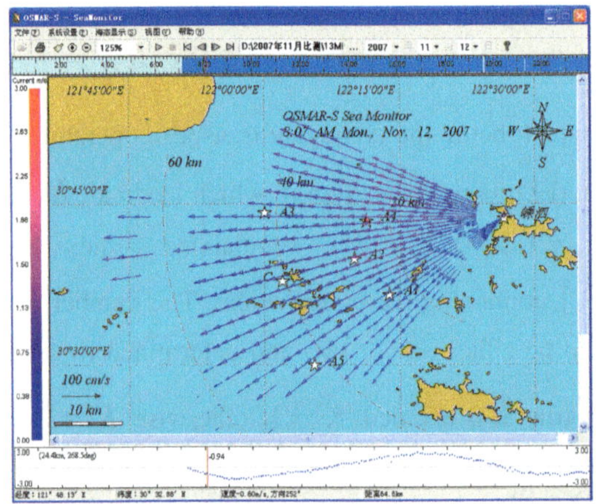

Fig. 4 Example of radial current velocity map by OSMAR-S

Step 4 - Forming the radial current map

A raw radial current velocity map is given by calculating the weighted or unweighted averages of the velocities falling on each radial angular grid point. There are often some vacancies on some grid points due to the randomness of scattering, so an extra interpolation is needed to give the complete radial current map. One frame of radar data leads to one radial current map, as shown in Figure 4.

Step 5 – Smoothing

Several maps within a given time period, e.g. half an hour, are combined together to give the final radial current map for a better performance.

4.3.2 Wind Direction Mapping

The wind direction can be achieved at the same time and in a similar manner as the current field mapping. In the direction finding process of the first-order spectral points, the positive-to-negative Bragg ratios (referring to (8)) are also recorded. The single-DOA spectral points are screened out and then the corresponding Bragg ratios are translated to the wind directions according to (9) to form a sparse wind direction field. Note that there are left-right ambiguities about the radial directions. Assuming a smooth and slowly-variant wind field, this ambiguity can be finally eliminated by seeking an averaging wind direction, $\theta_{w,0}$, that minimizes the total error in a given range of radial angles,

$$\theta_{w,0} = \arg\min_{\alpha} \sum_{\theta} \left| \min\{\varepsilon[\alpha,\theta_{w,+}(\theta)], \varepsilon[\alpha,\theta_{w,-}(\theta)]\} \right|^2 , \qquad (19)$$

where $\varepsilon[\alpha,\beta]$ denotes the acute angle between α and β, $\theta_{w,+}(\theta)$ and $\theta_{w,-}(\theta)$ are the two ambiguous wind directions given by (9) at radial direction θ, respectively. The final wind directions on the radial grid are then determined to be the ones closer to $\theta_{w,0}$.

4.3.3 Total Current Vector Mapping

At each time grid point, when two or more radial current maps from different radar sites are available, total vector current field can be synthesized. The spatial grid for vector current field usually consists of latitude and longitude lines, which does not coincide with the radial girds at all. An efficient method to estimate the total vector velocities is setting a circle around each vector grid point and then performing the least square method on all the radial velocities falling in it, as shown in Figure 5.

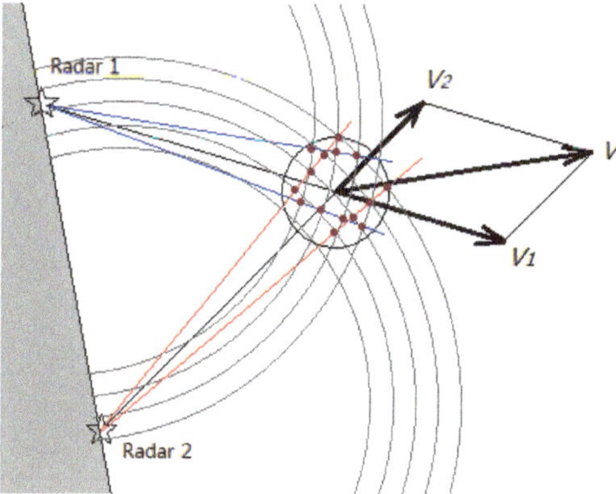

Fig. 5 Radial velocities falling inside the seeking circle are used to estimate the vector velocity on the origin of the circle

The geometric relation between the vector velocity and the radial projections can be denoted as

$$-\begin{bmatrix} v_1 \\ \vdots \\ v_Q \end{bmatrix} = \begin{bmatrix} \sin\theta_1 & \cos\theta_1 \\ \vdots & \vdots \\ \sin\theta_Q & \cos\theta_Q \end{bmatrix} \begin{bmatrix} v_x \\ v_y \end{bmatrix}, \tag{20}$$

where v_i and θ_i $(i=1,\cdots,Q)$ are respectively the radial velocities and directions falling in the circle, v_x and v_y are respectively the horizontal and vertical components of the vector velocity. The above equation can be rewritten in matrix form by

$$v_r = -bv \tag{21}$$

where $v_r = [v_1,\cdots,v_Q]^T$, $b = [\sin\theta \quad \cos\theta]$, $\theta = [\theta_1,\cdots,\theta_Q]^T$, and $v = [v_x \quad v_y]^T$. The least square solution of the total vector velocity is given by

$$v = -(b^T b)^{-1} b^T v_r. \tag{22}$$

The vector velocity can also be given in the polar form, i.e., the velocity is $V = |v| = \sqrt{v_x^2 + v_y^2}$ and the current direction is $\alpha = \mathrm{atan2}(v_y, v_x)$.

After the vector current field is extracted, a further two-dimensional median filtering is executed to correct some abnormal values and thus to guarantee a better performance, as shown in Figure 6.

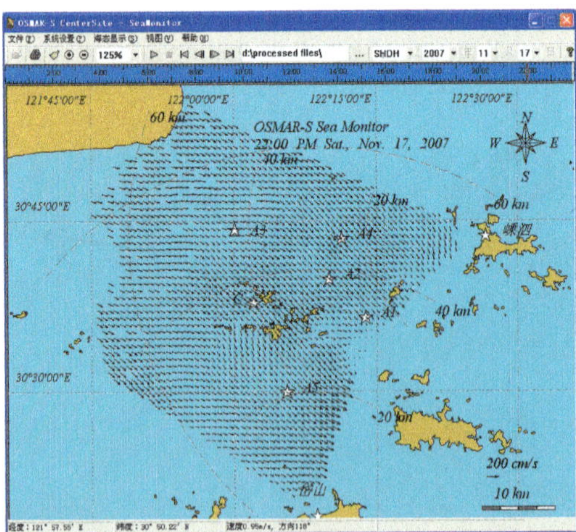

Fig. 6 Example of vector current map by OSMAR-S

4.4   Wave Height Estiamtion

OSMAR-S can provide high-quality sea echo spectra, usually having a more than 40dB SNR for the Bragg peaks at near ranges, which is shown in Figure 7. However, the large antenna beamwidth is an evident disadvantage for wave extraction. Spread and aliasing of the characteristic first- and second-order spectra occur frequently, as can be seen from Figure 7.

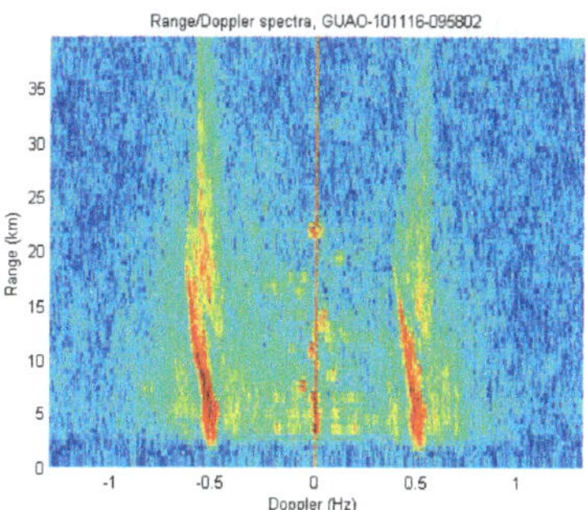

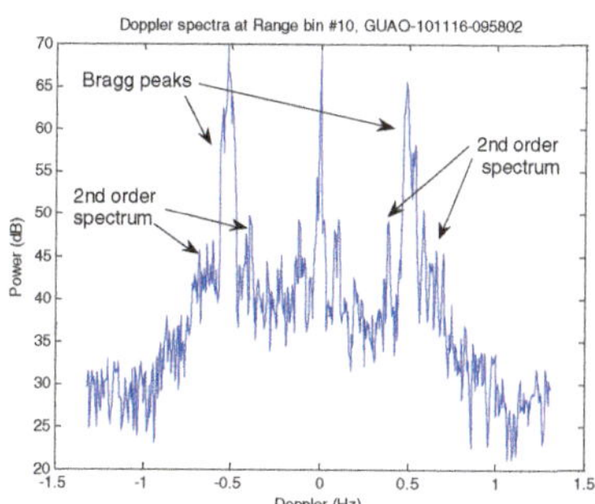

Fig. 7 Range/Doppler spectrum (left) and Doppler spectrum at the 10th range cell (right) recorded by OSMAR-S at 09:58 Nov 16, 2010. The radar frequency is 25MHz. The size of range cell is 500 m.

4.4.1   Beamforming and Power Spectral Estimation

Beamforming is a necessary step before wave extraction to provide a receive beam as narrow as possible. Specific weights are assigned to each antenna element and then the output signals from the antenna array are combined to form a beam with specific mainlobe and sidelobes.

4.4.1.1   Conventional Beamforming

The elemental patterns of the crossed-loop/monopole antenna are given in (18). The two loops have an equal half-power beamwidth, i.e., $\theta_l = \dfrac{\pi}{2}$. When the beam is directed to a direction of α the conventional beamforming (CBF) weights are given by

$$w_1 = \cos(\alpha + \frac{\pi}{4}), \ w_2 = \sin(\alpha + \frac{\pi}{4}), \ w_3 = 1 \qquad (23)$$

The synthesized beam pattern in the modulus sense by the CBF is

$$g(\theta) = \sum_{k=1}^{3} w_k a_k(\theta) = \cos(\alpha + \frac{\pi}{4})\cos(\theta + \frac{\pi}{4}) + \sin(\alpha + \frac{\pi}{4})\sin(\theta + \frac{\pi}{4}) + 1$$

$$= 1 + \cos(\theta - \alpha) = 2\cos^2(\frac{\theta - \alpha}{2}) \tag{24}$$

The CBF enables the beam pattern to be steered to any direction with a fixed beamwidth, $\theta_c = 2\arccos(\sqrt{2} - 1)(\text{rad}) \approx 131(\text{deg})$. The advantage of the CBF mainly lies in easy control of the main-lobe direction and suppression of the symmetric tail lobes. However, the directional resolution is quite low for wave extraction. In fact, the synthesized beamwidth by the CBF according to (24) is much larger than that of a single loop.

Assume the signals on the elements are $s_k(t)(k = 1,2,3, t = 0, \cdots, N - 1)$ with N being the number of the temporal samples, then the weighted sum of the array is given by

$$x(t) = \sum_{k=1}^{3} w_k s_k(t) \tag{25}$$

whose power spectrum, $P(f)$, is finally used for wave extraction. The Welch method is used to calculate the power spectrum [35]. The original sequence is divided into K subsequences of M points with an overlap of L points, which are denoted by $x_k(t)(k = 1, \cdots, K, t = 0, \cdots, M - 1)$. Then they are Fourier transformed and the periodograms, i.e., the squares of the moduli of the Fourier spectra $X_k(f)$, are averaged over all the subsequences, which leads to

$$P_{XX}(f) = \frac{1}{K} \sum_{k=1}^{K} |X_k(f)|^2 . \tag{26}$$

4.4.1.2 Improved Beamforming

To decrease the beamwidth while maintaining the electric scanning ability of the antenna, an alternative method is used. Firstly a two-sided cosine beam directed to α by combination of the signals on the loops is formed, that is,

$$y(t) = \sum_{k=1}^{2} w_k s_k(t), \tag{27}$$

with w_1 and w_2 are given by (22). Then the modulus of the cross-spectrum of $y(t)$ and $x(t)$ is used in place of the power spectrum, $P_{XX}(f)$, for wave extraction. In detail, $y(t)$ is divided into K subsequences in the same manner as $x(t)$, and their Fourier transforms are denoted as $Y_k(f)$. The cross-spectrum is thus given by

$$P_{XY}(f) = \frac{1}{K}\sum_{k=1}^{K}\left|X_k^*(f)Y_k(f)\right|, \tag{28}$$

with the superscript $*$ denoting complex conjugate. The corresponding synthesized beam pattern in the power sense is

$$G(\theta) = g(\theta)\cdot\sum_{k=1}^{2}w_k a_k(\theta) = [1+\cos(\theta-\alpha)]\cos(\theta-\alpha), \tag{29}$$

and the half-power beamwidth is $\theta_n = 2\arccos(\frac{\sqrt{5}-1}{2})(\text{rad}) \approx 104(\text{deg})$. Obviously there satisfies $\theta_n < \theta_c$.

Particularly, in the case that the look angles are within a range of less than 180 degrees, e.g. in a gulf or on the coast with a straight coastline, the monopole can be abandoned and (27) can be directly used to form a beam as a rotatable loop. The beam pattern in the modulus sense is given by

$$g_l(\theta) = \sum_{k=1}^{2}w_k a_k(\theta) = \cos\alpha\cos\theta + \sin\alpha\sin\theta = \cos(\theta-\alpha) \tag{30}$$

Although there exists strong back lobe (equal to the front lobe), the absence of sea echoes in the back directions guarantees the efficiency of such a method.

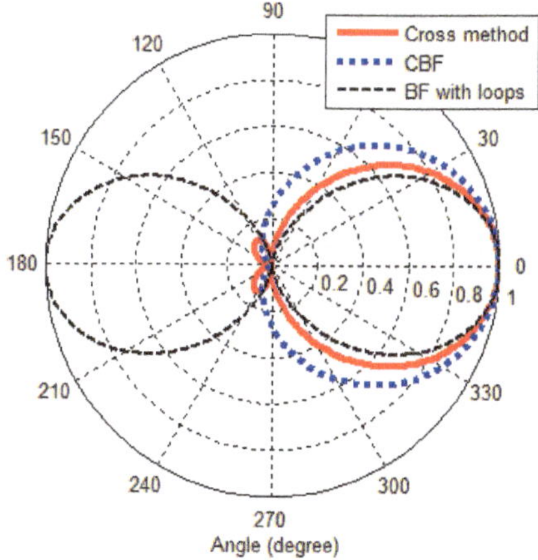

Fig. 8 Synthesized patterns in the power sense

The synthesized patterns mentioned above are shown in Figure 8. As can be seen, the cross-spectrum method can effectively improve the directional pattern at a price of slightly increased sidelobes, which generally have a smaller impact on the signal of interest than the large beamwidth of the mainlobe. The pattern synthesized with the loops only has the smallest beamwidth. So, when the sea echoes arrive from directions within a range of more than 180 degree, the cross-spectrum method is preferred, while the loop beamforming can give the best performance for a look angle constrained within a range of less than 180 degree.

4.4.2 *Wave Extraction*

4.4.2.1 Locating Second-Order Region

Firstly the nulls between the first-order and second-order spectral regions are located according to the spectral shape. Due to the spatial merging and the randomness of the sea surface scattering, there are often spectral splits. So, a relatively conservative strategy is used in OSMAR-S to avoid introduction of large errors from incorrect null positions.

In OSMAR-S, the algorithm for locating the spectral regions on the side with a stronger Bragg peak is described as follows. For simplicity the positive Bragg peak is assumed to be stronger than the negative one without loss of generality.

Initialization: Set the maximum radial velocity of the surface currents $v_{\max}$, the maximum wave spectral shift f_w, the Bragg peak SNR threshold η_B and the null-to-peak power ratio threshold η_n;

Step 1: Estimate the averaged noise power P_n, and accordingly the Bragg peak threshold $\gamma_B = P_n \cdot \eta_B$;

Step 2: According to the maximum radial current velocity, calculate the candidate Doppler interval of the first-order (Bragg) region; Find the strongest spectral peak within the interval, which gives $\{P_{\max}, f_{\max}\}$ and the null power threshold $\gamma_n = P_{\max} / \eta_n$;

Step 3: Find the null below the threshold γ_n between the strongest Bragg peak and the half maximum wave spectral shift, which gives $f_n \in (f_{\max}, f_{\max} + f_w / 2)$;

Step 4: Continue to find one more lower null on the right side of the previous null if possible as in Step 3; If the peak value between the successive nulls exceeds the Bragg peak threshold γ_B, the interval between the nulls is included into the first-order region and the new null is determined to be the valid null;

Step 5: Repeat Step 4 once more; The right-side null is determined to be $f_{n,r}$ and the second-order region is $[f_{n,r}, f_{n,r} + f_w]$;

Step 6: Determine the left-side null $f_{n,l}$ in the range of $(f_{max} - f_w/2, f_{max})$ according to the steps similar to Step 3 to 5.

The additional evaluation of the neighbor nulls to the conventional method according to Step 4 and 5 can effectively avoid the split first-order peaks to be mistakenly included into the second-order region.

4.4.2.2 Wave Height Estimation

In OSMAR-S, an empirical method is used to estimate the wave height, which is composed of two components. Although both the positive and negative spectral sides can give a wave height estimate independently, only the side where the stronger first-order peak locates is used in OSMAR-S for a better performance.

One component is an estimate obtained according to (15) and (16), but a little differently. It can be learned from Barrick's second-order RCS equation (11) that, besides the second-order spectral continuum, there are also integral singularities, namely spectral peaks, at Doppler frequencies $\sqrt{2}f_B$ and $2^{3/4}f_B$. The $\sqrt{2}$ singularity is also called second-harmonic peak due to the electromagnetic (EM) and hydrodynamic second-order effects [36]. The EM component is from sea waves of length $L = \lambda$ and the hydrodynamic component is from a second spatial harmonic with length $L = \lambda$. The $2^{3/4}$ singularity is due to corner reflector EM effect, which occurs when two sets of first-order sea waves pass through $45°$ with respect to the propagation direction. These peaks have been accounted for in Barrick's method by setting special weights at their frequencies, and they should also be excluded in the empirical methods using the un-weighted second-order to first-order spectral ratio. In real radar data, the second harmonic peaks are found to be generally much stronger than the $2^{3/4}$ peak and thus have a greater impact on the wave height estimation. So, only the neighborhood of $\sqrt{2}f_B$ is excluded from the integration, which is a 3-point interval with the center locating at $(\sqrt{2}-1)f_B$ from the maximum of the first-order spectra. The corresponding wave height estimate is given by

$$r_1 = \frac{\int_{\Omega_2'} \sigma^{(2)}(\omega)\,d\omega}{\int_{\Omega_1} \sigma^{(1)}(\omega)\,d\omega} \tag{31}$$

$$H_{S,1} = \alpha_1 (r_1)^{0.5}. \tag{32}$$

with Ω_2' being the corrected second-order region and α_1 being a calibration factor to be determined by fitting to the in situ data.

The other component is an index related to the ratio of the maximum of the second-order spectra to the maximum of the first-order spectra on the same spectral side. The second-order RCS equation (11) is a result from the narrow-beam assumption, which no longer stands in radars using compact receive antenna arrays such as the crossed-loop/monopole antenna. Spatial merging will occur and lead to big errors in the wave height estimation. Since the first- and second-order spectra have quite different properties, i.e., the former is discrete and the latter is continuous, the superposition effects in these spectra after being Doppler shifted by the sea currents are also different. Moreover, as can be seen from (12) and (15), the locations of the first- and second-order spectral regions, say Ω_1 and Ω_2, have crucial impacts on the wave extraction, but there is often no clear nulls between them. If some first-order peaks are mistakenly included into the second-order spectrum, the wave height will be greatly overestimated, and vice versa. Moreover, because the second-order spectral continuums are relatively weak, even a medium interference or clutter can lead to a significant rise in the wave height estimate. To account for the susceptibility to the noise and difficulty to locate the spectral regions due to the low spatial resolution, incorporation of the maxima-related index into the estimation can help to some degree. In fact, there do exist a quantitative relation between the integration and the maximum of the wave height spectrum for some particular wave models. Here two most popular spectral models are discussed for example, i.e., the Pierson-Moskowitz (P-M) spectrum [37] for a fully developed sea and the JOint North Sea WAve Project (JONSWAP) spectrum [38] for a not fully developed sea. The JONSWAP spectrum is given by

$$S(\omega) = \frac{\alpha g^2}{\omega^5} \exp(-\beta \frac{\omega_p^4}{\omega^4}) \gamma^a \tag{33}$$

where $\alpha = 0.0081$, $g = 9.8 \text{m/s}^2$, $\beta = 1.25$, $a = \exp[-\frac{(\omega - \omega_p)^2}{2\omega_p^2 \sigma^2}]$ with

$$\sigma = \begin{cases} 0.07, \omega \le \omega_p \\ 0.09, \omega > \omega_p \end{cases}, \quad \omega_p = 2\pi f_p$$ is the peak circular frequency, and γ is a peak enhancement parameter between 1 and 7. When $\gamma = 1$, it reduces to the P-M spectrum.

The overall power of the P-M spectrum can be analytically calculated by

$$E = \int_0^\infty S(\omega) \, d\omega = \frac{\alpha g^2}{5\omega_p^4}. \tag{34}$$

And the peak value is given by

$$S(\omega_p) = \frac{\alpha g^2}{\omega_p^5} e^{-\beta}. \tag{35}$$

The ratio of the overall integration to the modified peak value (RIP) is

$$k_1 = \frac{E}{[S(\omega_p)]^{0.8}} = 0.2(\alpha g^2)^{0.2} e = 0.517. \tag{36}$$

The overall power of the JONSWAP spectrum has no analytical expressions available but can be calculated numerically. The RIP in the JONSWAP spectrum with a given γ is also nearly a constant

$$k(\gamma) = \frac{E(\gamma)}{[S(\omega_p)]^{0.8}}. \tag{37}$$

So, with the nearly definite relationship between the overall integration and the peak value of the JONSWAP wave spectrum, the semi-empirical index based on the ratio of the maximum of the second-order continuum to the Bragg peak power (RSCB) for wave height estimation can be denoted as

$$r_2 = \{\frac{\max[\sigma_2(\omega)]}{\max[\sigma_1(\omega)]}\}^{0.8} \tag{38}$$

$$H_{S,2} = \alpha_2(r_2)^{0.5}, \tag{39}$$

with α_2 being another calibration factor.

The two components are then combined together to give the wave height estimate. Each model by itself may be more susceptible to certain kinds of external noise. However, a combination or fusion of these estimates will effectively decrease the error. The combined wave height estimate can be denoted as

$$H_S = w_{h,1} H_{S,1} + w_{h,2} H_{S,2}, \tag{40}$$

where $w_{h,1}$ and $w_{h,2}$ are two positive weights satisfying $w_{h,1} + w_{h,2} = 1$. In OSMAR-S, they are set $w_{h,1} = 0.6$ and $w_{h,2} = 0.4$ to place a more emphasis on the integration-based component.

To make a further improvement, smoothing is performed on the wave height time series to give the final result every hour on the hour.

4.5   Automatic Frequency Selection and RFI Suppression

In HFSWR detection, the predominant noise is from the external. The radio frequency interference (RFI), which frequently occurs, is a major source of error in the sea state estimates. The RFI mainly results from the signals transmitted by communications, broadcasts or other radio systems. When the interference signal received is with a constant frequency in the radar's coherent integration time (CIT), it is regarded as a stationary RFI and will produce a strip parallel to the range axis on the range/Doppler spectrum. However, the Doppler effect due to movement of the sources (e.g. communication signal transmits by a moving ship), the phase shift due to the change of the propagation path (such as an ionosphere-reflected RFI), the incoherence of the time clocks between the interference and the radar, or the modulation by voice or music signals, which is the worst case, may all lead to a time-varying frequency resulting in a non-stationary RFI. The non-stationary RFI may have a Doppler spread up to one Hertz, thus it will form a much wider strip and mask much larger useful spectral regions. Due to the directionality of the RFI, the radial current map extracted will shown some wild values in certain angular sectors accordingly. And because the second-order spectral continuums are relatively weak, even a moderate interference can lead to a significant rise in the wave height estimate. So, the radar should have an anti-interference ability to make it more robust and reliable. In OSMAR-S, there are two measures to guarantee the signal qualities. One is the automatic frequency selection (AFS) system to sense the EM environment and select a quiet frequency band for the radar to work, and the other is the RFI suppression software module.

4.5.1   Automatic Frequency Selection (AFS) System

The waveform used in OSMAR-S is the frequency modulated interrupted continuous wave (FMICW) as shown in Figure 9. Each complete sweep period is broken into a number of time slots by the transmit/receive (T/R) pulse train. Between successive sweeps, there are small intervals conserved for the radar to finish the signal processing such as the FFT and get prepared for the next sweep. Just by use of this conserved interval, the radar is able to measure the external noise spectrum, accomplish a real-time statistic analysis, and adjust the radar's work frequency accordingly [39].

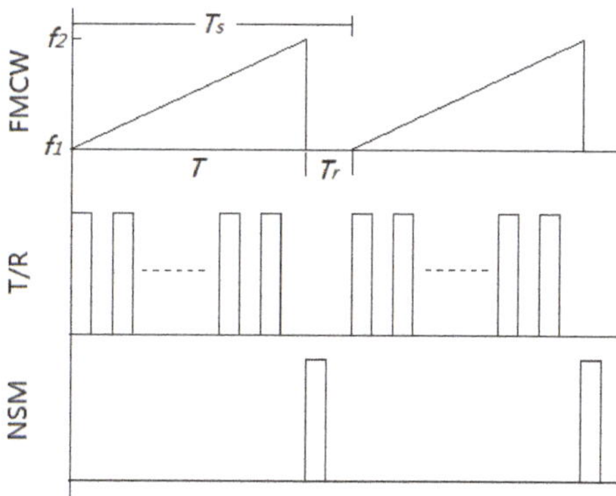

Fig. 9 Timing diagram of the receiver

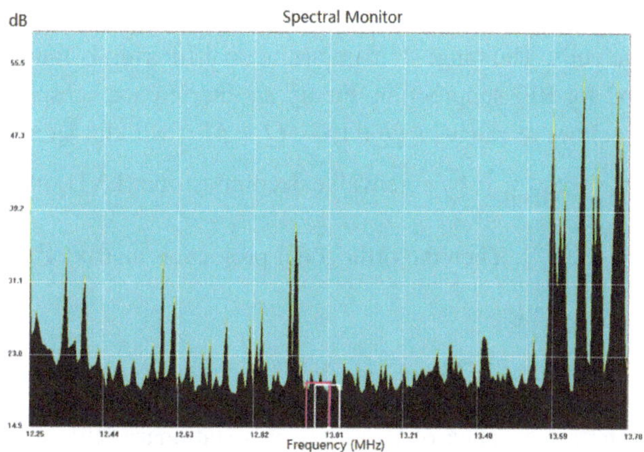

Fig. 10 Example of noise spectrum collected by OSMAR-S

Each radar frame period, T_s, can be divided into the sweep period, T, and the noise spectrum monitoring (NSM) period, T_r, with $T_s = T + T_r$. After each regular sweep period, a NSM pulse is triggered. During the noise monitoring period T_r, the radar transmitter doesn't work, while the local oscillator (LO) keeps running. The external noise is bandpass sampled, with the bandwidth determined by the frontend bandpass filter of the radar receiver. The frequency of the LO during T_r is changed every frame so that the noise spectrum in a rather wide frequency band can be achieved. The noise time series sampled in each frame is FFT processed to give a noise spectrum of a small bandwidth, which is accumulated to the existing noise spectrum. Sampling and accumulation in a number of frames are required to produce a complete noise spectrum over a wide frequency band, as shown in Figure 10.

After the noise spectrum is obtained, the most suitable frequency for the radar is calculated. If the noise level at this frequency is continuously lower than the current noise level and the difference exceeds a given threshold, the radar frequency will be changed. The AFS system has provided a low-level guarantee to enhance the radar's adaptability in complex EM environments.

4.5.2 RFI Suppression

Even equipped with the AFS system, the radar will inevitably encounter the RFI frequently. The RFI suppression is a necessary step in HFSWR to guarantee the signal quality.

The theoretical analysis of RFI in an HFSWR with a linear frequency-modulated continuous wave (FMCW) can be found in [40, 41]. It has been shown that, the RFI generally forms a stripe parallel to the range axis on the range/Doppler spectrum and the interference time series are highly correlated across the range cells. The range domain orthogonal filtering is found to be a very efficient method for RFI suppression. Firstly an interference subspace $\mathbf{V}$ is constructed using echoes at ranges ($Q+1,\cdots,Q+M$) without signals of interest, say $\mathbf{X} = [\mathbf{s}_{Q+1},\cdots,\mathbf{s}_{Q+M}]$, by eigenvalue decomposition (EVD) of its correlation matrix, $\mathbf{R} = \dfrac{1}{M}\mathbf{XX}^{H}$. Then the orthogonal projection matrix $\mathbf{P}$ is formed and the final cleaned signal is

$$\mathbf{s}_{i,\mathrm{sup}} = \mathbf{Ps}_i = \mathbf{s}_i - \mathbf{VV}^{H}\mathbf{s}_i \tag{41}$$

with i being the range index of interest. For a stationary RFI, strong range domain correlation is observed, and therefore the range domain orthogonal filtering does a nearly perfect work. However, when the RFI tends to be non-stationary, the correlation is weakened, which may degrade the performance of, or even fail, the global cancellation method.

To solve the non-stationary problem, range domain cancellation is done in a short-time mode [42]. The original slow time sequences are broken into a number of segments. Then detection and cancellation of the RFI are performed on each segment respectively, that is, in a short time manner. Signals at some far range bins are used as reference for the RFI suppression. Firstly, the average Doppler power spectrum at the reference range bins is calculated. The noise floor is estimated and the RFI is detected with a threshold of interference-to-noise ratio (INR). Then, if there is any RFI, the reference signals are used to estimate the interference subspace and orthogonal projection filtering is performed on the signals at effective range bins. Finally, the output segments are combined together for further processing. By such measures, interference in each segment now can be considered to be almost stationary, and the strength of correlation is recovered, which ensures the performance of cancellation in both stationary and non-stationary circumstances. Moreover, the computation time is greatly reduced since the EVD of a large matrix is much slower than multiple EVDs of small matrix. So, the short-time method also has much lower computation cost. Figure 11 shows an example of the RFI suppression.

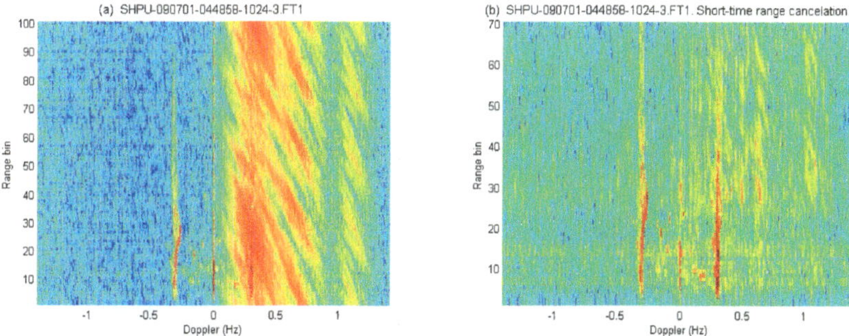

Fig. 11 Example of RFI suppression. (left) Original range/Doppler spectrum; (right) after cancelation.

4.6 Results of Field Comparison Experiments

A number of experiments have been done to verify the sea state monitoring capability of OSMAR-S. Here three examples are shown to demonstrate the surface current velocity and the wave height measurements respectively.

4.6.1 Hangzhou Bay Experiment

In November 2007, a comparison experiment was performed in Hangzhou Bay, East China Sea to verify the current measuring ability of OSMAR-S. Two 13MHz OSMAR-S sites, namely DAIS and SNSI, were constructed and jointly worked to map the surface current field of the bay area, which is shown in Figure 12. Both radar sites have worked for longer than one month. The radial current velocities recorded are shown in Figure 13, from which the solar and lunar tide periods can be clearly seen. Five locations in the common coverage region of the two radars, say A1 through A5, were chosen for the contact instruments such as the current meter and acoustic Doppler current profiler (ADCP) to achieve in situ measurements at the same time. These locations were particularly selected to form two crossed profiles for a more comprehensive verification. At each location a SLC9-2 direct reading current meter was towed by a ship and placed one meter below the sea surface, while at A4 a RCM-9 Aanderra current meter and a SONTEK-1M ADCP were used alternatively due to the obviously large error of the SLC9-2 current meter originally placed there. The in situ measurements began at 10:00 Nov 12 and ended at 14:00 Nov 13. The comparison curves of the current velocities

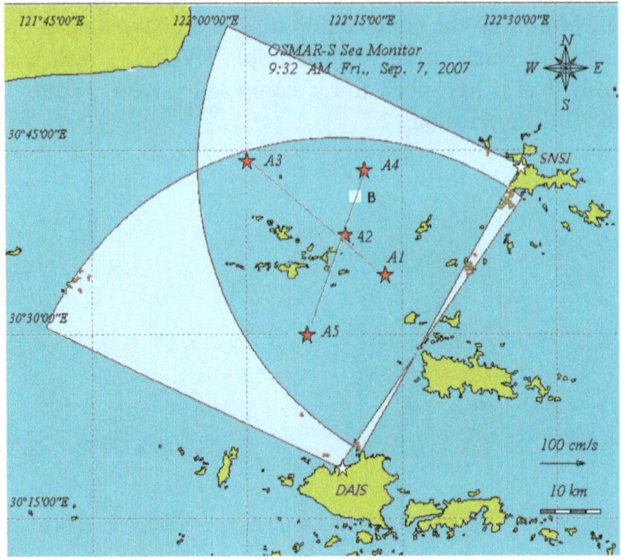

Fig. 12 Map of Hangzhou Bay Experiment in 2007

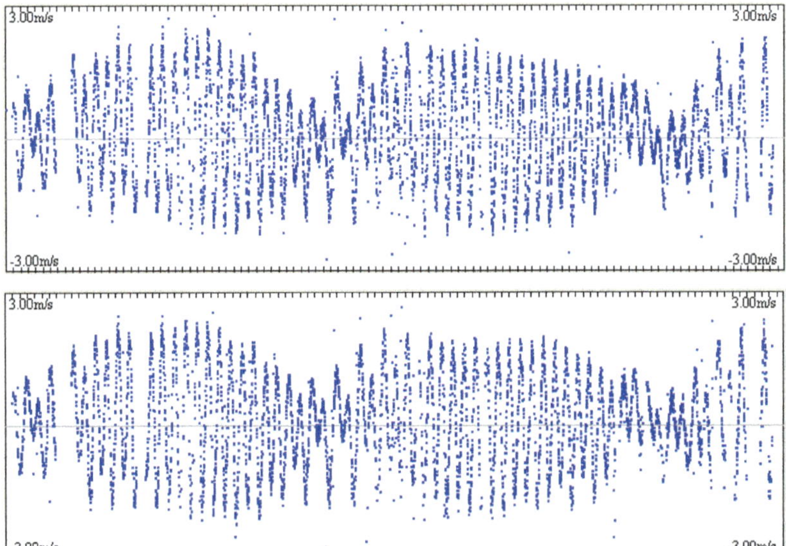

Fig. 13 Radial current velocity recorded per 20 minutes by the two OSMAR-S sites from 0:00 Aug 13 to 23:40 Sep 26, 2007. (Up) the DAIS site, the range is 20 km and the direction is 320 degree from the north; (down) the SNSI site, the range is 40 km and the direction is 270 degree from the north.

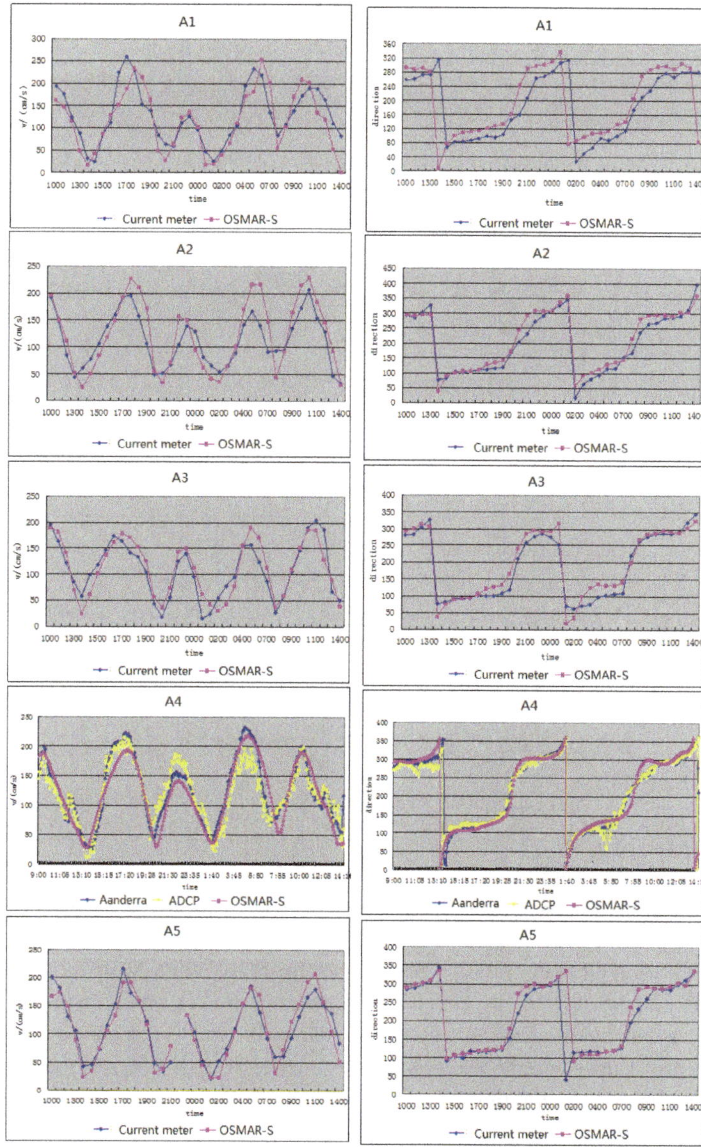

Fig. 14 Current velocities (left) and directions (right) by OSMAR-S versus in situ instruments

Table 1 Error analysis of the comparison results

		Mean Error	RMSE	Correlation Coefficient
A1	Vel (cm/s)	-11.81	36.77	0.87
	Dir (deg)	37.00	48.42	0.84
A2	Vel (cm/s)	9.80	32.16	0.82
	Dir (deg)	10.12	69.17	0.95
A3	Vel (cm/s)	2.58	23.66	0.91
	Dir (deg)	8.04	24.20	0.97
A4 Aanderra	Vel (cm/s)	-4.27	19.96	0.94
	Dir (deg)	6.28	17.05	0.99
A4 ADCP	Vel (cm/s)	-0.57	29.53	0.84
	Dir (deg)	7.70	21.74	0.99
A5	Vel (cm/s)	-3.23	19.11	0.94
	Dir (deg)	4.12	20.36	0.74

and directions obtained by radar versus in situ instruments at different locations are shown in Figure 14. The statistic performance of the comparison is given in Table 1. It can been seen that, generally the radar measurements coincide well with the in situ instruments. Except for the current direction at A5, the correlation coefficients in both velocities and directions are very high. Taking the difference of the scales in time and space into account, it can be concluded that OSMAR-S performs well in current measurements.

4.6.2 Shanwei Experiment

During the 16th Asian Games Sailing Competition held in Guangdong Province, China in November 2010, OSMAR-S was used as an important equipment in a large monitoring network to provide oceanographic parameters for the weather forecasting of the Shanwei sea area. The geographic map of the observation system is shown in Figure 15.

The two OSMAR-S sites are GUAO (22°41.2′N, 115°29.8′E) and ZHEL (22°39.5′N, 115°32.2′E) at 25 MHz, and there are two buoys, namely Buoy A (22°39.5′N, 115°32.2′E) and C (22°36.0′N, 115°32.9′E) providing in situ wave parameters. Buoy A is about 5km far from GUAO site and C is about 11.1km far. The wave heights and periods measured by the GUAO site are shown in Figure 16. The two radar sites gave similar wave parameters, showing good internal coincidence of the radar system. The correlation coefficient between the wave heights measured by the radars and the buoys are above 0.8. This experiment result has proved the wave capability of the OSMAR-S system.

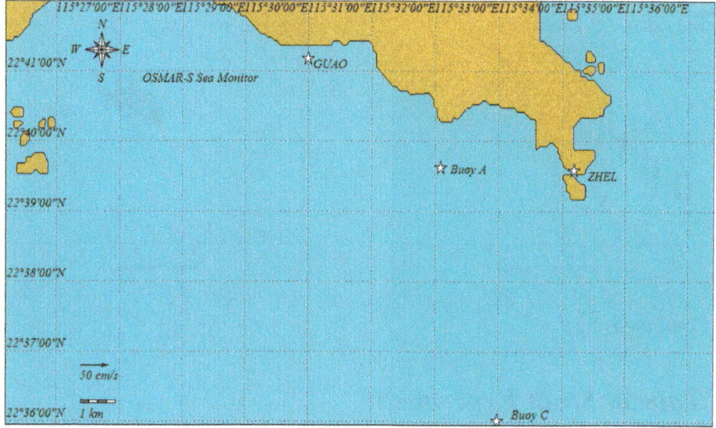

Fig. 15 Geographic map of the Shanwei observation system

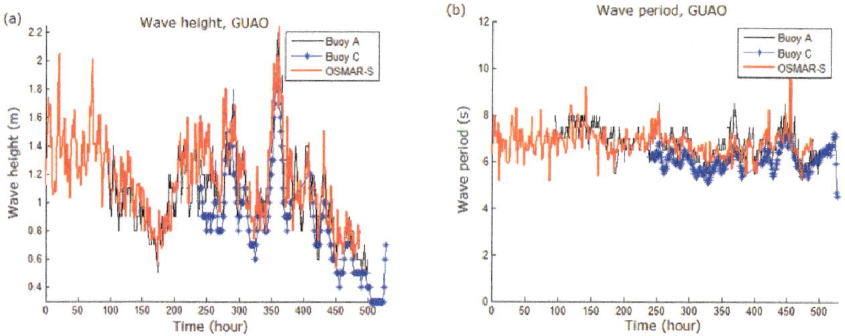

Fig. 16 Comparisons of wave height and period estimated by the GUAO radar and *in situ* buoys. (a) Wave height; (b) Wave period; (c) Scatter plot of the wave height versus Buoy A; (d) Scatter plot of the wave period versus Buoy A; (e) Scatter plot of the wave height versus Buoy C; (f) Scatter plot of the wave period versus Buoy C.

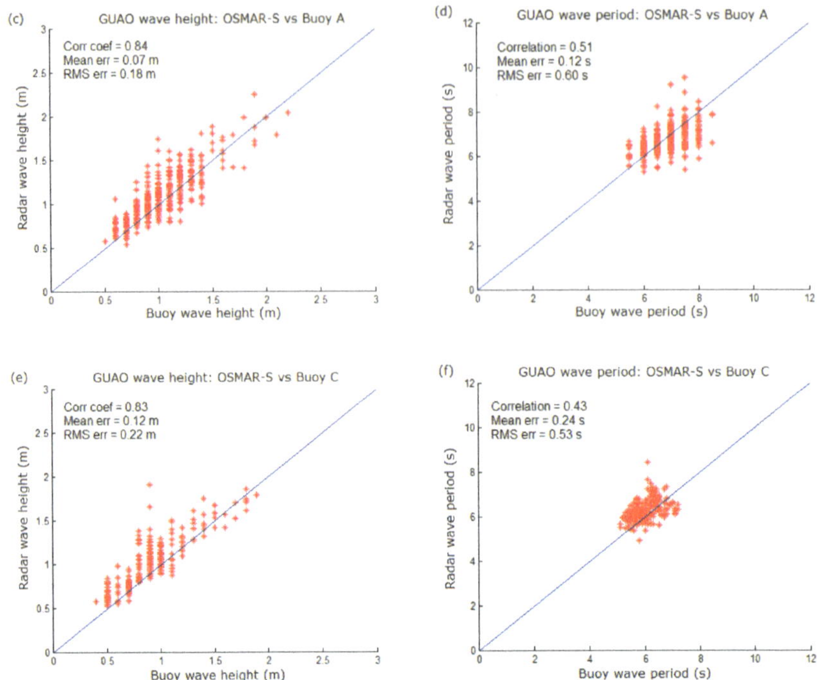

Fig. 16 (*Continued*)

4.6.3 Taiwan Strait Experiment

To further verify the radar performance under a wider range of wave heights, another experiment of OSMAR-S was performed at XIAN (23°44.60′N, 117°36.29′E) in the Taiwan Strait, China from Jan 10 to Mar 31, 2013. The geographic map is shown in Figure 17, where the OSMAR-S is marked by a pentagram. During this experiment, the wind predominantly blows from the northeast with a speed up to 20 m/s, and a maximum wave height above 5 m is recorded correspondingly. The OSMAR-S works at 13 MHz with a bandwidth of 60 kHz, and all the other parameters are the same as those in the 25 MHz experiment.

Till now, there is no in situ buoy data available for comparison, but the wave parameters output from GROW2012 (Global Reanalysis of Ocean Waves, 2012) [43] developed by the Oceanweather Inc. can be used as the "ground truth" data. The wave output is available at an hourly time step on a grid of 0.5 degrees (corresponding to about 55 km). In consideration of the coverage of the XIAN site, the No. 111879 (23°30′N, 118°0′E) grid point is the most suitable for comparison, which is about 48 km far from the radar as indicated by a 4-pointed star in Figure 17.

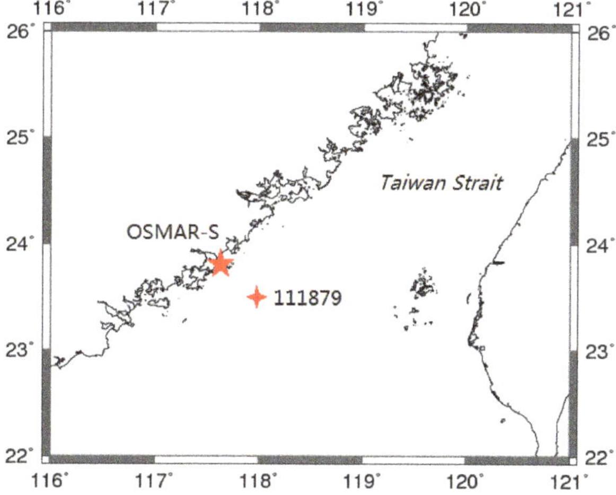

Fig. 17 Geographic map of the Taiwan Strait experiment

The significant wave heights are calculated on the range cell of 10 km in the direction of the grid point where comparison data is available, and the comparison results are shown in Figure 18. The final correlation coefficient between the radar and GROW2012 model results is 0.74 and the RMS error is 0.77 m. This has once more proved the wave extraction capability of OSMAR-S.

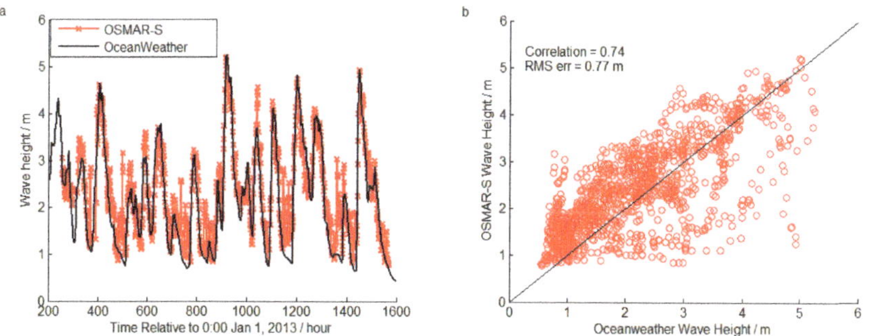

Fig. 18 Significant wave height by OSMAR-S versus Oceanweather Inc. (left) time sequences; (right) scatter plot.

4.7 Conclusion

OSMAR-S is a portable HFSWR that can be setup in a rather wide variety of situations. As a main characteristic of the portability, the extremely compact crossed-loop/monopole antenna is a key to the success of OSMAR-S, but it also brings challenges to the subsequent signal processing and data processing. Several measures have been taken in both the hardware and software levels to account for the disadvantage of use of such a small-aperture array. OSMAR-S is a useful oceanographic instrument and will play a more and more important role in monitoring of the sea states.

References

[1] http://en.wikipedia.org/wiki/Sea_current

[2] Teng, C.: Wave measurements from NDBC buoys and C-MAN stations. In: OCEANS 2002 MTS/IEEE, vol. 1, pp. 517–524 (2002)

[3] Nieto, J.C., Reichert, K., Dittmer, J.: Use of Nautical Radar as a Wave Monitoring Instrument. Coastal Engineering 37, 331–342 (1998)

[4] Dobson, E., Monaldo, F., Goldhirsh, J., Wilkerson, J.: Validation of GeoSAT altimeter-derived wind speeds and significant wave heights using buoy data. J. Geophys. Res. 92(C10), 10719–10731 (1987)

[5] Paduan, J.D., Graber, H.C.: Introduction to High-Frequency Radar: Reality and Myth. Oceanography 10(2), 36–48 (1997)

[6] Barrick, D.E.: Remote sensing of sea state by radar. In: IEEE International Conference on Engineering in the Ocean Environment, Ocean 1972, pp. 186–192 (1972)

[7] Gill, E., Huang, W., Walsh, J.: On the Development of a Second-Order Bistatic Radar Cross Section of the Ocean Surface: A High-Frequency Result for a Finite Scattering Patch. IEEE Journal of Ocean Engineering 31(4), 740–750 (2006)

[8] Wyatt, L.R., Thompson, S.P., Burton, R.R.: Evaluation of high frequency radar wave measurement. Coastal Engineering 37(3-4), 259–282 (1999)

[9] Zhou, H., Wen, B., Wu, S., et al.: Sea States Observation with a Portable HFSWR During the 16th Asian Games Sailing Competition. Chinese Journal of Radio Science 27(2), 293–300 (2012)

[10] Wu, X., Li, L., Li, Y., et al.: Experimental Research on Significant Waveheight Detecting with HFSWR OSMAR071. Oceanologia Et Limnologia Sinica 43(2), 210–216 (2012)

[11] Crombie, D.D.: Doppler spectrum of sea echo at 13.56 mc/s. Nature 175, 681–682 (1955)

[12] Gurgel, K.-W., Antonischki, G., Essen, H.H., et al.: Wellen Radar (WERA): a new ground-wave HF radar for ocean remote sensing. Coastal Engineering 37(3), 219–234 (1999)

[13] Wu, S., Yang, Z., Wen, B., et al.: Test of HF ground wave radar OSMAR2000 at the Eastern China Sea. In: OCEANS 2001. MTS/IEEE Conference and Exhibition, vol. 1, pp. 646–648. IEEE (2001)

[14] Ponsford, A.M., Sevgi, L., Chan, H.C.: An integrated maritime surveillance sys-tem based on high-frequency surface-wave radars. 2. Operational status and system performance. IEEE Antennas and Propagation Magazine 43(5), 52–63 (2001)

[15] Barrick, D.E.: 30 Years of CMTC and CODAR. In: IEEE/OES 9th Working Conference on Current Measurement Technology, CMTC 2008, pp. 131–136. IEEE (2008)

[16] Wen, B., Li, Z., Zhou, H., et al.: Sea surface currents detection at the Eastern China Sea by HF groud wave radar OSMAR-S. Acta Electronica Sinica 37(12), 2778–2782 (2009)

[17] Wyatt, L.R., Ledgard, L.J.: OSCR wave measurements - some preliminary results. IEEE Journal of Oceanic Engineering 21(1), 64–76 (1996)

[18] Zhou, H., Wen, B.: Calibration of antenna pattern and phase errors of a cross-loop/monopole antenna array in high-frequency surface wave radars. IET Radar, Sonar & Navigation (2014), doi:10.1049/iet-rsn.2013.0141

[19] Helzel, T., Hansen, B., Kniephoff, M., et al.: Introduction of the compact HF radar WERA-S. In: 2012 IEEE/OES Baltic International Symposium (BALTIC), pp. 1–3. IEEE (2012)

[20] Wu, X., Li, L., Shao, Y., et al.: Experimental determination of significant wave-height by OSMAR071: Comparison with results from buoy. Wuhan University Journal of Natural Sciences 14(6), 499–504 (2009)

[21] Barrick, D.E.: First-order theory and analysis of mf/hf/vhf scatter from the sea. IEEE Trans. Antennas Propagation 20(1), 2–10 (1972)

[22] Barrick, D.E.: Remote Sensing of the Sea State by Radar. In: Derr, V.E. (ed.) Remote Sensing of the Troposphere. NOAA/Environ Mental Research Laboratories, ch. 12, pp. 1–46. U.S. Government Printing Office, Washington, DC (1972)

[23] Longuet-Higgins, M.S., Cartwright, D.E., Smith, N.D.: Observations of the Directional Spectrum of Sea Waves using Motions of a Floating Buoy. In: Ocean Wave Spectra. Prentice-Hall, Englewood Cliffs (1963)

[24] Long, A.E., Trizna, D.B.: Mapping of north Atlantic winds by HF radar sea backscatter interpretation. IEEE Trans. Antennas Propagation 21, 680–685 (1973)

[25] Lipa, B.: Derivation of directional ocean-wave spectra by integral inversion of second-order radar echoes. Radio Science 12(3), 425–434 (1977)

[26] Howell, R., Walsh, J.: Measurement of ocean wave spectra using narrow-beam HE radar. IEEE Journal of Oceanic Engineering 18(3), 296–305 (1993)

[27] Hisaki, Y.: Nonlinear inversion of the integral equation to estimate ocean wave spectra from HF radar. Radio Science 31(1), 25–39 (1996)

[28] Wyatt, L.R.: Limits to the inversion of HF radar backscatter for ocean wave measurement. Journal of Atmospheric & Oceanic Technology 17(12), 1651–1665 (2000)

[29] Barrick, D.E.: Extraction of wave parameters from measured HF radar sea-echo Doppler spectra. Radio Science 12(3), 415–424 (1977)

[30] Maresca, J.W., Georges, T.M.: Measuring RMS wave height and the scalar ocean wave spectrum with HF skywave radar. Journal of Geophysical Research: Oceans 85(C5), 2759–2771 (1980)

[31] Heron, M.L., Dexter, P.E., McGann, B.T.: Parameters of the air-sea interface by high-frequency ground-wave Doppler radar. Marine and Freshwater Research 36(5), 655–670 (1985)

[32] Gurgel, K.-W., Essen, H.H., Schlick, T.: An empirical method to derive ocean waves from second-order Bragg scattering: prospects and limitations. IEEE Journal of Oceanic Engineering 31(4), 804–811 (2006)

[33] Cohen, L.: Time-frequency distributions-a review. Proceedings of the IEEE 77(7), 941–981 (1989)

[34] Schmidt, R.O.: Multiple Emitter Location and Signal Parameter Estimation. IEEE Trans. Antenna and Propagation 34(3), 276–280 (1986)

[35] Welch, P.D.: The use of fast Fourier transform for the estimation of power spec-tra: a method based on time averaging over short, modified periodograms. IEEE Trans. on Audio and Electroacoustics 15(2), 70–73 (1967)

[36] Barrick, D.E., Weber, B.L.: On the nonlinear theory for gravity waves on the ocean's surface. Part II: Interpretation and Applications. Journal of Physical Oceanography 7(1), 11–21 (1977)

[37] Pierson, W.J., Moskowitz, L.: A proposed spectral form for fully developed wind seas based on the similarity theory of SA Kitaigorodskii. Journal of Geophysical Research 69(24), 5181–5190 (1964)
[38] Hasselmann, D.E., Dunckel, M., Ewing, J.A.: Directional wave spectra observed during JONSWAP 1973. Journal of Physical Oceanography 10(8), 1264–1280 (1980)
[39] Wen, B., Shen, W., Zhou, H., et al.: Method to realize frequency monitoring by the HF radar receiver. Chinese Patent No. 200610018499.4 (2006)
[40] Zhou, H., Wen, B., Wu, S.: Target Detection in the Environment of Radio Fre-quency Interference by HF Radar. Modern Radar 26(9), 29–32 (2004)
[41] Zhou, H., Wen, B., Wu, S.: Dense Radio Frequency Interference Suppression in HF Radars. IEEE Signal Processing Letters 12(5), 361–364 (2005)
[42] Zhou, H., Wen, B.: Radio Frequency Interference Suppression in Small-Aperture High-Frequency Radars. IEEE Geoscience and Remote Sensing Letters 9(5), 788–792 (2012)
[43] Hope, M.E., Westerink, J.J., Kennedy, A.B., et al.: Hindcast and validation of Hurricane Ike (2008) waves, forerunner, and storm surge. Journal of Geophysical Research: Oceans, 2013 118(9), 4424–4460 (2013)

Chapter 5
Using Motion Sensor for Landslide Monitoring and Hazard Mitigation

K.-L. Wang, Y.-M. Hsieh, C.-N. Liu, J.-R. Chen,
C.-M. Wu, S.-Y. Lin, and H.-Y. Pan

Abstract. Owing to tectonic collision, about 70% of Taiwan's land are mountains and foothills. With several unfavourable conditions such as typhoons, strong rainfalls, and earthquakes, the landslide hazard is a critical issue in Taiwan. This issue is increasingly important because the use of hillsides for residence are growing in recent years. To address the issue and to protect its citizens, Taipei City Government initiated a pilot project to monitor landslide displacements in 2010. The project involves a selected hillside community with dip slope in Taipei City. The community had a major landslide in 1983.The monthly monitoring with non-automated devices is not helpful for residence to react to landslides. To help improve the situation, we use motion sensors to develop fully automated devices to infer displacements. We also used numerical analyses to help determine the best

K.-L. Wang · C.-N. Liu · J.-R. Chen
Department of Civil Engineering, National Chi Nan University,
No. 1, University Road, Puli, Nantou 545, Taiwan

Y.-M. Hsieh
Department of Construction Engineering,
National Taiwan University of Science and Technology, No.43,
Sec. 4, Keelung Rd., Da'an Dist., Taipei 106, Taiwan

C.-M. Wu
Construction and Disaster Prevention Center, Feng Chia University,
No. 100, Wenhwa Rd., Seatwen, Taichung 407, Taiwan

S.-Y. Lin · H.-Y. Pan
Geotechnical Engineering Office, Public Works Department,
Taipei City Government, Taiwan
e-mail: ymhsieh@mail.ntust.edu.tw

© Springer International Publishing Switzerland 2015

H. Leung and S.C. Mukhopadhyay (eds.), *Intelligent Environmental Sensing,*
Smart Sensors, Measurement and Instrumentation 13, DOI: 10.1007/978-3-319-12892-4_5

locations to install these devices. They were mounted to the ground surface to sense the tilt. The tilt converts to displacements by assuming slip surfaces. Small displacements were detected during a typhoon event in July of 2013.With the developed monitoring device, standard operation procedures were developed accordingly. Community residents has been educated how to react to unfavourable conditions. These conditions are defined by displacements exceeding certain thresholds.

Keywords: landslide, monitoring, triaxial accelerometer, decision-making process.

5.1 Introduction

5.1.1 Location of Study Site

The monitoring of landslide usually adopted visual inspection or manual monitoring per month. Real time landslide displacement monitoring system is including in-hole extensometer, ground extensometer, building crack meter. However, in-hole extensometer requires boring hole and costs a lot of funding. Ground extensometer has to be cross landslide crack on the surface, where is not easy to define before slip happens. Building crack meter has to be set up on the position where crack happens. Thus a landslide displacement tracking sensor with high efficient and economically is required. Fig. 1 illustrates the location of research area and the buildings are located in the community.

Fig. 1 Location of community and monitoring station

The 3D building was acquired from Google Earth 3D database. The sliding direction is from the community to down slope. The dip angle of the rock formation underneath the community is about N80°~90°E/30°~40°S. Cracks of brick wall at station 2(Fig. 2) was observed and the tilt direction is consistent with the direction of dip slope.

5.1.2 Geological Condition

The potential landslide area is about 0.089 hectare. The primary geological formation is dark grey to black shale or sandstone and shale interlayer with some local thick sandstone. The boring holes shows that underground conditions are:

(1) Surface soil:a thickness of 3 meter from ground surface was documented. The soil is brown clayey sand with small rocks. In-situ density of surface soil is about 1.98~2.00 ton/m^3.

(2) Rocks: underneath surface soil to 30meter from ground surface is rock. In-situ density of rock is about 2.36 ton/m^3.

Fig. 2 Cracks of brick wall at station 2

5.2   Landslide Numerical Analysis

The numerical simulation of landslide is adopted FLAC program to derive Factor of Safety, landslide depth and displacement distribution on profiles. FLAC is a two-dimensional explicit finite difference program for engineering mechanics computation. This program simulates the behavior of structures built of soil, rock or other materials that may undergo plastic flow when their yield limits are reached. Materials are represented by elements, or zones, which form a grid that is adjusted by the user to fit the shape of the object to be modeled. [1]

Two slope profiles were adopted for simulation as shown in Fig. 1. The triggering factor to initiate landslide is assumed as rainfall induced high ground water level. Thus the Factor of Safety of these two profiles were simulated to 1.0

from rising ground water level and weaken geological strength parameter. Parameters for simulation are as shown in Table 1. The simulation result is as shown in Fi. 3 and Fig. 4. As shown in Fig. 3, displacement is derived near ground surface with approximately 5m sliding depth. However, a small crack about 0.5cm was observed about 10m behind the crest of slope, which means the displacement measured could be affected by two sliding depths. The sliding simulation of profile 2 is as shown in Fig. 4. Two major displacement groups are simulated, which means two sliding depth could be existed as profile 1. The sliding depths of profile 2 simulated are about 3m and 10m, respectively. Thus the location of monitoring stations are selected from the largest displacement location as shown in Fig. 3 and Fig. 4.

Table 1 Parameters for landslide numerical simulation

Type	Unit Weight (kN/m^3)	Cohesion (kPa)	Friction Angle (degree)
Sandstone	20	0.1	32
Siltstone/Shale	20	0.1	27

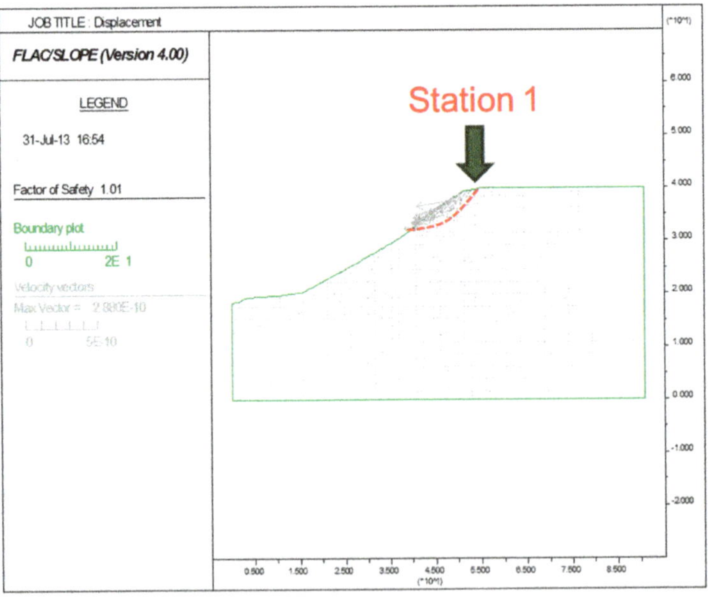

Fig. 3 Landslide displacement simulation of profile 1

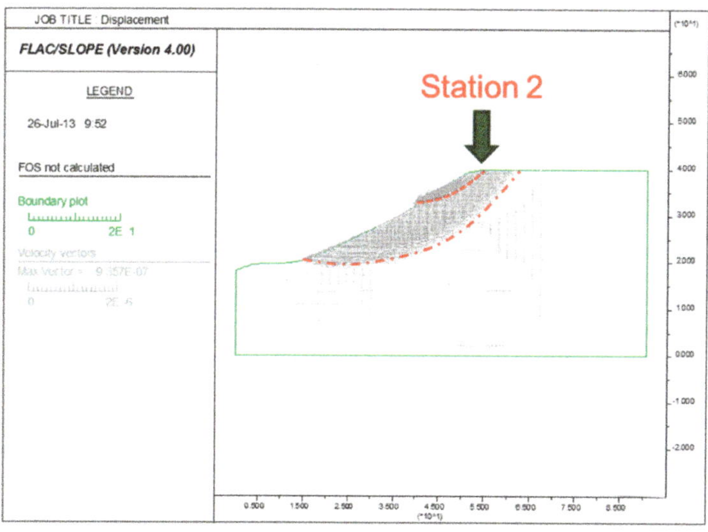

Fig. 4 Landslide displacement simulation of profile 2

5.3   Tilt Measuring Station

The tilt measurement station is composed of four major components: 1) power source, 2) GPRS module, 3) system board, and 4) triaxial accelerometer.The system schematic is shown in Fig. 5. Each component is discussed in subsequent sections.

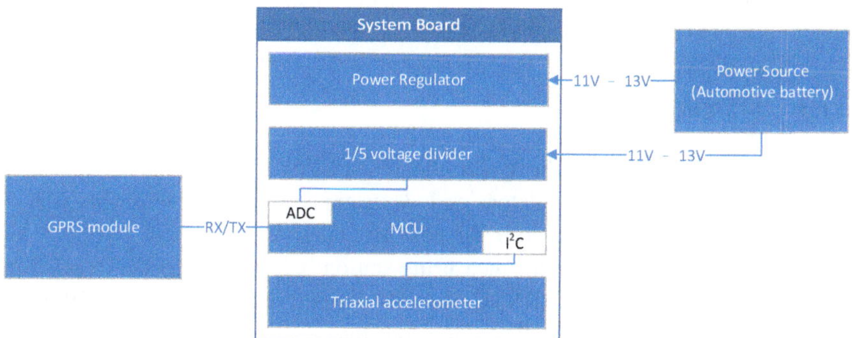

Fig. 5 System schematic of the tilt measuring station

5.3.1 Power Source

The power source of the tilt measurement station is a standard lead-acid batter commonly found in cars.It provides 10 – 13V direct current to the power regulator on the system board. The power regulator coverts the power down to 5V and feed to MCU module and GPRS module.The power source also feeds a tiny current to a 1/5 voltage divider consists of resistors with high resistivity.The purpose of the 1/5 voltage divider is to bring voltage down to the range that ADC (analog-digital conversion) on MCU (micro-controller unit) can measure.Nominal voltage output from the power source is between 10V – 13V, and measured voltage after this 1/5 voltage divider is thus between 2.0V – 2.6V.Measuring the voltage output of from the power source gives maintainers an indication when the battery should be replaced.Due to the very low power consumption of the tilt-measuring station we built, one fully charged 60AH (amperage-hour) lead-acid battery can operate for more than three months.

5.3.2 GPRS Module

The GPRS module we used is SIM900 [2] based GPRS module from EFCom [3]. According to its specification, it needs a five volts input, and maximum of two amperages during transmission.Three pins of the GPRS modules were used in the GPRS module.The RX pin of the GPRS module receives AT commands and data from the system board.These AT commands controls states of the GPRS module. The TX pin of the GPRS module sends responses and data from the GPRS module to the system board.Finally, there is a PWR pin that controls the power state of the GPRS module.This pin is initially low.When this pin is pulled high for more than one second and then pulled low again, the module will be powered up.At idle, the module is turned off completely through AT commands to save power.

5.3.3 System Board

Other main components on the system board includes 1) MCU (micro-controller unit) module, 2) power regulator, and 3) triaxial accelerometer. The triaxial accelerometer will be introduced in the next section.

The MCU adopted for the tilt measuring station is STM32F303VCT6 [4] from STMicroelectronics.It is an ARM Cortex M4 based solution with 256 Kilobytes flash memory and 12-bit ADC.There are four reasons for using this MCU.Firstly, the module is commercially available at a very reasonable price.Secondly, it features very low power consumption at sleep and standby modes.This helps to reduce the overall maintenance cost.Thirdly, the MCU features enough number of analog and digital inputs/outputs to interface with sensors and measure analog signals.Finally, the MCU module that we used has triaxial accelerometer built-in, and thus reduces the required soldering work and other interfacing issues.

Power regulator converts input voltages (of around 12V) to desired output voltages for different modules that we use to build the tilt measuring station.By design, both the GPRS module and the MCU module takes 5V input, and therefore we need a single power regulator to convert the output from the power source to 5V.We used KIS3R33S [5] to obtain needed 5V.The input voltage of KIS3R33S ranges between 7.0 and 24.0V.The output voltage is 5V.It is capable to deliver 2.5A continuous current, and it can peak at 4A for short duration. KIS3R33S has a nominal conversion efficiency of 96%, and thus reduces wastage of electricity.

5.3.4   Triaxial Accelerometer

The triaxial accelerometer we adopted is LSM303DLHC [X5] from STMicroelectronics. It is capable of measuring gravity in three orthogonal directions.When the sensor is placed horizontally, the ideal reading out of the sensor is (0.0, 0.0, -1.0g).In which, g (9.81m/s^2) stands for acceleration due to gravity load.Once the gravitational accelerations are measured, tilt in x and y directions can be calculated using angles θ and ψ obtained from equations shown in Fig. 6 [7][8], with coordinate systems and angles are defined in Fig. 7.The accelerometer LSM303DLHC also features I^2C digital communication interface and selectable full-scale range of ±2g, ±4g, and ±8g.The full-scale range of ±2g is used because only steady-state tilt is of interest, and this selection offers the highest sensitivity at 1mg.This 1mg sensitivity allows us to measure changes in tilt of 1/1000.In addition to gravity condition, the LSM303DLHC has also built-in temperature sensor to measure the temperature of the accelerometer.It is known sensed gravity varies with temperature [9].Due to this dependency, the calculated tilt varies with temperature. Readings from the temperature sensor can be used to correct or eliminate such thermal drift.

5.4   Information System

The overall system is illustrated in Fig. 8. It consists of three roles of devices: 1) tilt measuring station, 2) central server and backup server, and 3) client devices. The following sections introduce these roles of devices, and discusses software running on each devices.

$$\theta = \tan^{-1}\left(\frac{A_{x,out}}{\sqrt{A_{y,out}^2 + A_{z,out}^2}} \right)$$

$$\psi = \tan^{-1}\left(\frac{A_{y,out}}{\sqrt{A_{x,out}^2 + A_{z,out}^2}} \right)$$

Fig. 6 System schematic of the tilt measuring station [7]

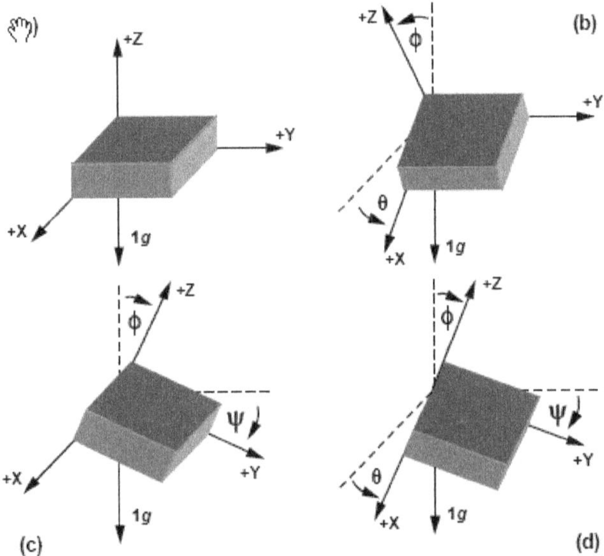

Fig. 7 System schematic of the tilt measuring station [7]

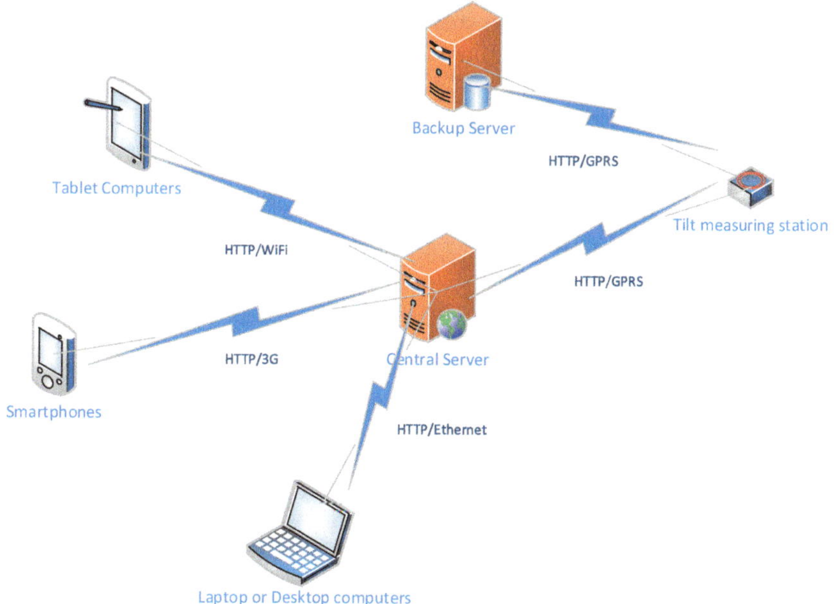

Fig. 8 Software system architecture

5.4.1   Tilt Measuring Station

The tilt measuring station measures tilt at installed location.Its components are discussed in the previous section, this section mainly discusses its software.The station runs in-house built firmware developed fully in C language.The main body of the firmware is an infinite loop performing the following tasks:

1. Initialize ADC, accelerometer and setup sampling rate.
2. Average 1000 accelerometer outputs and take temperature sensor reading.This averaging helps reduce noise from readings of accelerometers.Also, read the ADC value corresponding to the input voltage from the power source.
3. Compute tilt using the averaged accelerometer readings with equations listed in Fig. 6.
4. Prepare a message to be sent to central and backup servers in JSON [X9] format.The message contains a timestamp obtained from the GPRS module, measure accelerations in three directions, the temperature, converted tilt, and ADC value corresponds to the voltage of input power source.It should also be noted the timestamp obtained from GPRS module is obtained from the GPRS stations setup by telecom.
5. Store the prepared message in a message buffer.

6.   If the number of messages stored in the buffer is less than a user-specified count (communication interval), then go to step 13.

7.   Turn on the GPRS module and configure it for Internet access.

8.   Send all messages in the buffer to the central server and the backup server through GPRS module using the POST method defined in HTTP protocol [11].

9.   Turn off the GPRS module.

10.   If transmission is not successful, go to step 13.

11.   Clear message buffer.

12.   Parse returned message from central server to obtain and set new sampling interval and communication interval.

13.   Compute how many seconds T_{delay} should MCU sleep before the next measurement. Then, setup the alarm provided by MCU to wake up after T_{delay} seconds.Finally, turn off all sensors and ADC, and then goes into standby mode.

14.   Go back to step 1 after waking up.

It is noted the station communicates with central and backup servers through REST-style web services through HTTP/Post request.It is important to use standardized communication protocols so that the servers can be developed using any technology (PHP, ASP.NET, JSP, etc.)The current setup shown in Fig. 8 has limited scalability.This is because the one single server has limited capability that can only provide services to a limited amount of devices.It is possible to use distributed computing technology or cloud computing concept to build monitoring systems with unlimited scalability, high reliability and highly accessible systems [12].The unlimited scalability of such system is achieved by adding more network servers.The reliability of such system can be assessed by using simple probability concepts [13].It is proven that such system is as reliable as a single machine setup, and networking failure can be compensated by retrial mechanisms.

5.4.2 *Central Server and Backup Server*

The central server serves three purposes.First, it stores all monitoring data sent by tilt measuring stations.Second, it shows or visualizes monitoring data to connected client devices.Finally, it sends alarms to interested parties through email or SMS (short message service).In the project described in this chapter, the server runs the following free and open-source software: 1) Linux operating system, 2) Apache web server software, 3) MongoDB database management system.Application software are developed using PHP and JpGraph (for charting).The backup server runs Windows operating system, and web service for storing monitoring data is developed using Node.JS with 20 lines of code, as listed in Fig. 9.

In this project, we attempted two relative new technologies: MongoDB and Node.JS.They are very effective, and can be recommended for similar projects.MongoDB [14] is a document-oriented database management system.It was developed with high performance and scalability in mind.Because of its document orientation, complex data structures can be easily stored in MongoDB.Developers usually find it is easy to store data defined in objects or structures can be stored easily in MongoDB without no or little data conversion effort.This is not true for conventional relational databases, which needs more effort to map data into tabular formats. The other technology Node.JS [15] is a server-side technology aiming to develop web servers or web services using JavaScript.Before Node.JS, JavaScript was mainly used for client-side programming for developing user interfaces in browsers. Now, Web servers and web services can be easily developed using Node.JS.We used Node.JS to program our backup server for evaluation.The Node.JS code is shown in Fig. 9, the 20 lines of code in Node.JS creates a web service listening on port 3068 at line 20 on our backup server.When a HTTP request reaches the web service, the request is first parsed and the body of the request is convert to a JSON formatted string and stored into file "backup.dat" at line 6.It is seen in this example that using Node.JS to create web services is very simple and effective. Node.JS has a module repository [16] which currently has roughly 64,800 modules to help create web servers, access various database management systems, etc. In summary, both MongoDB and Node.JS are recommended for developing similar projects.

```
 1 var http = require("http");
 2 var qs = require('querystring');
 3 var fs = require('fs');
 4
 5 function writeData(data) {
 6 fs.appendFile("backup.dat", JSON.stringify(data) + "\n");
 7 }
 8
 9 var server = http.createServer(function(request, response) {
10 request.on('data', function(data) {
11 var d = qs.parse( data.toString() );
12 writeData(d);
13 });
14
15 request.on('end', function() {
16 response.end("275");
17 });
18 });
19
20 server.listen(3068);
```

Fig. 9 Backup server code written with Node.JS

5.4.3   Client Devices

The client devices require only standard web browsers and URL to the central server.The main information is shown in Fig. 10.From the status shown, the following information can be obtained: 1) battery status (full, enough, need replacement), 2) last data recipient time, 3) signal strength (strong, moderate, marginal, and weak), 4) sensors temperature, and 5) slope status (unknown, green, yellow, red).The battery status is determined by the measured voltage from the output of the battery, and the measured voltage and its corresponding status and displayed color are listed in Table 2.So the battery status helps maintainers to determine whether the battery should be replaced.The last recipient time is used to determine whether the system's operation is normal or not.If the last recipient time is more than 30 minutes ago, then the status will turn yellow; if it is more than 60 minutes, then it will turn red.This helps to catch attentions of maintainers.Signal strength refers to the sensed signal strength of GPRS module at the tilt measuring station.By using the RSSI (radio signal strength indicator) that can be obtained using AT commands issues to GPRS module, the signal strength can be displayed.Table 3 lists the RSSI values and displayed signal strength and color.If the signal strength is always weak, then there is a high probability of data loss.It is better to relocate the tilt measuring station, or to change or adjust the antenna in order to get better signal strength.The sensor temperature is mainly used for reference only.It should be noted the temperature inside the station can be as high as 50 degree centigrade if the station is under direct exposure of sunlight.Finally, the slope status is determined by the measured tilt.Averaged tilt comparing to the averaged obtained some time ago is used to infer the status.Different criterions were used to determine the status and discussed in other sections.If the needed average quantity is not available, the status "unknown" is shown.

Stations	#1	#2
Battery status	Satisfactory	Satisfactory
Last data reception time (min. ago)	11.9	26
Signal strength	Excellent	Good
Sensor temperature	31.6	29.6
Slope condition	Green	Green

Fig. 10 Monitoring status

In addition to the status shown in Fig. 10, which is mainly developed for maintainers.

Users through client devices can also display monitoring data at user-specified interval.Fig. 11 shows monitored tilt over one month period.It is interesting to observe that tilt varies with daily temperature changes, as well as some long-term trends in development.

Table 2 Mapping between measured batter voltage, status, and displayed color

Measured voltage	Status	Color
12.2$^+$	Full	Green
12.0 – 12.2	Enough	Yellow
12.0$^-$	Need replacement	Red

Table 3 Correspondence between RSSI and displayed status and color

RSSI	Status	Color
RSSI > -73	strong	Green
-73 >= RSSI > -83	moderate	Green
-83 >= RSSI > -93	marginal	Yellow
-93 >= RSSI	weak	Red

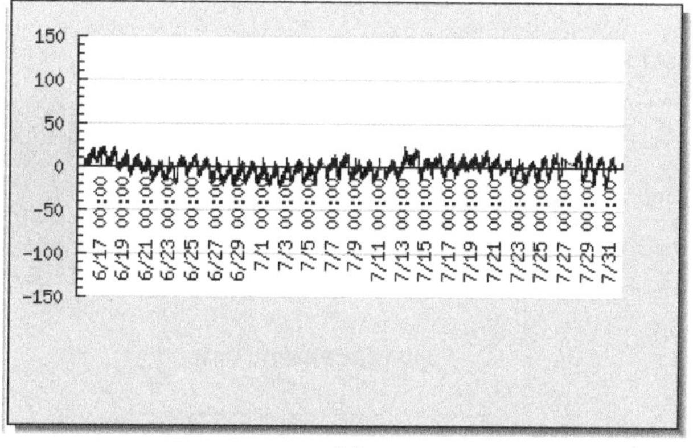

(a)

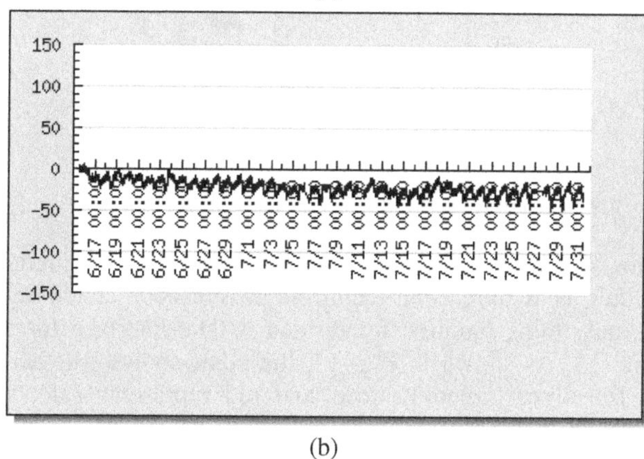

(b)

Fig. 11 Monitored tilt variations over time in (a) x-direction and (b) y-direction

5.5 Landslide Management with Motion Sensor Monitoring System

For practical usage of landslide management, a warning threshold such as displacement is required. However, displacement is not easy to define whether to act. Considering how many people are in the building and possible landslide crack is very close to building, a standard for landslide warning is proposed as shown in Table 4. Three levels are defined – Notice, Alarm, and Action. To prevent confusing data noise, an extra requirement to change action level is the displacement keeps moving in the same direction. The last step is to transfer tilt measured from previous mentioned device. A simple process is adopted in this system. A tilt to displacement concept is as shown in Fig. 12. Displacement (L) means the multiplication of sliding depth (D) and tilt angle (θ). Sliding depth comes from the numerical simulation and other in-situ observations. Tilt angle comes from the sensors aligned in study site. The main purpose for this project is providing a direct signal for residence. Thus despite of Notice, Alarm, and Action definition, signal color green, yellow, and red are set to each level, respectively.

Table 4 Standard for landslide warning

	Notice	Alarm	Action
Landslide displacement detector	2mm / month & same direction	2mm / 2day & same direction	2mm / 8 hrs & same direction
Light	Green	Yellow	Red

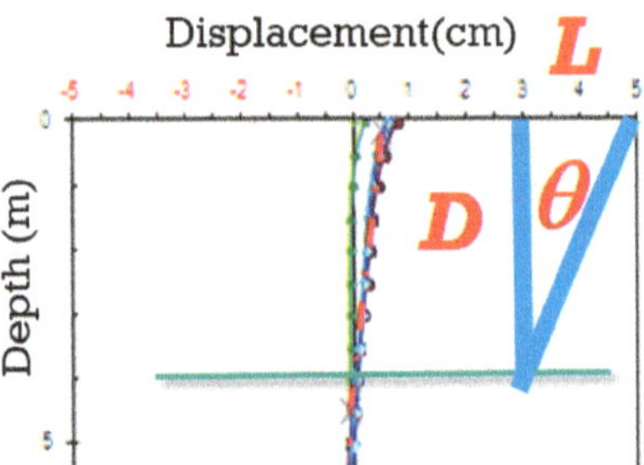

Fig. 12 Tilt to displacement illustration

A website is created for residence and engineers simultaneously. The displacement/tilt is required for engineers to consider action level. However, signal is the only thing required for residence. The interface for residence is as shown in Fig. 13. As shown in Fig. 13, the color shows whether the residence should act. The signal green, yellow, and red represents safe, stay alert, and evacuate.The web interface can be access by every residence without password. Moreover, a monitoring panel is set up in the community committee office for quick alarm service.

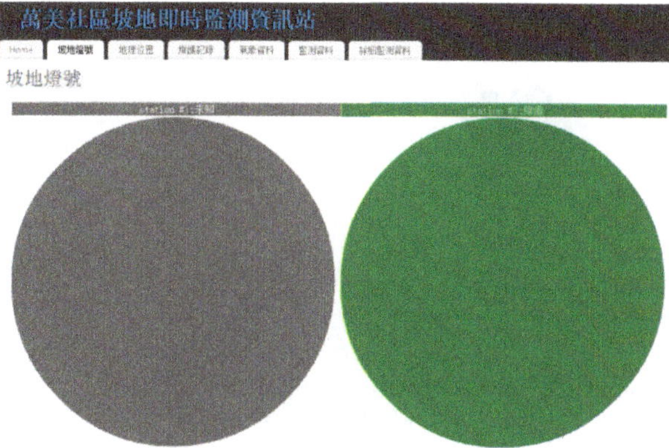

Fig. 13 Real time monitoring system (signal for residence)

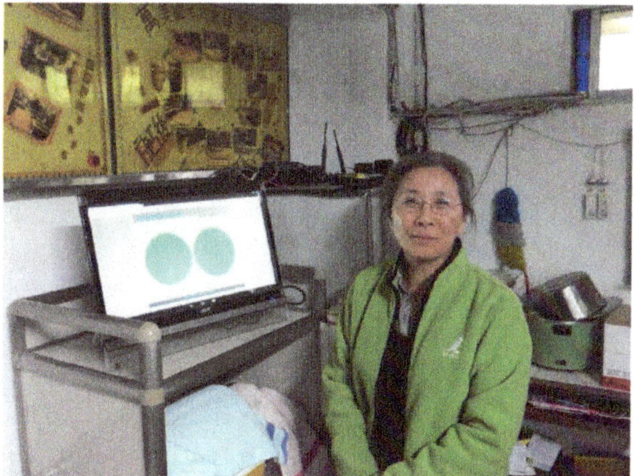

Fig. 14 Monitoring panel set up in community committee office with chief director

5.6   Conclusion Remarks

An efficient and economic design of landslide monitoring station was developed in this research and works fine during past months. Small displacement was observed during typhoon Soulik in July, 2013. However, the displacement was small and cannot find trend from time series display. Thus a scatter plot of two sliding directions is as illustrated in Fig. 15. The observation show that tilt was documented at Station 1 and the tilt happened during typhoon after check time data. To provide useful landslide data for hazard mitigation purpose, long term monitoring is necessary and displacement for slip of slope mass requires further study. However, the monitoring website can provide local residence a quick and clear signal to face possible landslide situation and provide decision making reference.

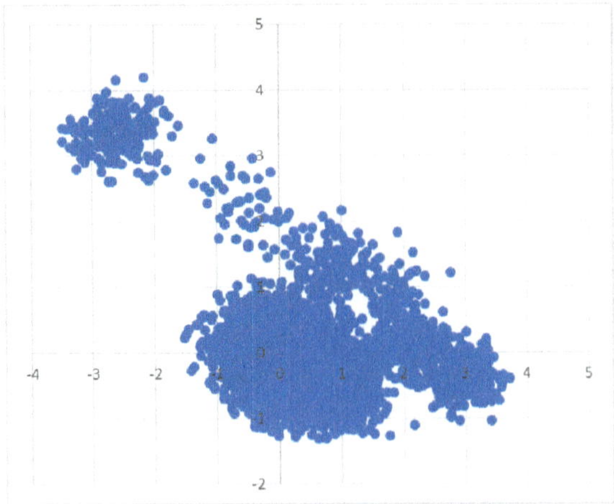

Fig. 15 Combined tilt direction of Station 1 in July, 2013

Acknowledgment. The authors would like to give their grateful thanks to Taipei city government to provide funding for this test project. The staff in the landslide monitoring community also helps to indicate potential landslide crack locations to help select monitoring sites.

References

[1] http://www.itascacg.com/flac/overview.html

[2] SIM900, http://wm.sim.com/producten.aspx?id=1019 (accessed on: March 25, 2014)

[3] EFCom GPRS/GSM Shield,
    http://www.elecfreaks.com/wiki/index.php?title=EFCom_GPRS/
    GSM_Shield (access on: March 25, 2014)

[4] STM32F303VC, Analog and DSP with FPU ARM Cortex-M4 MCU with 256 Kbytes
    Flash, 72 MHz CPU, MPU, CCM, 12-bit ADC 5 MSPS, PGA, comparators,
    http://www.st.com/web/en/catalog/mmc/FM141/SC1169/
    SS1576/LN1531/PF252054 (accessed on: March 25, 2014)

[5] Hobbypower Kis3r33s Dc-dc 7-24v to 5v 3a Usb Power Converter Buck Module,
    http://www.amazon.com/Hobbypower-Kis3r33s-Dc-dc-
    Converter-Module/dp/B00I04T3A6 (accessed on: March 25, 2014)

[6] LSM303DLHC, Ultra compact high performance e-compass: 3D accelerometer and
    3D magnetometer module,
    http://www.st.com/web/catalog/sense_power/FM89/
    SC1449/PF251940 (accessed on: March 25, 2014)

[7] Fisher, C.J.: Using an Accelerometer for Inclination Sensing. AN-1057 Application
    Note, Analog Devices, http://www.analog.com/static/imported-
    files/application_notes/AN-1057.pdf (accessed on: March 25, 2014)

[8] Pedley, M.: Tilt Sensing Using a Three-Axis Accelerometer. Freescale
    Semiconductor Application Note, Document Number: AN3461, Rev. 6 (2013)

[9] Pedley, M.: High Precision Calibration of a Three-Axis Accelerometer. Freescale Semiconductor Application Note, Document Number: AN4399, Rev. 1 (2013)

[10] Introducing JSON, `http://www.json.org/` (accessed on: March 25, 2014)

[11] Fielding, R., Gettys, J., Mogul, J., Frystyk, H., Masinter, L., Leach, P., Berners-Lee, T.: Hypertext Transfer Protocol – HTTP/1.1 (1999),
`http://www.w3.org/Protocols/rfc2616/rfc2616.html`
(accessed on: March 25, 2014)

[12] Hsieh, Y.M., Hung, Y.C.: A scalable IT infrastructure for automated monitoring systems based on the distributed computing technique using simple object access protocol Web-services. Automation in Construction 18(4), 424–433 (2009)

[13] Hsieh, Y.M., Chen, Y.H.: Critical reliability assessments of distributed field-monitoring information systems. Automation in Construction 26, 21–31 (2012)

[14] MongoDB, `http://www.mongodb.org/` (access on: March 25, 2014)

[15] Node.JS, `http://nodejs.org/` (access on: March 25, 2014)

[16] Node Packaged Modules, `https://www.npmjs.org/` (access on: March 25, 2014)

Chapter 6
Distributed Intelligent Monitoring System for Water Environment

Yuhao Wang, Junle Zhou, Hongyang Lu, Xiaolei Wang, and Henry Leung

Abstract. As more concern is attached to water environmental protection, the need for continuous water quality monitoring is highlighted. This paper proposes a real-time online water environment monitoring terminal which contains the acquisition, storage, processing and transmission of water environment parameters and control instruction. The sensing part of the terminal includes the novel planar electrode sensor, commercial available sensors, GPS and IP camera. The data acquisition board provides compatible interfaces and stores water quality information. Mesh network is used for transmission. The performance shows the terminal can be effectively applied to distributed water environment automatic monitoring.

6.1 Introduction

According to a report of United Nations, about 25 million people die of drinking the polluted water every year, and billions of people cannot get access to the clean drinking water in the world. Additionally, it is reported by the World Health Organization (WHO) that nearly 75 percent of the diseases are related with the water pollution. In this context and faced with the increasingly serious water

Yuhao Wang · Junle Zhou · Hongyang Lu
School o Information Engineering, Nanchang University, China, 330031

Xiaolei Wang
School of Electronic Information, Wuhan University, China, 430072

Henry Leung
Department of Electrical and Computer Engineering,
University of Calgary, Canada-T2N 1N4
e-mail: leungh@ucalgary.ca

© Springer International Publishing Switzerland 2015
H. Leung and S.C. Mukhopadhyay (eds.), *Intelligent Environmental Sensing,*
Smart Sensors, Measurement and Instrumentation 13, DOI: 10.1007/978-3-319-12892-4_6

resources shortage as well as the water pollution problem, it becomes a necessity for the water environment regulators and water industries to adopt the advanced process controlling system to strengthen the monitoring of the water environment, under the pressure of laws and promoted by the modern market[1-3].

Online automatic monitoring system of the water environment is a comprehensive online automatic monitoring network which takes the online automatic analysis equipment as the core and applies the modern sensor technology, automatic measurement technology, computer applications technology and the related professional software as well as the communication network. The application of such system can realize the real-time continuous and the long-distance detection, which can make it possible to make the correspondent analysis based on the monitored situation of water quality to take fully advantage of the water resources[4-6]. For its part, it still remains a larger gap and many problems in our water quality monitoring station compared with the foreign ones, hence making it urgent for the construction of the water quality monitoring net in China for the time being now, to improve the research and development (R&D) ability, enhance the theoretical analysis and break through the bottlenecks of the related technology of the online automatic detection system.

With the development of integrated circuit and wireless sensor network, the online water quality analysis system has made a rapid development for its convenience and real-time advantages compared with the artificial sampling and laboratory analysis method, which waste both time and energy. In this way, such online method and the corresponding application technology improved the stability the reliability of the analysis method, consequently, the water quality parameters monitored online become much more and the functions of water monitoring system tend to be perfect[7-9]. In 2003, the international organization for standardization (ISO) created a set of standards code named as ISO15839-2003, namely Water quality - On-line sensors/analyzing equipment for water - Specifications and performance tests. This standard defines the performance characteristics of the online water quality analysis of the analysis instruments and establishes the test program to evaluate and test the parameters of the performance characteristics. And also, such general standard provides evidence and support for the development, production as well as the acceptance of the online water quality analysis instruments. Based on the different intended purposes, online water quality monitoring system can be categorized into two types, which are the monitoring analysis and the processing analysis[10]. In this paper, we propose a newly designed water quality monitoring terminal. The terminal contains not only novel planar electrode sensor buy also some commercial detecting and monitoring devices while providing all kinds of required interfaces. A low power consumption MCU is employed for data acquisition, data storage and other managements. Data and instructions are transmitted by a mesh-type network, grouping of the monitoring nodes before transferring the sensor data to higher levels. The following part will first describe of the entire system then detail the each part of it.

6.2   The Overall Design of the Water Quality Monitoring Terminal

After the study of the trend and technology of the online water quality monitoring system, it can be easily figured out that it faces generally three issues[11-13]:

1) Firstly, the system is not highly intellectualized, and the price of the water sensors are at a rather high level which adds the total cost and lower the cost-performance;
2) Secondly, the geographical distribution of the water quality monitoring stations is necessarily dependent on the data transmission network which requires high productive, low cost and high coverage;
3) Thirdly, the operation requirements and the dispersion of the water quality monitoring station makes it much more complicated to manage and maintain the system.

Focused the above issues, we put forward an overall design scheme which is consisted of the following four parts:

1) The application of diversified sensors to make an overall analysis of the water quality. Sensors we employed includes sensors which can detect the temperature of water, PH, salinity, turbidity dissolved oxygen and other conventional parameters; IP camera which is used to monitor the floaters on the water; GPS employed to locate the spatial locations of the water quality monitoring stations; and the planar electrode sensors which are self-developed and applied as the detection of the inorganic salt ions in the water;
2) The design of the data acquisition board which is of high performance, low power consumption and can be compatible of varied sensor interfaces[14]. This data acquisition board is based on the STM32 and μC/OS-II real time multi tasks operation system, taking advantage of the embedded technology, this terminal integrates part of the sensor interfaces on the data collecting board according to the IEEE1451 intellectual sensor interfaces regulations and standards[15].
3) The design of distributed wireless transmission platform based on the Mesh technology. To meet the requirements of the uniqueness of the operation environment of the terminal as well as the communication of data, the distributed data wireless transmission modules employ the self-designed open source router hardware NCU-Mesh solution, which can provides multi general interfaces with plug and play, and can be well extended. In wireless Mesh network, every node can send and receive signals, and every one of them can directly communicate with one or more peer nodes, in this context, the transmission bandwidth and transmission distance can be reachable at 12Mbps and thousands miles respectively, which fully meet the requirements that the sensors of ecological perceptual layer should sense information and video information and transfer the spatial information[16].

4)   The adoption of LabVIEW, a virtual development tool of equipment and software, on the interface of the terminal. LabVIEW has a strong ability of network communication and completely support the TCP/IP transport protocols, and at the meantime, it applies the graphic programming mode and the visualized compiling environment, and inherits the advantages of structuration and modularization of the coding language. In this design, the water quality data are based on the TCP/IP wireless transmission protocols, the information integration and the interface programming are on the basis of the LabVIEW, and the information of the spatial locations of checkpoints, the video information of the surface of the water and the water quality data are displayed in the PC interface.

Based on these considerations,the structure diagram of the water quality monitoring terminal as shown in the figure 1.

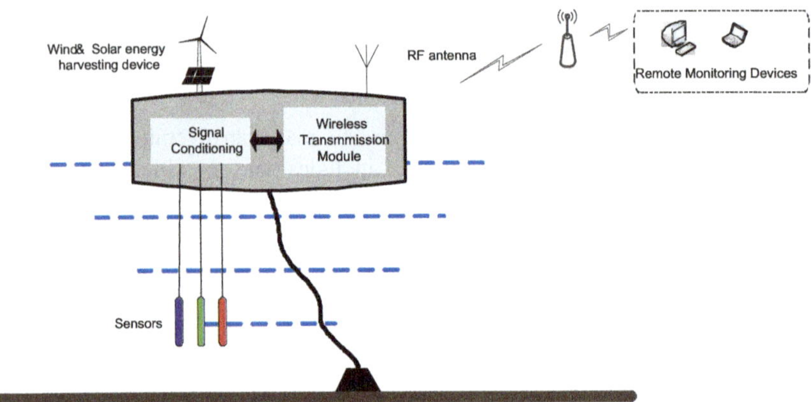

Fig. 1 Structure diagram of the water quality monitoring terminal

6.3   Sensors for Water Quality Detection

In order to accomplish current legal requirements as well as environmental and social concerns we need to build water quality monitoring programs. This chapter provides an available lab based and real-time methods of water pollutants monitoring for monitoring of nitrates. From an environmental perspective, nitrate fertilizers are increasing used in agriculture, the quantity of nitrate in rivers and ground waters is increasing cumulatively. Nitrate is an important analyte for environmental, food and human health monitoring and thus its detection and quantification is essential .The severity of the water quality environment has triggered to need for new methods and systems which enable the monitoring of chemical and biological pollutants in real time.

Environmental water monitoring include measurements of physical characteristics such as PH, temperature, conductivity, turbidity as well as chemical parameters such as dissolved oxygen, , nitrogen and phosphorus compounds and abundance of certain biological taxa. There is an urgent need in on-line monitors that are able to detect the excess of pollutants established by the official water quality regulations. A more sensitive sensor are required as well as standards are often associate to the modern limits of detection. There is multi-parameters water quality monitors used in wastewater monitoring.

Most commonly measured water parameters and associated sensing technologies are described as follow: The content of heavy metal parameters is mainly obtained by colorimetric sensing technology as well as other methods such as Atomic Absorption pectrometry and Ion selective electrode. The conductance can be detected by conductivity cell, annular ring electrode, nickel electrode, titanium and noble metal electrode. The dissolved oxygen of water can be got by means of Membrane electrode, 3-electrode voltametric method or optical sensor. Ions (e.g.Cl-, NO3-, NH4+) can be measured by Ion-selective electrodes. There are many methods monitoring PH such as titration with Sodium Hydroxide, ion sensitive field effect transistor (ISFET) and proton selective glass bulb electrode, proton selective metal oxide. Generally, the temperature can be measured by thermistor and the turbidity can be acquired via optical sensors and nephelometric method.

6.3.1   IP Camera and the GPS Module

With the increase of the monitoring targets, floaters of the water has much more aroused the attention of people, however, in the present market, restricted by the transmission bandwidth, the water quality monitoring analysis instruments almost cannot transmit the high definition video information of the surface of the water. Under such circumstances, the adoption of IP Camera in this design make it possible for the real time observation of the floaters on the surface of the water. What the major difference between the IP Camera and the general camera is that the IP Camera is actually a combination of the video server and the camera. This camera possesses of all the general functions of the traditional cameras for example, the picture capture function, and also, it is embedded a digital condensed controller and an operation system based on WEB, hence making it possible to transmit the data information to the client users via Internet of Wi-Fi after the data are condensed and encrypted.

However, the dispersion of the spatial locations of the water quality monitoring points is one of the difficulties of the online automatic water quality monitoring system. On one hand, the exact location technology provides the convenience of locating the spatial locations of the mobile buoy monitoring system, while on the other hand, it also makes it convenient to obtain the exact locations of the pollution sources in time. Global Positioning System is a satellite navigation and positioning system constructed by America, and it can provide all-weather, continuous and real-time three-dimensional navigation positioning and velocity measurement for the users, and additionally, it also can provide the high precision time transfer and the accurate positioning service. The GPS module we implied is HOLUX GR – 213 intellectual satellite receiver, a complete satellite position receiver, which is embedded the satellite receiving antenna and employs the Third Generation satellite receiving chip which is developed by SIRF Company in America. The all-round functions of this GPS module is its most favored advantage, and in this receiving chip, 200 thousand satellite tracking arithmetic units are embedded which are capable of tracking 20 satellites and can fully satisfy the requirements of the strict professional positioning and the personal consuming.

Fig. 2 GPS receiver and the Camera module

6.3.2   Sensors of 5 Conventional Parameters of the Water Quality

The five conventional parameters of the water quality monitoring system is a set of the economic multi-parameters water quality monitoring system. As shown in figure 3, the standard system consists of multi-channels data recorder and five 4-20mA water quality sensors which are used to monitor the temperature of the water, PH, salinity turbidity and dissolved oxygen. Such multi-channels data recorder includes seven simulation channels and two digital channels which are used for data recording, a USB and a RS232 communication interface. The principal theories of sensors and the technique index displayed in the table 1.

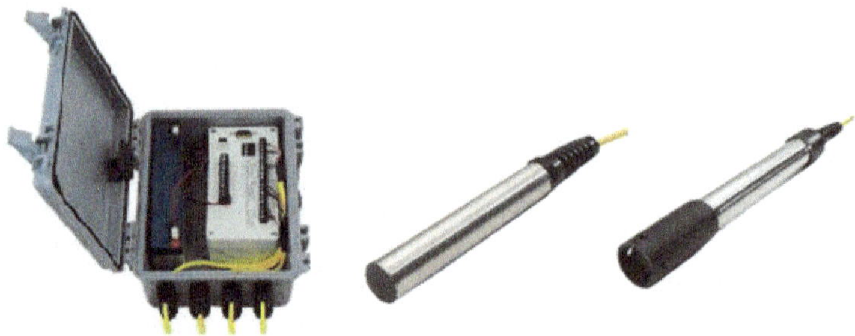

Fig. 3 Part of sensors of 5 conventional parameters of the water quality

The Temperature Sensor

Temperature measurement of thermal resistance via temperature sensor is based on the characteristic that the thermal resistance of metals rises with the increase of the temperature, according to this theory, the change of the temperature can be measured by the change of the value of thermal resistance, because it can transform the thermal signals of the objects into the digital ones. In this course, it adopts Pt100 platinum resistance as the measurement element and its resistance value and the variation of the temperature are under the unified standard

$$R_t = R_0[1 + \alpha(t - t_0)] \tag{1}$$

and its original temperature is related with the temperature coefficient of the materials used.

The PH Sensor

PH is another important index to decide whether the water environment is good or not, and according to the measurement theories, there are different methods to monitor PH: one is colorimetric method based on the color change of the PH reagent and the other is potentiometric method based on the measurement of the voltage on the both side of the electrodes. In the recent market, a complex PH electrode is generally used on the PH sensor, and it is composed of indicator electrode made of glass electrode and reference electrode made by calomel electrode.

The Conductivity Sensor

In physics, conductivity means the ability that liquor conducts electricity, and it is an important index to monitor the condition of the water environment. In the pure water, there are few impurity ions and its conductivity is relatively low, when the content of inorganic acid, alkali or the salt in the water increases, the conductivity of the water will increase correspondently, therefore, the varies of the conductivity can roughly reflect the pollution condition of the water. In this article, the WQ-COND conductivity sensor is primarily employed, it makes use of the level four electrode measurement technology and can provide a wide-range of indication of the conductivity and the temperature, and it also adopts 2%/°C automatic temperature compensation.

The Dissolved Oxygen Sensor

The dissolved oxygen means the amount of the molecular oxygen in the water. When the probe of the dissolved oxygen sensor is in the water, it can produces the voltage gap between the two electrodes when it is under the impressed voltage, and in this context, the metal ions will go through the membrane on the anode and restore on the cathode. The speed of electricity current produced in this course is in proportion to the transmission speed of the oxygen in the water, and it can be monitor via monitoring electricity current. In this article, the dissolved oxygen sensor applied is WQ401 type made in America and the electricity output is in the range of 4-20mA and the accuracy can be reachable at about 0.5%.

Table 1 The main detection parameters of the water quality monitoring system

The primary detection sensor of the system		
Parameters Detected	The Principal theories and the methods	The technique index
The temperature sensor	Thermistor	The output electricity current : 4-20mA The range of the measurement : -50~+50°C The accuracy : ±0.1°C The operation temperature : -50~+100°C
The PH sensor	proton selective glass bulb electrode, proton selective metal oxide	The output electricity current : 4-20mA The range of the measurement : 0-14pH The accuracy : 2% of the full range of the measurement; The operation temperature : -5 to +55°C
The conductivity sensor	Titanium or noble metal electrode	The output electricity current : 4-20mA The range of the measurement : 0-10,000μS/cm The accuracy : 1% of the full range of the measurement; The operation temperature : -40 to +55°C
The dissolved oxygen sensor	optical sensor	The output electricity current : 4-20mA The range of the measurement : 0-100% of the saturation , 0-8mg/L(temperature compensates to 25 °C) The accuracy : ±0.5% of the full range of the measurement; The operation temperature : -40 to +55°C The range of the voltage : 10-36VDC
The turbidity sensor	nephelometric (light scattering) method	The range of measurement : 0-50NTU and 0-1000NTU Accuracy : ±1% of the full measurement Source of the light : infrared LED (880nm) The output electricity current: 30mA+ output by the sensor

Table 1 (*continued*)

		The highest resolution ratio : 1280*720pix The compression method : H.264 The frame frequency : 30 frames highest; Internet protocol : TCP/IP, HTTP, ICMP, Etc. Signal-to-noise ratio : >50db
IP Camera	Via the internet or LAN (local area network) to transmit the video and the audio	
GPS location module	Adopt the third generation SIRF high performance chip, which can quickly located and possesses of exact searching and computation abilities.	The accuracy of location : less than 10 meters; Time used in location : 0.1s Controller : ARM7TDMI GPS chip : SIRF star III

The Turbidity Sensor

A large number of methods can be adopted to measure the turbidity, however, either the chemical method or the sensor detection method are largely related with varies of the character of the light in the water. And based on that, it can be categorized into the three following ways: spectrophotometry method, gravimetric method and nephelometer method. The turbidity sensor used in this article is well consistent with the USEPA method of 180.1, and it adopts the 90° optical scattering method, the focused incident beam will be oriented on the water monitored, and then the particles in the water will reflect the beam, and the light intensity will be measured by the light detection instrument located at the 90° orientation to the beam. And the light intensity detected will be accordant with the turbidity of the water.

6.3.3   *Planar Electrode Sensors for Water Detection*

Planar electrode sensors are widely used in various fields of modern technology and engineering such as industrial areas, agriculture, and engineering/scientific applications. There are three types of sensors according to their impedance characteristics. Sensors of inductive characteristic have been used as detector for nondestructive testing of the integrity of conductive and magnetic materials. Another type is capacitive type which is commonly with interdigital electrode. This kind of sensor has many applications such as moisture content measurement in pulp, monitoring the change in impedance caused by the growth of immobilized bacteria, human health confirmation based on the content of water in human skin, humidity sensors, chemical sensor, food inspection for human safety and estimation of material dielectric properties such as food, saxophone reed and leather. The third type is the combination of inductive and capacitive planar elements which is categorized as a passive sensor and operates on remote query basis (wireless). An example for it is real-time monitoring of water content in civil engineering materials. Unlike solid and air, liquid has its own character when the sensor is put in, and the most obvious contrast is ionic adsorption.

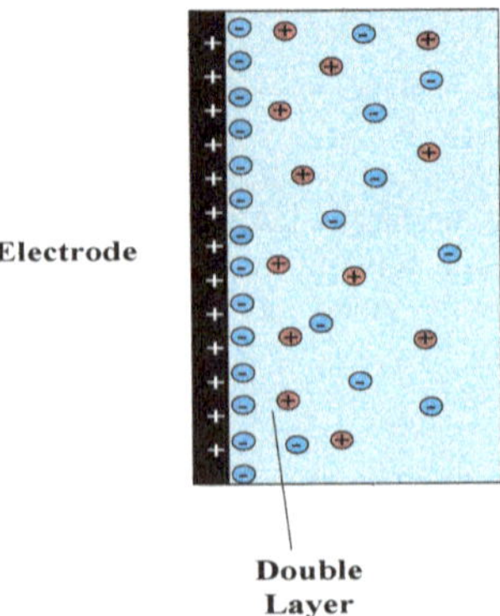

Fig. 4 The schematic diagram of schematic diagram

Figure 4 is the schematic diagram of water sample and sensor electrodes in which adsorption effect can be distinctly observed. Charged ions are attracted at the interface between coated electrode and water solution. This phenomenon is usually called electrical double layer (EDL)[19,20].

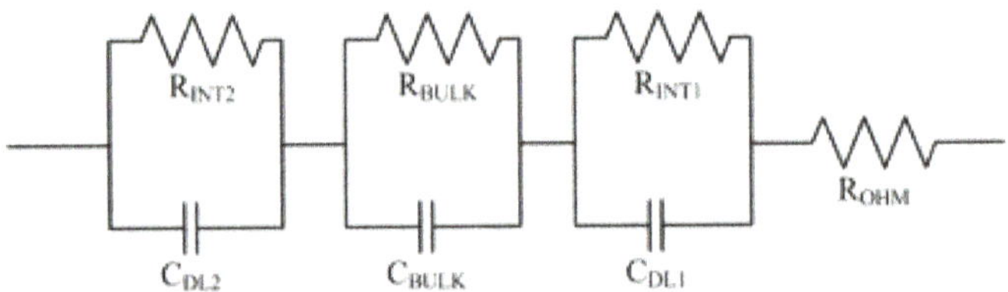

Fig. 5 The equivalent circuit of water sample and sensor electrodes

Figure 5 is the equivalent circuit of schematic diagram, C_{DL} is double layer capacitor, R_{INT} is interface resistances. R_{BULK} is the resistance of solution, C_{BULK} is the capacitance of the sensor and R_{OHM} is the resistance of cable and electrode.

As double layer capacity can be separate into a series of diffuse-layer and Stern-layer capacitors, the relationship between the double-layer, Stern-layer, and diffuse-layer capacitances can be written in the form

$$\frac{1}{C_{DL}} = \frac{1}{C_S} + \frac{1}{C_D} \tag{2}$$

C_S is only a minimal dependency on electrolyte concentration; therefore, it is customary to assume that C_S is constant at low electrolyte concentration which is often the case for the concentration of pollutants. On the other hand, C_D shows strong dependency on electrolyte concentration. This capacitance has a minimum value when the surface charge density is zero and increases exponentially with the surface charge density and electrolyte concentration. As a result, when concentration is low, C_S is much large than C_D, while under high electrolyte concentration, the value of C_D is much larger than that of C_S. Because the overall capacitance is always determined by the smaller of two capacitors in series and the pollutant concentration is relatively low, C_{DL} can be equated to C_D in this case[17]. The thickness of diffuse layer can be determined from the formula for Debye length λ DEBYE which is the measure of the distance in the electrolyte over which a small potential perturbation decays by $1/e$–effectively a region of space charge with an excess of charges where electroneutrality no longer holds, derived from the Poisson-Boltzman equation[17]. In equation 2 " is relative permittivity of material",0 is constant permittivity of a vacuum, K_B is Boltzmann constant, T is thermodynamic temperature, e_0 is elementary charge, z_i is charge of species type i (number of electrons), C$*$ is concentrations of charged species. Corrected by the applied voltage V, C_D (F=cm2) can finally express as [18]

$$\lambda_{DEBYE} = \sqrt{\frac{\varepsilon\varepsilon_0 K_B T}{8\pi e_0^2 \sum z_i^2 C_i^*}} \tag{3}$$

$$C_D = \frac{\varepsilon\varepsilon_0}{4\pi\lambda_{DEBYE}} \cosh\frac{z_i e_0 V}{4K_B T} \tag{4}$$

The interfacial resistance R_{INT} is predominantly caused by sorption resistance representing the impedance to current resulting from kinetics of specifically adsorbed or desorbed species in the electrochemical double layer. A derivation for electron-transfer kinetics in electroactive monolayers with an amount of adsorbed species Γ (mol/cm2) resulted in expressions for adsorption resistance R_{ADS} as[18,19]:

$$R_{INT} = R_{ADS} = \frac{2R_G T}{F^2 A\Gamma K_f} \tag{5}$$

Where R_G is the gas constant, T is thermodynamic temperature, F is Faraday's constant, A is the electrode surface area, and K_f is the rate constant for the adsorption-driven kinetics. In order to simplify the calculation of R_{BULK} and C_{BULK}, the model in Figure 4 can be seen as a parallel plate electrode. f in a bounded sample area with the surface area of the electrodes exposed to the analyzed sample A, and thickness f the sample between the electrodes d, according to Ohm's Law, The "bulk" media resistance is defined as

$$R_{BULK} = \frac{1}{\sigma}\frac{d}{A} = \frac{d}{AF}\frac{1}{\sum z_i \mu_i c_i^*} \tag{6}$$

The bulk resistance of a homogeneous material depends on the bulk concentration c^* of conducting species, their mobility u, charge z, sample temperature T, and the electrode geometry of the area A in which current is carried. According to the definition of capacitance, C_{BULK} can be express as

$$C_{BULK} = \frac{\varepsilon\varepsilon_0 A}{d} \tag{7}$$

Since each part of the equivalent circuit is obtained, the total impedance of the circuit is

$$Z = R_{OHM} + \frac{R_{BULK}}{1+\omega^2 R_{BULK}^2 C_{BULK}^2} + j\frac{\omega R_{BULK} C_{BULK}}{1+\omega^2 R_{BULK}^2 C_{BULK}^2} + \frac{R_{INT}}{1+\omega^2 R_{BULK}^2 C_{DL}^2} + j\frac{\omega R_{BULK} C_{DL}}{1+\omega^2 R_{BULK}^2 C_{DL}^2} \tag{8}$$

And two cut-off frequencies can be defined according to the frequency spectrum characteristics of the impedance.

$$f_L = \frac{1}{2\pi R_{INT} C_{DL}} \tag{9}$$

$$f_H = \frac{1}{2\pi R_{BULK} C_{BULK}} = \frac{\sigma}{2\pi\varepsilon\varepsilon_0} \tag{10}$$

The reason why the frequency concerned with solution is higher than that concerned with interface is because C_{DL} is usually much larger than C_{BULK}. When applied frequency is lower than f_L, the impedance can be simplified as

$$Z = R_{OHM} + R_{BULK} + \frac{R_{INT}}{1+\omega^2 R_{INT}^2 C_{DL}^2} + j\frac{d}{\omega\varepsilon\varepsilon_0 A} \tag{11}$$

The bulk impedance is simplified into bulk resistance, while all the imaginary part of the impedance is contributed by interfacial impedance. It is obvious that the interfacial impedance is not ignorable at this frequency range. As interfacial impedance is concern with ion species, it can provide a parameter in inorganic material identification. When the frequency is higher than f_H , the impedance can be simplified as

$$Z = R_{OHM} + j\frac{1}{\omega C_{BULK}} = R_{OHM} + j\frac{d}{\omega\varepsilon\varepsilon_0 A} \tag{12}$$

In this case, the impedance shows a strong capacitive character. As the capacitive impedance is decreasing with an increasing frequency, it is probably that only coated electrode remains at high frequency. When the frequency is between f_L and f_H, bulk resistance dominates the impedance, and can be used to evaluate conductivity and concentration according to equation 5. Taking the coating electrode in to consideration, R_{OHM} is set as 104ohm, some typical values of very low concentration solution are $R_{BULK} = 10^7 ohm, C_{BULK} = 10^{-10} F, R_{INT} = 10^6 ohm,$

$C_{DL} = 10^{-6} F$ .According to equations above impedances of the 10 times and 100 times concentration of the very low solution can be calculated. Figure 6 illustrates the spectrum of 3 concentrations.

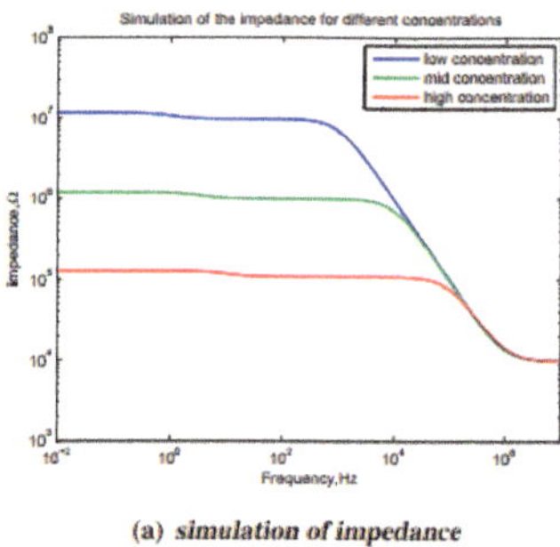

(a) *simulation of impedance*

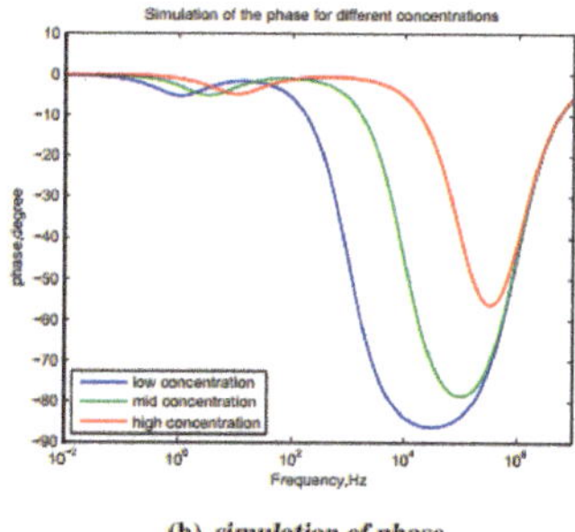

(b) *simulation of phase*

Fig. 6 Simulation of spectrum at 3 concentrations

Figure 6, the impedances can be divided into three parts by the two cut-off frequencies. Although the value of interfacial impedance is relatively smaller than bulk impedance, the decreasing process of interfacial impedance can be seen clearly in the figure of phase. The altering of impedance characteristic from resistance to capacitance is obvious too. As discussed above, the best region for interfacial impedance measurement is below f_L, the frequency between f_L and f_H is suitable for material conductivity and concentration of conducting species C^*, while the ε can be calculated at frequency above f_H. The influence of concentration is distinct, too. The impedance and at higher concentration is evidently smaller than that of lower concentration, while the cut-off frequencies of higher concentration is larger than that of lower concentration. The differences are mainly caused by larger resistor in purer water.

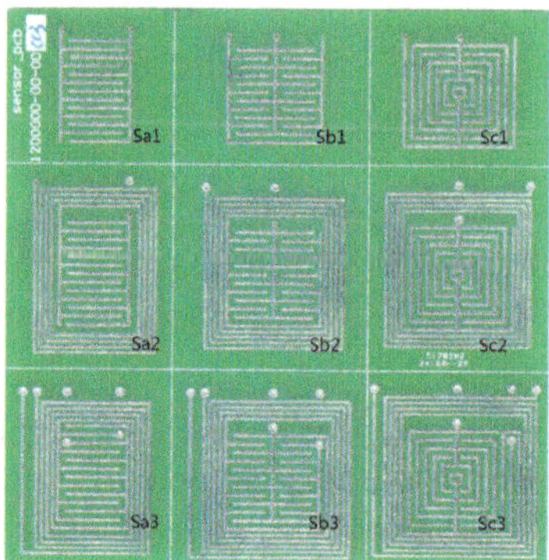
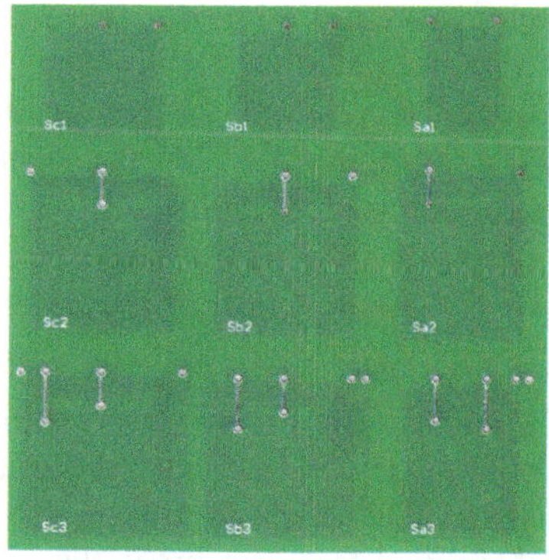

Fig. 7 Top and bottom layers of the planar electrode sensor

Figure 7 shows the fabricated sensors. Sa, Sb, Sc represent different kinds of interdigital electrodes, and the following number indicates the coils type. The sensors are fabricated by PCB technology, and the thickness of the board is 1.6mm.Copper track is covered by stannum so that the thickness of the metal is increased. The width of the copper and the pitch distance are 0.5mm. The whole PCB is 99mm ×99mm× 1.6mm, and the area of each sensor is about $1100mm^2$ All the sensors are coated with silicon rubber to form a coating layer in order to prevent any direct contact with water sample.

6.4 Design of the Data Acquisition Board

With the development of the technology of integrated circuit and the improvement of the sensor design technology, it becomes more and more important to improve the data colleting ability and the extensibility of the data acquisition system. However, in the present market, the open source hardware which can meet the standards and requirements to provide the convenience of interfacing the diversified data collecting board are in a large shortage, in this context, we made a research on the types of the interfaces and the data types of recent different kinds of sensors. According to IEEE1415 interfaces of intellectual sensors standards and regulations as shown in Figure 8, a general water quality data acquisition system is self-designed and this system can meet the following requirements:

1) The solution of high performance and low power consumption: the selection of the overall design of the hardware is strictly controlled to lower the power consumption in order to adapt to the bad wild environment, hence improving the reliability of the system. At the meantime, the 32 bit microcontroller can handle with all kinds of data and algorithms.
2) The play-and-plug interfaces which can be used in different kinds of water quality sensors. The design of the circuit of the interface accords with the IEEE1451 interface of intellectual sensors standard and regulations, which has a strong compatibility[21].
3) The rationalization of the PCB design of the circuit board: the design of the circuit accords with the general principal of the design of the PCB, and it has well anti-jamming performance, which can obtain the best performance and leave the possibility of modification and extending.
4) The transplantation of the operation system to improve the multi-tasks processing ability. Since it requires the hardware possesses of multi-tasks processing ability of the interface of different kinds of sensors and the ability of processing and transferring for the diversified data, it is of necessity to transplant the lightweight system μC/OS-II completely.

The component of data acquisition board is shown as shown in Figure 9. Firstly, the self-designed data acquisition board downloads the control programs via J-LINK; Under the control of STM32MCU, on one hand, AD modules collect the information from sensors in the real time, while on the other hand, the internet modules will be in charge of transferring the data after some necessary operation programs; Additionally, the internet modules will transfer the data package to the next node, supported by TCP/IP protocol; finally, the client program will receive the data package and display them in real time. And then, we will make a detailed description of the water quality data acquisition system from the hardware and the software[22].

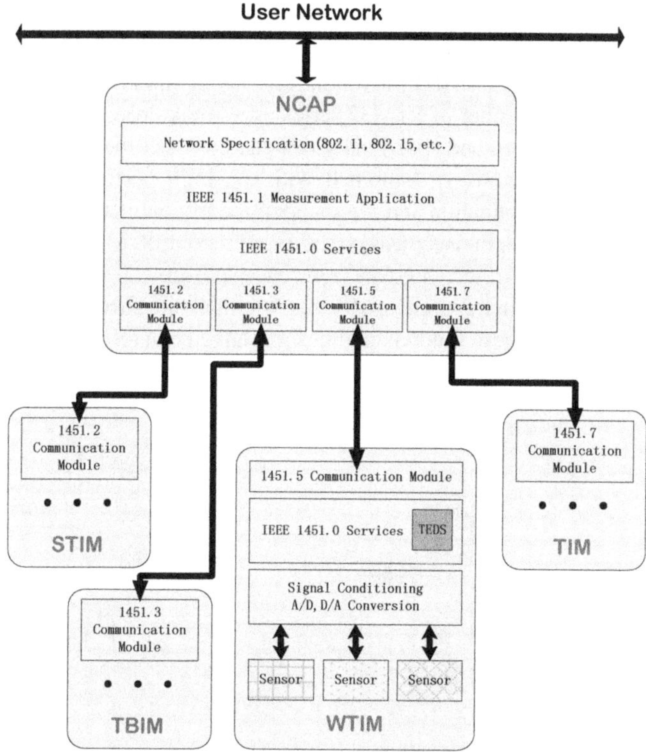

Fig. 8 The IEEE1451 interface of intellectual sensors standard

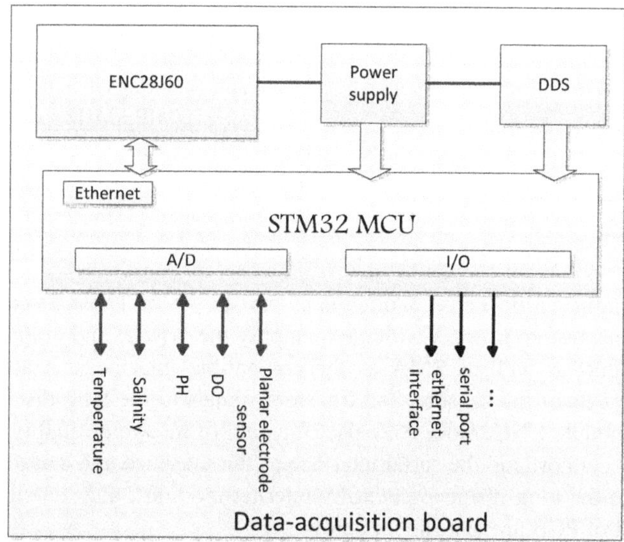

Fig. 9 The component of data acquisition board

6.4.1   The Hardware Design of Data Acquisition Board

This system employs STM32f103 microcontroller as the main chip, and such 32bid flesh memory microcontroller of STM32 series uses breakthrough Cortex-M3 kernel which is produced by Arm and specifically designed to meet the requirements in the field of embedded technology, combined with the high performance, low-power dissipation, real-time application and the competitive price together. The whole data collecting hardware system is consisted of master control MCU-STM32, J-Link debug circuit, power circuit, regulate reference power, DDS signal generator, integrated power amplifier circuit, and other internet interface circuit and sensor interface circuit. All of these fundamental designs have been tested to be reliable and stable and they all reach the requirements of the simulations[23].

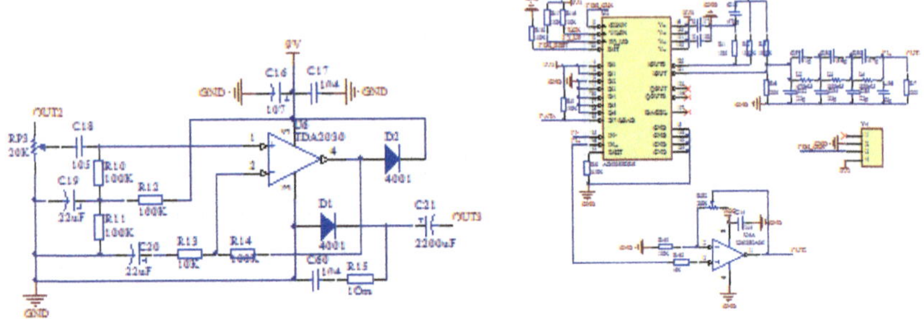

Fig. 10 Power amplifier circuit and DDS signal generator circuit

Table 2 Basic information of the Components

Basic information of the Components	
STM32f103Z	maximum operating clock frequency 72Mhz,1.25DMIPS/MHZ,
AD9850	maximum clock frequency 25MHZ
DDS signal generator	sinusoid frequency 10Khz–1Mhz
ENC28J60	clock frequency 25Mhz,28pin

Basic information of the Components as shown in Table 2.As for the power supply, this system uses 5V voltage to supply the power, and employs serial port, Ethernet port as well as SMA as the compatible interfaces, it also uses AD to collect signals of the sensors, and transfers signals to the long distance clients via internet module ENC28j60. And also, the data of the water quality sensors which take Ethernet port and the serial interface as the interface and can transmit the data directly to the long-distance client by internet modules, and the circuit board and the layout of PCB are just shown below.

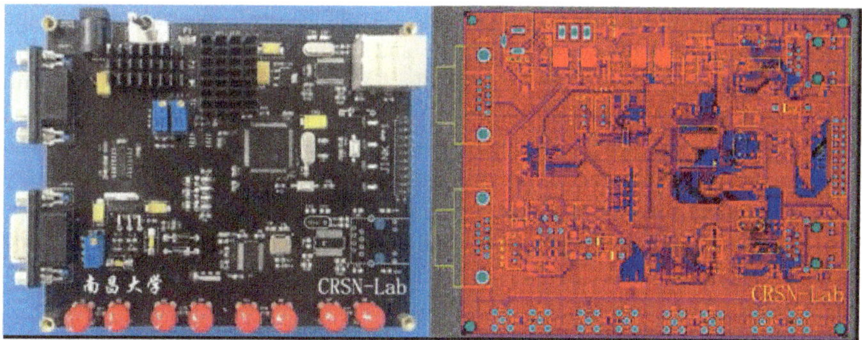

Fig. 11 Circuit board and the layout of PCB

ENC28J60 is an independent Ethernet controller with the standardized serial peripheral interface (SPI), which can be used as the Ethernet interface for any application equipped with the SPI controller. ENC28J60 meets all the rules and regulations of IEEE 802.3, and employs a series of packet filtering mechanisms to limit the input data package. And also, it provides an internal DMA module to realize the IP calibration and calculation supported by the hardware and the rapid data throughput. As for the communication with the main controller, it is realized by SPI and two interrupt pins, whose transmission speed can be reachable to 10Mb/s, and such two dedicated pins are used to connect LED to instruct the state of network activities.

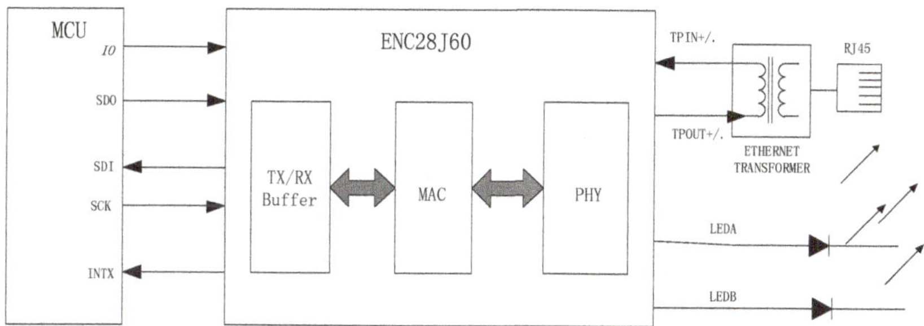

Fig. 12 Control system element of ENC28J60

6.4.2   *The Software Design of Data Acquisition Board*

The purposes of hardware initializing process are mainly for the clock configuration of the controller itself and the initialization of the peripheral equipment. As a matter of fact, to realize the execution of multi-tasks, an operation system is in need, since µC/OS-II has the perfect information box mechanism, it can well meet the requirements of the communication between the two main tasks in this design. As a preemptive real-time multi-tasks kernel, μC/OS-II is upgraded from μC/OS, whose code is open-sourced and can be clipped, it is written by ANSIC language and includes a small part of assembly language code, hence making it available for different micro-controllers with diverse architectures.

Since μC/OS-II is just a real-time operation kernel, it only includes some basic functions such as task scheduling, task management, time management, memory management and the communication and synchronization among tasks, however, other additional services like input and output management, file system and network are not included.

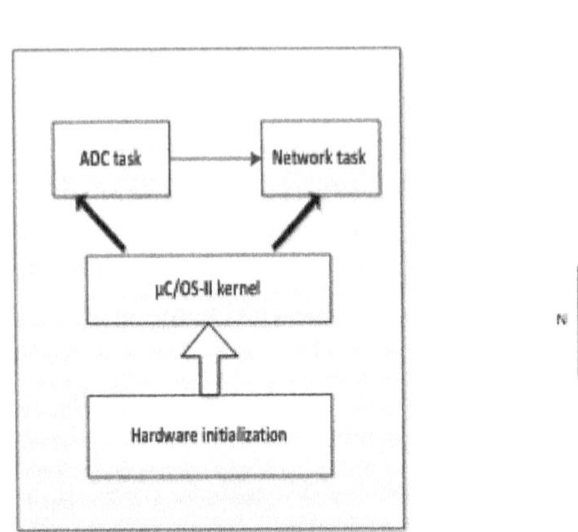

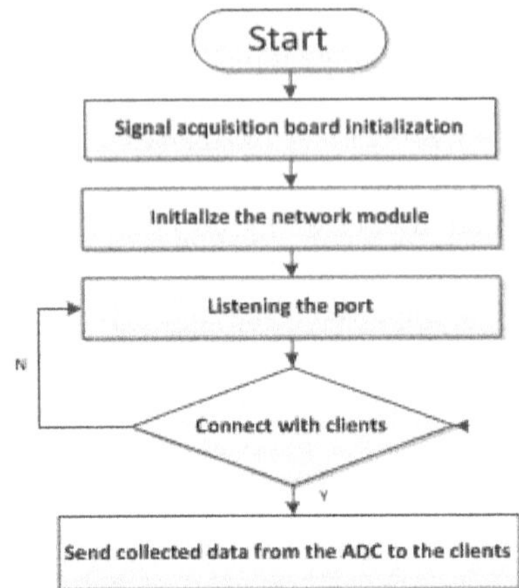

Fig. 13 The flowchart of the program

The transplantation operations of μC/OS-II involves five files,namely, includes.h ,os_cpu.h ,os_cpu_a.asm ,os_cpu_c.h and os_cfg.h. The detailed transplantation procures have been introduced in many books and references both online and offline, and we have consulted a lot to them when it comes to the study of transplantation methods, therefore, we just give the following flowchart of the operations of transplantation and you'll ignore the detail description from this part.

6.5 The Distributed Data Wireless Transmission

To make the water quality intellectual sensor terminal adapted in the complicated environment, the whole equipment should be portable and the process of the operation is supposed to be independent, so that they can minimize the independence on the external devices. There are some alternatives which are shown in the following:

1)   The generalization of the GPRS technology makes it possible to be the alternative with a high coverage of the telecommunication network. However, this technology has a big disadvantage which are respectively the lower transmission rate and the high dependence. Theoretically, the transmission speed is 171.2Kbps, which can no meet the requirement of the specific situation, moreover, the transmission speed for the present cannot reach that. Meanwhile, it is necessary for the GPRS to run on the carrier network, especially in the application of the earthquake relief work and the investigation activities in the field. Due to the low coverage, the damage of the base station, congestion issues of the public network and the other problems, the transmission is not of good quality and reliable which may cause the failure of the transmission.

2)   It can be accessible to the relatively reliable connection and the high-speed transmission rate by the WLAN technology, but it still remains some drawbacks. First of all, the transmission distance is short, and if we enhance the transmit power of the antenna by force, it will damage the human health and the hardware equipment, which cannot meet the design requirements. And secondly, such technology is independent on AP, and if the users will communicate with each other, they first should connect a fixed AP. And also, this technology cannot be well accessible in the remote areas or when the fundamental facilities are damaged.

3)   The transmit speeds of the four 3G communication programs which are CDMA2000, WCDMA, TD-SCDMA and WiMAX, all meet the requirements of transmission, but same with GPRS technology, 3G mobile communication technology is also of great independence of the base station, and they share the disadvantages of the constraints of the coverage and the malfunction of the equipment of the base station. And at the same time, restricted by the long distance of the communication in the field and the vast amount of the video data, all of the technologies, such as the Bluetooth, Zigbee, the infrared technology and other close quarters technologies, cannot reach the requirements of the design, due to the constraints of the bandwidth and the distance.

Based on the requirements of the different kinds of wireless technology analysis and the data transmission ability of water quality sensor terminal, the distributed data wireless transmission modules employ the self-designed open source router hardware MCU-Mesh solution as shown in Figure 14, which can provides multi general interfaces with plug and play, and can be well extended. In wireless Mesh network, every node can send and receive signals, and every one of them can directly communicate with one or more peer nodes, in this context, the transmission bandwidth and transmission distance can be reachable at 12Mbps and thousands miles respectively, which fully meet the requirements that the sensors of ecological perceptual layer should sense information and video information and transfer the spatial information. Some intellectual advantages of wireless Mesh technology such as ad-hoc network, self repairing, and self administration, have been fully taken in the distributed transmission modules, which can highly transfer the data processed by the upper level to the intellectual client.[24]

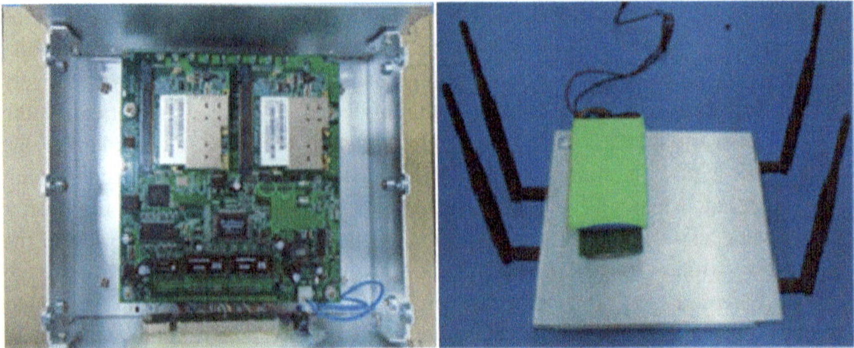

Fig. 14 Hardware structure of NCU-Mesh

6.5.1   Mesh Network

Wireless Mesh network, which is also called multi-hop, is a new wireless network technology different from the traditional wireless network. In the traditional wireless local area network (WLAN), each client gets access to the internet via a wireless link connected with AP, users need to visit a fixed access point (AP) to realize the communication with others, and such network structure is defined as single-hop network. While in wireless Mesh network, any node of the wireless applications can be used as both an AP and a router at the same time, and every node in the network can send and receive signals, and communicate with one or more peer nodes directly. Compared with the traditional communication method, such as GPRS, CDMA, and Bluetooth and so on, Mesh has a great advantage over these wireless technologies as follows[25].

> **Broadband Transmission**
> The transmission bandwidth of GPRS, CDMA and Bluetooth is not large, and generally, all the bandwidths of the above three are simply large enough to transfer data and audio, while the maximum bandwidth of Mesh technology can reach about 12Mbps, which can completely meet the requirements of the communication both in audio and video.

> **Be Available to Connect Wherever and Whenever Without Infrastructures**
> Both GPRS and CDMA need to be supported by the infrastructures, where there is no base station, there is no communication. In the case of emergency, it will be bound to affect the quality of communication if the fixed communication infrastructure is destroyed. However, as for Mesh technology, infrastructure is not the necessity, and it only takes three vehicles to make up an ad-hoc network to realize the wireless communication without getting access to GPRS or CDMA .

> ## The Long Transmission Distance
>
> Compare with the Bluetooth Technology, the transmission distance of Mesh system can be reachable at hundreds of thousands miles. Although Bluetooth technology does not need the support of infrastructures, its communication distance is only about 10 miles, hence making it not suitable to the large distance communication.

> ## The Strong Network Survivability
>
> As long as it is in the range of communication, Mesh system can realize the communication, and it is irrelevant with the number of nodes. Also, the join and exit of the network for the communication nodes have little influence on the data transmission.

> ## Great Confidentiality
>
> Since such network is a network of its own, other applications which are not in the range of the same frequency band will not be available to communicate. At the same time, other communication applications which are out of the range cannot communicate with it, because such network only exists in the range of vehicles equipped with such apparatuses.

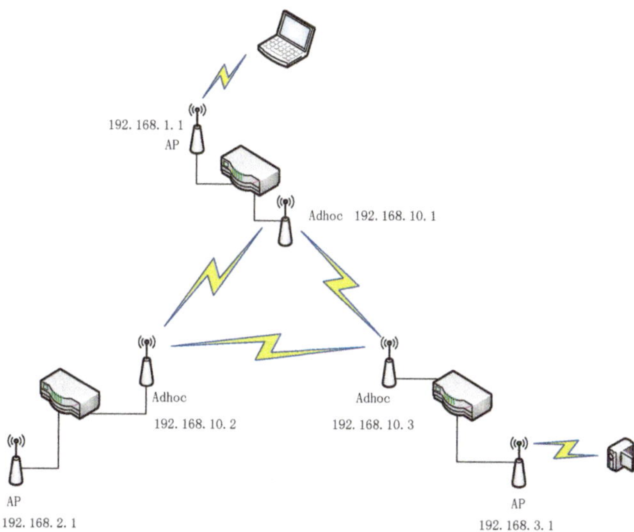

Fig. 15 The application map of Mesh network

> **Low Cost**
> Supported by the data bases, it takes a lot extra communication fees for GPRS and CAMA, while it does goes for the Mesh communication technology.

Taking Figure 15 as an example, only four Meshes can constitute a temporary emergency network to realize the tasks of rescuing and monitoring.

6.5.2 Constitution of NCU-Mesh Hardware System and Software System

The design of the NCU-Mesh hardware adopts the AR7161 wireless network controller chip solution developed by Atheros as shown in Figure 16. The AR7161 is Atheros' first-generation high performance, cost effective and scalable wireless network processor that enables efficient design of solutions that address triple-play services like voice, video and data. When combined with the Atheros IEEE 802.11a, 802.11b, 802.11g and 802.11n wireless chipsets, the AR7161 provides customers with bestin-class WLAN solutions for home and enterprise access points, routers and gateway applications. To ensure the good extendibility and the enough number of the interface, such as serial port, USB2.0 interface and four LAN ports and so on. The detailed hardware configuration is as follows.

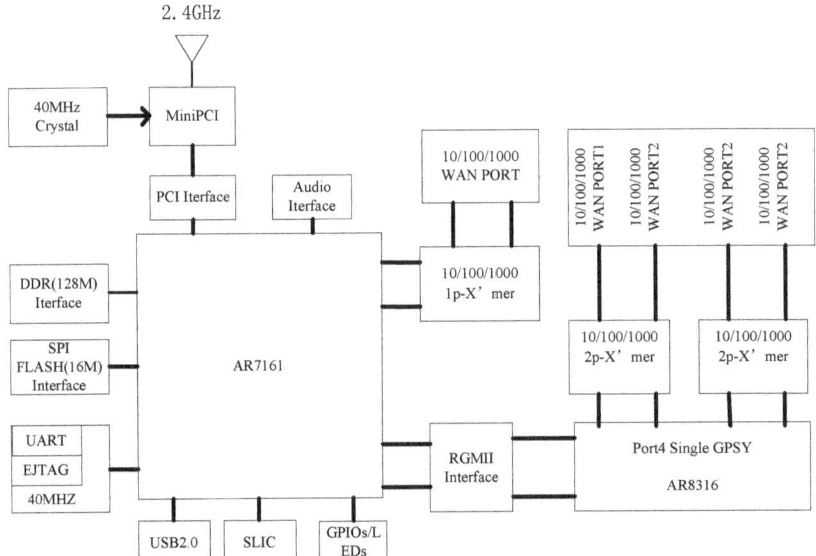

Fig. 16 Constitution of NCU-Mesh hardware system

1) Mainboard: RT210W of Askey
2) CPU: 4702 of BROACOM, 125MHZ
3) Memory: SD with a memory of 16M
4) One FLASH: 4M
5) One Ethernet interface: under the standard of 10/100Base
6) One Wireless Network Card: 4318mini PCI of BROADCOM
7) A piece of 4318 Card
8) One Antenna of 2db
9) And one jumper with shield magnet ring uf.1 to rp-sma

Table 3 The parameters and the technical indexes of the hardware system

parameters	technical index
Supply voltage and current	12V±0.5V ; 200mA
Single-hop transmission distance	Ordinary cases, 300-500 meters, up to 15 km
the number of Supported users	100-200
maximum throughput of Physical layer	300megabits per second

The software diagram of the system as shown in Figure 17.To fully takes the advantage of the broadband transmission of NCU-Mesh, for example, the convenient connection wherever and whenever, without need of infrastructures, the relatively long transmission distance, the strong survivability of network, the low cost as well as the great confidentiality, it is of great importance to code the controlling program and the configuration files, after well configure the NCU-Mesh hardware circuit. And the functions of the NCU-Mesh applied in the water quality sensor terminal is well shown in the above chart. The bottom layer of the system supports 802.11g protocol, infrastructure and ad-hoc(peer network) operation mode, and it works in the frequency band of 2.4GHz. As for the routing layer and transport layer, they respectively adopts REAR router protocol and BCRT transmission control protocol. After the configuration, Mesh network is built to realize the real-time data transmission to the clients and the real-time monitoring of the water conditions. And the detail configuration process is shown in the following:

1) Transplant the openwrt system, which means downloading the official firmware of openwrt to the open source hardware via TFTP protocol.
2) Download the related software, which means connecting the wan interface of the open source router card to the Ethernet, and then downloading some necessary software and x-driver by logging into the routers via telnet protocol.
3) Configurate the openwrt files, which means entering the configuration files of the open source routers such as wireless, network, firewall, olsrd. config and so on.

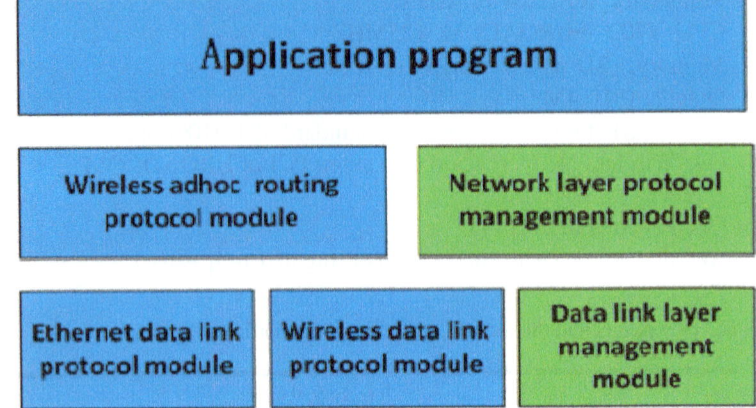

Fig. 17 The software diagram of the system

6.6 The Experiment Setup and the Performance Testing of the Water Quality Monitoring Terminal

6.6.1 The Testing of the Sensor and the Performance Analysis

Several experiments were conducted to test and evaluate sensors. The experimental setting is shown in figure 18. A function waveform generator which can create 10Vpp standard sinusoidal waveform was set as the input signal for the sensors. The frequencies arrange from 1 kHz and 1 MHZ. A beaker is used for the sample container and the sensor was immersed into the water sample. The Agilent 9404A mixed signal oscilloscope was used to record the output signals. The impedance of absolute value, real part and imaginary part can be calculated by equation.

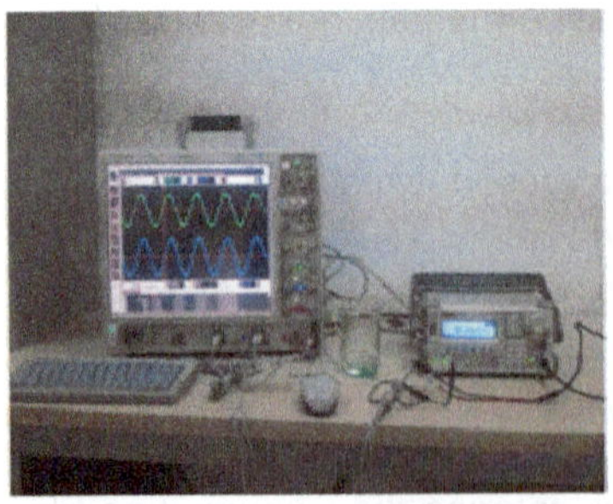
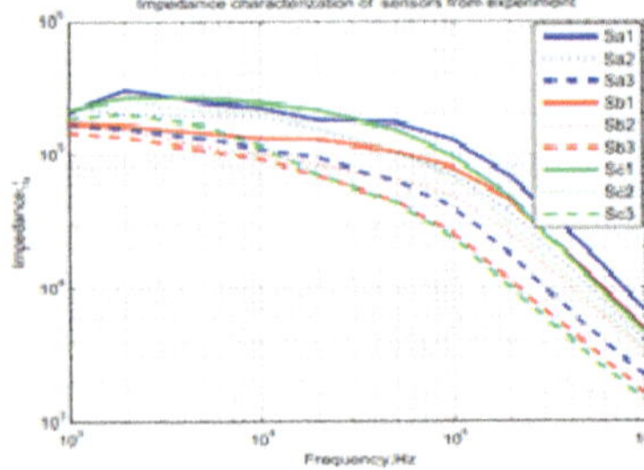

Fig. 18 The testing platform and impedance spectrum of sensor

For the same type interdigital electrode, the type with no meander has the highest impedance, and the only one meander type has medial impedance, and the same interdigital with double meanders has the lowest impedance .For the same meander, traditional interdigital has the highest impedance while the impedance of hooked interdigital and double interdigital are similar. Unlike the stationary status at low frequency in simulation, the impedance of sensors in experiment decreases when frequency is increasing.

The aim of the quantitative experiment is to testify the respondent reactions of the planar electrodes sensors in the two different inorganic salt solutions of different concentrations. In the experiment, the two different solutions are respectively the 5mg/l sodium nitrate and Sodium dihydrogen phosphate solution and the 10mg/l sodium nitrate and sodium dihydrogen phosphate solution, and the frequency the tests ranges from 20HZ to 1MHZ. It can be figured out by the following data of the figure 19 that the planar electrode sensor has a good resolving capacity in all of the four solutions, especially when it is in a lower range of frequency. The stability in fixed water of planar electrode sensor directly determines the effectiveness of the sensor, therefore we select sc2 sensor to test stability at solution concentration of 1mg/L,10mg/L and 100mg/L, and the sampling rate is 10min/sample. The results data demonstrate that a good sensor reliability as shown in figure 20.

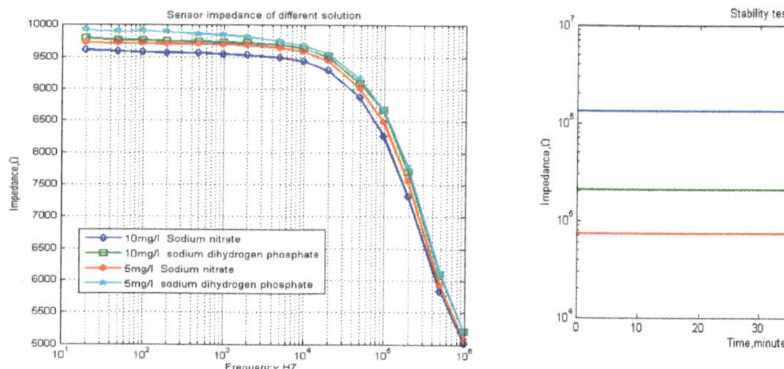

Fig. 19 The impedance spectrum in the two solutions of different concentrations

Fig. 20 The stability of planar electrode sensor in water monitoring

6.6.2   The Performance Testing of Data Acquisition Board

What the data acquisition board collects in the real time are the planar electrode sensor data of the five different structures and the five conventional sensor data. Several experiments were conducted to test and evaluate data acquisition board. In the experiment, the five conventional sensors are located in the same water situation, and the data monitored in the different time are shown in the following picture, and it shows that the stability is great. To testify the effectiveness of the data acquisition board applied in collecting the data of planar electrode sensors sensor sb2 is selected in the 10mg/L sodium nitrate solution to make two contrasted experiment, one adopts the exact signal generator and the oscilloscope to collect the data, and the other is the self-designed data acquisition board. It can be worked out in the following figure contrast that the maximal error is about 4.24% in the two different conditions.

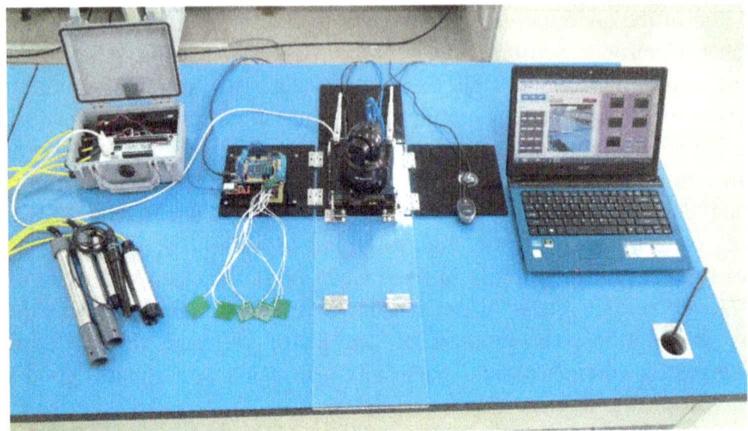

Fig. 21 The water quality monitoring terminal

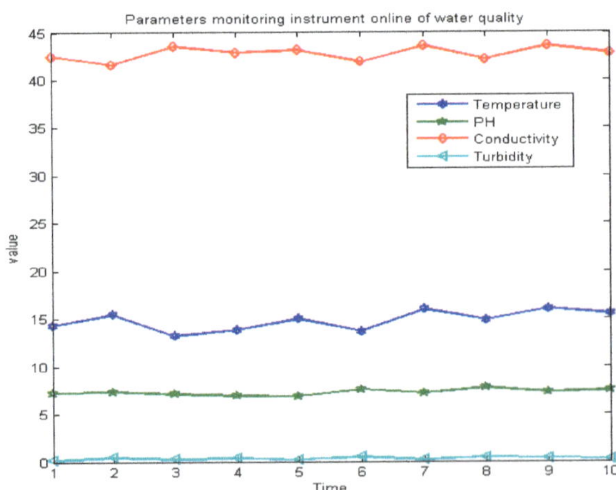

Fig. 22 The four conventional parameters monitored in the different time

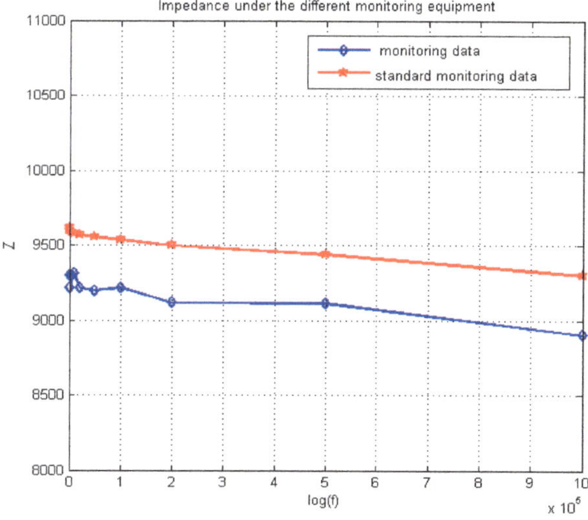

Fig. 23 The measured data of the data collecting board and the standard instrument

6.6.3   *The Testing Results of the NCU-Mesh*

After the configuration of the NCU-Mesh, the system will run the REAR route protocol, and transfer the video data packet collected by the IP Camera. In the testing process, the water quality sensor terminal will be located on the river bank, which is used to testify the data transmitting speed of 10m, 50m, and 200m away from the terminal. And the result shows that the video can be played normally in the process while the data transmit speed can be reachable at least for 2.58 Mbps.

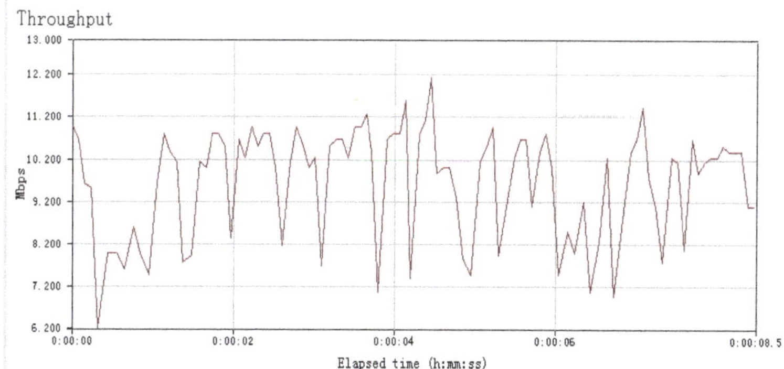

Fig. 24 The data transmit speed of the 10m away from the terminal. The maximal data transmit speed is 12.121Mbps and the minimum is 6.202 Mbps, and the average is 9.443Mbps.

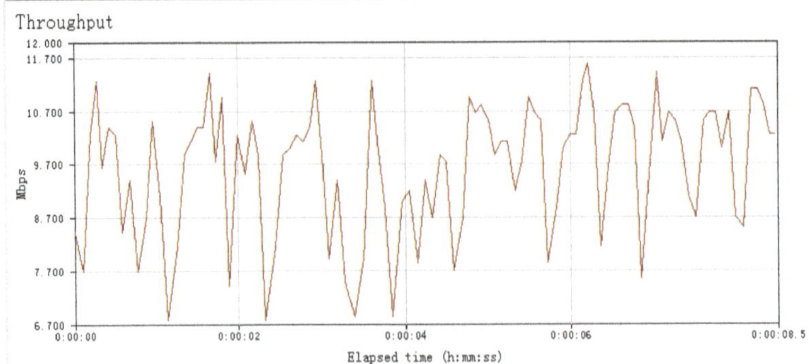

Fig. 25 The data transmit speed of the 50m away from the terminal .The maximal data transmit speed is 11.594Mbps and the minimum is 6.780 Mbps, and the average is 9.431Mbps.

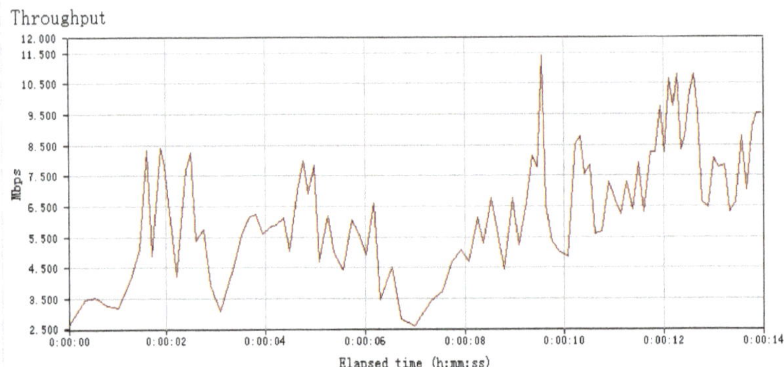

Fig. 26 The data transmit speed of the 200m away from the terminal .The maximal data transmit speed is 11.429Mbps and the minimum is 2.589 Mbps, and the average is 5.763Mbps.

6.7 Conclusion

In this paper, based on five conventional parameters of water quality monitoring sensors, IP cameras and GPS module positioning technology space, and combined with the excellent performance of planar electrode sensor which used for inorganic ions monitoring, as well as self-designed multi-channel signal acquisition which using a distributed Mesh networking technology achieve high bandwidth long-distance data transmission, we propose a newly designed water quality monitoring terminal. The terminal contains not only novel planar electrode sensor but also some commercial detecting and monitoring devices while providing all kinds of required interfaces. A low power consumption MCU is employed for data acquisition, data storage and other managements. Data and instructions are transmitted by a Mesh network, grouping of the monitoring nodes before transferring the sensor data to higher levels. Our planar electrode sensor and wireless Mesh technology intelligent terminal water quality is successfully developed, and has certain reliability and validity, achieve the desired overall performance.

References

[1] Kassal, P., Steinberg, I.M., Steinberg, M.D.: Wireless smart tag with potentiometric input for ultra low-power chemical sensing. Sensors and Actuators, pp. 254–259 (2013)

[2] Yue, R., Ying, T.: A Novel Water Quality Monitoring System Based on Solar Power Supply and Wireless Sensor Network. Procedia Environmental Sciences, 265–272 (2012)

[3] Chang, N.-B., Pongsanone, N.P., Ernest, A.: A rule-based decision support system for sensor deployment in small drinking water networks. Journal of Cleaner Production, 152–162 (2013)

[4] Mukhopadhyay, S.C., Mason, A. (eds.): Real-Time Water Quality Monitoring. SSMI, vol. 4. Springer, Heidelberg (2013)

[5] Ribero, H.M.C., Almeida, A.C., Rocha, B.R.P.: Water Quality Monitoring in Large Reservoirs Using Remote Sensing and Neural Networks. IEEE Latin America Transactions 6(5) (September 2008)

[6] Capella, J.V., Bonastre, A., Orsa, R., Peris, M.: In line river monitoring of nitrate concentration by means of a Wireless Sensor Network with energy harvesting. Sensors and Actuators, B 177, 419–427 (2013)

[7] Zhuiykov, S.: Solid-state sensors monitoring parameters of water quality for the next generation of wireless sensor networks. Sensors and Actuators, B 161, 1–20 (2012)

[8] DeRouin, A.J., You, Z., Hansen, M., Diab, A.: Development and Application of the Single-Spiral Inductive-Capacitive Resonant Circuit Sensor for Wireless, Real-Time Characterization of Moisture in Sand. Journal of Sensors 2013, Article ID 894512, 7 pages

[9] Schuster-Wallace, C.J., Grover, V.I., Adeel, Z., Confalonieri, U., Elliott, S.: Safe Water as the Key to Global Health, The United Nations University, pp. 4–7 (2008)

[10] Li, Y., Migliaccio, K.: Water Quality Concepts, Sampling, and Analyses, p. 113. Taylor and Francis Group (2011)

[11] Colin, R., Quevauiller, P.: Monitoring of Water Quality, pp. 26–29. Elsevi (2009)

[12] Schlicker, D., Washabaugh, A., Shay, I., Goldfine: Inductive and capacitive array imaging of buried objects. Insight Nons-Destructive Testing and Condition Monitoring 48(5), 302–306 (2006)

[13] George, B., Zangl, H., Bretterklieber, T.: A Combined Inductive-Capacitive Proximity Sensor and Its Application to Seat Occupancy Sensing. In: I2MTC-International Instrumentation and Measurement, Singapore, pp. 13–17 (2009)

[14] Guo, Z., Yuan, J.: Design of Ocean Intelligent Sensor Based on STM32. In: 2009 International Conference on Test and Measurement, pp. 124–127 (2009)

[15] Song, E.Y., Lee, K.B.: Sensor Network based on IEEE 1451.0 and IEEE p1451.2-RS232. In: IEEE International Instrumentation and Measurement Technology Conference, I2MTC 2008, May 12-15 (2008)

[16] Rabbi, M.F., Rahman, M.T., Uddin, M.A., Salehin, G.M.A.: An Efficient Wireless Mesh Network: A New Architecture

[17] Zhao, C.L., Qin, M., Huang, Q.A.: A Fully Packaged CMOS Interdigital Capacitive Humidity Sensor With Polysilicon Heaters. IEEE Sensors Journal 11(11), 2986–2992 (2011)

[18] Justin, B.O., Zhanping, Y., Julian, M.B., Ee, L.T., Brandon, D.P., Keat, G.O.: A Wireless, Passive Embedded Sensor for Real-Time Monitoring of Water Content in Civil Engineering Materials. IEEE Sensors Journal 8(12), 2053–2058 (2008)

[19] Bhadra, S., Bridges, Thomson, Freund, M.S.: Electrode Potential-Based Coupled Coil Sensor for Remote pH Monitoring. IEEE Sensors Journal 11(11), 2813–2819 (2011)

[20] Yunus, M.A.M., Mukhopadhyay, S.C.: Novel Planar Electromagnetic Sensors for Detection of Nitrates and Contamination in Natural Water Sources. IEEE Sensors Journal 11(6), 1440–1447 (2011)

[21] Lee, K., Song, E.: Wireless Sensor Network Based on IEEE 1451.0 and IEEE 1451.5-802.11. In: The Eighth International Conference on Electronic Measurement and Instruments, pp. 5–802 (2007)

[22] Hu, J.: The Design of Wireless Data Acquisition System Based on STM32 and virtual instrument. IEEE (2012) 978-1-61284-683-5/12

[23] He, G., Pan, X., Liang, S.: Research on the Designation of Diesel Engine Data Acquisition. IEEE (2011) 78-1-4244-8165-1/11

[24] Takahashi, J., Yamaguchi, T.: Communication Timing Control and Topology Reconfiguration of a Sink-Free Meshed Sensor Network With Mobile Robots. IEEE-ASME Transactions on Mechatronics 14(2), 187–197 (2009)

[25] Kim, Y.S., DeBruhl, B., Tague, P.: MeshJam: Intelligent Jamming Attack and Defense in IEEE 802.11s Wireless Mesh Networks, IEEE 10th International Conference on Mobile Ad-Hoc and Sensor Systems. IEEE (2013) 978-0-7695-5104-3/13

Chapter 7
Application to Environmental Surveillance: Dynamic Image Estimation Fusion and Optimal Remote Sensing with Fuzzy Integral

Zhongliang Jing, Han Pan, and Gang Xiao

Abstract. Image fusion has been increasingly important in environmental surveillance. It is well-known that multi-scale decomposition methods lack of spatial-temporal adaptability. To tackle this problem, based on Kalman filtered compressed sensing, a dynamic image estimation fusion framework is provided. A parametric spatial-temporal fusion method is proposed for modelling fusion pattern. This method is capable of learning model parameters adaptively in state space. Furthermore, an optimal fusion method for remote sensing images is presented, which exploits statistical property of the decomposition coefficients and utilizes the fuzzy integral. The numerical experiments on the ground-truth datasets indicate the efficiency and effectiveness of the proposed methods.

7.1   Introduction to Image Fusion

The term image fusion is commonly referred to a universal procedure for combining multi images into a single image [1][2]. This process can make the information, obtained from different sensors, more visual perceptible, processing and understandable. It is a natural way to take advantage of the multi-sensors to efficiently monitor the environment. Particularly, it is clear that the imaging sensors become more and more available for environment sensing, e.g., visible and infrared cameras, laser and other spectral sensors. This process has a significant impact on the field of environmental surveillance. More exactly, it is very important to merge, detect, and discriminate the relevant information simultaneously

Zhongliang Jing · Han Pan · Gang Xiao
School of Aeronautics and Astronautics,
Shanghai Jiao Tong University, China
e-mail: zling@sjtu.edu.cn

© Springer International Publishing Switzerland 2015
H. Leung and S.C. Mukhopadhyay (eds.), *Intelligent Environmental Sensing,*
Smart Sensors, Measurement and Instrumentation 13, DOI: 10.1007/978-3-319-12892-4_7

when measurements from different sensors are available. This issue recently has attracted considerable research attention due to its wide range of potential applications. Moreover, the main purpose of information fusion is to reduce the uncertainty representation of the information and provide universal representational formation in term of the spatial layout and the spectral characteristics of the observed scene. In this chapter, the problem of dynamic image fusion is explored for merging more salient features by taking the space-varying sensing process of multiple sensors into account.

7.1.1 *Limitations of Single Sensor System*

Single sensor cannot provide a compact and complete representation of the observed scene. The common imaging sensors in various fields are the charge couple device (CCD) and complementary metal-oxide semiconductor (CMOS). These sensors are usually band-limited and provide complimentary information, depending on the field of view (FOV). In the practical applications, the sensors capture the actual scene at different times, views, frequency-bands and exposure times because of the different operation situations. These sampling procedures can lead into a huge requirement on the storage and computational cost to process the resulting captured images. More exactly, some algorithms are required to deal with the problems of multi-modality, multi-temporal, multi-view and multi-exposure image pairs or sequences. In addition, some considerations should be account for removing the effect of the inherent limitation of different conditions, such as the sensor noise, blocky artifacts and miss-registration. Image sequences, usually in the form of spatial-temporal volumes, would increase the workload for displaying and processing the redundant and complementary information. Therefore, an alternative approach should be developed for preserving the frequency characteristic of these images and integrating multi-source information in a single view. In the overall procedure, it is important to identify the technological possibilities and issues for determining how the information management system of the environment sensing must be designed to meet with these challenges and providing a comprehensive perspective on the true scene. In summary, some mechanism should be explored for handling these issues and challenging regarding the process of the environmental sensing and surveillance. Image fusion can be a common scheme for these problems within a dynamic environment. Moreover, some methods, depending on the application fields or environment, should be promising balance among the memory requirement, computational cost and complexity.

7.1.2 *Advance of Image Fusion*

Multi-sensor image fusion is capable of favorable development because of its importance in environmental surveillance and monitoring. The increasing environmental sensors not only provide an uncertainty and redundancy representation of the observation scene, but also offer a potential to learn more about several low level mechanisms of the system. More exactly, this scheme can improve the coverage of the systems of environment sensing and the robustness to various operations situations. For example, this approach can combine geometrical features of the observed scene, improve the capability to discover, extract and classify the desired objects in the application of remote sensing. Moreover, it can save a huge amount of storage space and computation time. Obviously, image fusion provides an alternative solution to these problems arising from the application of environmental sensing, at least to some extent.

Intuitively, environmental sensing can be viewed as a procedure to distinguish and learn the complex condition of the environment. Moreover, environmental sensing not only requires the collection of the information from a group of the sensors, but also it is desired to process it and discover the pattern of its internal behavior. Fortunately, image fusion offers an improved comprehension on the underlying structure or behavior of the environment. It can integrate pertinent geometrical information into a single informative image, e.g., a combination of high spatial and spectral or spatial-temporal information. This technology can be categorize into three classes broadly, i.e., pixel-level, feature-level and decision-level. Significant attention has been given to this topic. In this chapter, we focus on the pixel-level and optimal image fusion because of their excellent effectiveness on the construction of the complete representation of the perceived scene.

7.1.3 Related Works

Classical image fusion methods [1] include multi-resolution analysis (MRA) based schemes, principal component analysis (PCA) and artificial neural networks (ANNs) and etc. Among these methods, the MRA based schemes are proved to be more preferable, including pyramid based and wavelet-like transformations. In a nutshell, most MRA based fusion algorithms concentrate on two primary aspects. On the one hand, numerous methods, especially wavelet-like decomposition methods, are developed for decomposing the image into different levels according to user-defined design criteria. These image representation methods provide various alternative solutions to model the local features of image content. On the other hand, another important aspect is to develop fusion rules to merge more salient features into the fused image. Some common fusion rules [1] involve the selection of the maximum pixel intensity, weighted averaging, windows-based or region-based rules. These rules can capture fine details of the scenes efficiently. It should be noted that the couple of these key aspects leads to the success of MRA based schemes.

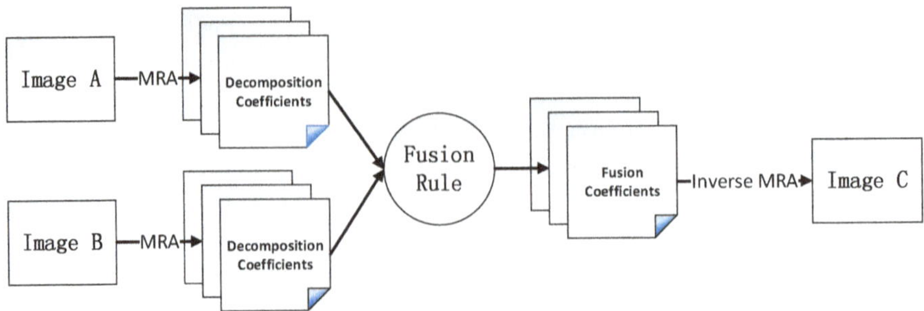

Fig. 1 An illustration of multi-resolution analysis image fusion scheme

A general description of general MRA based image fusion procedure is depicted in figure 1. However, classic MRA based fusion methods suffer from some shortcomings due to the redundant representation of the input signal and the separation of frequency band, such as shift-variant and the temporal stability or adaptability in some degree. More exactly, these issues would bring in false-information or flickering features in the final results which may destroy the local image structure. There are numerous MRA-variant based fusion methods. The shift invariant discrete wavelet transform [3] (SIDWT) can be used to remove the side effect of the down-sampling processing in DWT. The dual tree-complex wavelet transform (DT-CWT) presented in [4] can fuse the directional information. To overcome the redundancy representation of wavelet transform, Yang [5] proposed a lower storage and nearly shift-invariant image fusion method. These fusion methods solve the above described problems to some extent. Some multi-resolution geometric analysis (MGA) methods are applied for combining global features and structures, e.g. ridgelet[6], shearlet[7], contourlet[8] and nonsubsampled contourlet transform [9][10][11] (NSCT). A key feature of these fusion methods usually is the combination of the MGA methods and some data analysis tools. Another efficient approach for video fusion is surfacelet transform [12], which can detect and combine the temporal information naturally. However, this fusion method also suffers from some practical limitations. First, it is not easy to tune the parameters and reduce computation complexity and memory requirement because of its extremely redundant representation on local features. Second, although these schemes may provide a common framework for video fusion, they sometimes lack of temporal adaptability. The important visual information or the temporal changes has not exploited enough. Third, there are few research efforts focus on modeling the underlying temporal structure of the fusion pattern. At last, the fusion rules, originated from the static image fusion method, cannot produce a regional consistency fusion results in terms of the visual quality.

Recently, numerous methods have been proposed to address the above problems. One approach is sparse representation (SR) based methods [13][14], which can provide an analytic solution to represent the natural image efficiently. It improves the fusion performance by taking advantage of dictionary learning or others. The main advantage of this method lies in modelling the geometrical cha-racteristics of the same scene efficiently. One emerging technology for solving this problem is the theory of compressed sensing (CS)[15-18]. It has been applied to some application fields, e.g. astronomical imaging and image fusion [19-21]. Although these methods tackle some aspects of these problems, various numerical examples indicate that the computational effort to improve the fusion performance would increase the computational complexity and storage requirement significant-ly. Meanwhile, some research efforts has been devoted to dynamic image fusion because the temporal changes or object motion highlight visual perception [39].

7.1.4 Dynamic Image Estimation Fusion and Optimal Remote Sensing

The goal of this chapter converts the problem of the environmental surveillance into a unified framework using Kalman filtered compressed sensing [27] (KFCS) and fuzzy integral separately, which can make the overall procedure more auto-matic and intelligent using the explicit and implicit information.

This chapter is arranged into two parts. In the first part, a dynamic image esti-mation fusion method, characterized by the spatial-temporal fusion method with a discriminative measure, is proposed. It directly exploits the spatiotemporal redun-dancy, which is available in dynamic imaging because parts of the adjacent im-ages remain static over time. They can provide a unified framework to model the fusion pattern of the multi-model image through assuming that the fusion process has a simple parametric form. These considerations can lead to the appearance of more high frequencies in the visual fusion results. A remarkable feature of the proposed fusion method is that the introduction of spatial and temporal state space. This procedure allows us to separate the different components of image se-quence according to the underlying property fields. The proposed parametric fu-sion method can describe the spatial-temporal alternation of the fusion results. This approach provides a mechanism for constructing and learning the weighted coefficients adaptively.

In the second part, fuzzy integral is investigated for automatizing feature-level image fusion optimally. It can take advantage of the broad similarities by calculat-ing the correlations of different decomposition coefficients across the different domains. Fuzzy integral can provide a simple and general mathematical tool to identify the optimal fusion weights. This unified framework can greatly avoid the effect of the color distortion introduced by IHS transform. More exactly, it can make the physical scene more available for visual examination and improve the time efficiency. The numerical results illustrate the effectiveness of the proposed methods in terms of visual quality and the objective evaluation of fusion perfor-mance. The presentation of this chapter is based on the works [22-23, 41, 42].

7.2 Dynamic Image Estimation Fusion with Kalman Filtered Compressed Sensing

7.2.1 Kalman Filtered Compressed Sensing

7.2.1.1 Some Notations and Assumptions

In this subsection, some notations and assumptions about the presented methods are presented. An arbitrary vectors defined in the vector space for the length of the vector). Symbol $S = \{s_1, s_2, ..., s_i, ..., s_M\}$ stands for the input image matrix. And the constant index M indicates the column size of S. Then it can be noted that the size of S is $N \times M$. For the vector s_i, in this chapter, it is assumed to be compressible and can be projected or represented in a group of orthogonal basis $\psi = \{\psi_1, \psi_2, ..., \psi_i, ..., \psi_M\}$. The $N \times N$ basic matrix need to be designed or chosen, which also is referred as a linear transformation matrix. This procedure can be formulated as follows,

$$s = \psi u, \tag{1}$$

where u denotes the transform vector against the basis matrix ψ. The coefficient u_i can be obtained by calculating $u_i = \langle s_i, \psi \rangle$. For some special basis matrix such as the wavelet basis, the coefficients u can be assumed to be sparse. The sparsity of u is defined as follows,

$$\|u\|_0 = K \ll N, \tag{2}$$

where K denotes the number of the non-zero elements. In this chapter, index set T for extracting the desired entries in u or corresponding to the non-zero elements in u, is defined. The size of T is K. Similarly, the complement of T is denoted as T^c. For simplicity, the symbols $\cup, \cap, /$ indicate the usual set operations.

7.2.1.2 Compressed Sensing

The fundamental principle of compressed sensing [15-17] is presented in this subsection. The key difference to other theories is the introduction of the measurement matrix ϕ. The overall procedure can be expressed as follows,

$$y = \phi s = \phi \psi u = \Phi u, \tag{3}$$

where y denotes the measurement vector, which is assumed to be non-adaptive and random. Compressed sensing involves two key aspects, the design of the measurement matrix and the algorithm for recovering underlying signal s from y. In practice, the choice of ϕ is very important because it has an influence on the computational cost and the reconstruction accuracy. There are some common measurement matrices such as Gaussian random matrix [17], Rademacher matrix [24] and Hadamard circulant matrix [25]. The main goal of measurement matrices is to protect the salient features from being damaged. It can be noted that Eq.(3) is an underdetermined system or ill-posed problem. This problem usually can be solved through a l_0 or l_1 minimization problem. Some recovery methods have been proposed, such as matching pursuit-like methods and alternative direction multiplier method [26] (ADMM). In order to preserve the original structure of signals, the measurement matrix should satisfy the restricted isometry property (RIP) [17], which can be expressed mathematically as follows,

$$1 - \varepsilon \le \frac{\|\Phi\theta\|_2}{\|\theta\|_2} \le 1 + \varepsilon,$$

where ε is a small and positive constant, and θ is an arbitrary vector. Under the RIP conditions, the solution to l_1 minimization problem would be exact and stable [15][16].

7.2.1.3   Kalman Filtered Compressed Sensing (KFCS)

This subsection will briefly describe the KFCS by N. Vaswani [27]. The measurement process of the CS is conceived as a dynamic observation process. It is assumed that the support of signal S in the basis ψ is time-varying slowly for video. N. Vaswani considered the temporal correlations between frames and extended the static CS to a non-stationary case. Given a sparse state vector u_t and the observation vector y_t, the state equation of KF-CS is defined as follows,

$$u_t = Au_{t-1} + v_t,\ A = I,\ v_t \sim \mathcal{N}(0, Q_t), \tag{4}$$

where v_t denotes the state noise which follows a zero-mean Gaussian distribution. $\mathcal{N}(\cdot,\cdot)$ denotes the Gaussian distribution, Q_t is covariance matrix, . I stands for identity matrix. Matrix Q_t is assumed to be diagonal with different or same entries. The initialization of matrix Q_t is presented in [28].

The measurement equation of the KFCS is expressed as follows,

$$y_t = Hu_t + \omega_t,\ H = \phi\psi,\ \omega \sim N(0, \sigma^2 I), \tag{5}$$

where ω_t denotes the measurement noise and is assumed to be i.i.d. $N(0, \sigma^2 I)$.

The main workflow of KF-CS is illustrated in figure 1. The Eq.(6)-(9) are defined accordingly as follows,

$$\tilde{y}_t = y_t - \Phi\hat{u}_{t,T_{t-1}}, \tag{6}$$

where $\tilde{y}_t$ denotes the filtered error. The reconstruction process, for restoring the sparse entries from filtered error $\tilde{y}_t$, can be expressed as follows,

$$\hat{\alpha} = \arg\min_{\alpha} \left\| \tilde{y}_t - \Phi\alpha \right\|_2^2 + \gamma \left\| \alpha \right\|_1 , \tag{7}$$

where γ is the positive regularization parameter. To update the estimation state vector, it can be expressed as follows,

$$\hat{u}_{t,CS} = \hat{u}_t + \hat{\alpha}. \tag{8}$$

In order to extract the temporal information, it is important to detect and preserve the sparse pattern of the state vector, i.e., the indices set T_t :

$$T_t = \left\{ i \in \{i,...,m\} : \left| \hat{u}_{t,CS} \right|_i > \beta \right\}, \tag{9}$$

where β is a threshold to select the desired coefficients. As shown in figure 2, The combination of Eq.(8) and (9) can add or delete the desired entries from the estimate error. In another word, the KFCS can recursively maintain the entries set of the state vector u_t and provide a dynamic representation on the temporal change of the video.

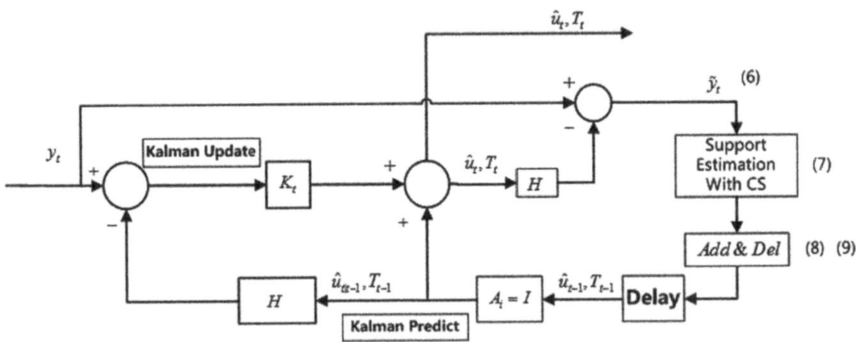

Fig. 2 The work flow of the KFCS

7.2.2 *Dynamic Image Estimation Fusion*

7.2.2.1 Challenge of Dynamic Image Fusion

Traditional fusion strategies may not be suitable to dynamic image fusion. There are some common fusion rules [29] for data fusion or tracking fusion including the best linear unbiased estimation (BLUE), weighted least squares rule (WLS) and etc. Beyond these strategies, various more implementable and feasible fusion rules for static image fusion are developed, e.g., linear weighted fusion and maximum pixel selection rule. It also should be noted that their conceptual simplicity indicates some practical limitations [30][31] for video fusion. Some practical examples indicate clearly that these fusion rules may generate drastic temporal changes in the fusion results when applying it to image sequences. Although, these fusion rules provide some perspective on the fusion pattern, some forms of the model constraints of these rules restrict their robustness and flexibility to some degree. Or more exactly, additional structures of the spatial or temporal information should be investigated and exploited within some of the individual disciplines.

The nature of correlation of the related frames and the potential temporal structure may provide alternative solution to dynamic image fusion. Based on this, more challenging problems may occur. First, it is needed to model the spatial-temporal dependence of the image sequences. Unfortunately, it is not easy to capture and model the potential structure of the temporal and spatial information simultaneously because of the complex image content in video. Careful consideration may be taken to integrate temporal information in a natural way. Second, intrinsic geometrical structure of the fusion pattern should be considered carefully. Third, another potential issue is that the model parameters for estimating the process of spatial-temporal fusion cannot be recognized adaptively. In summary, it is needed to fully explore the relationship of the spatial-temporal information which make it better-posed by using the prior knowledge about the special structures.

7.2.2.2. Definitions and Assumptions

In this subsection, the fusion pattern of the image sequence is assumed to be separable in spatial and temporal direction. Both directions for real-world images fusion are fused independently by exploiting the statistical property of both directions. The proposed method involves striking a balance between the effort of measuring the similarity between different components in state space and the computation power for obtaining the adaptive fusion coefficients. The parametric fusion method aims to extend the original problem from a stationary process into a non-stationary one. The separation of the original state space can directly factor out the temporal information. It is possible that the assumption can intrinsically model the underlying fusion structure and produce an intellectual fusion result through measuring the similarity in both directions.

In this chapter, multi-source image sequences are assumed to be spatially registered. For simplicity, we consider two input image sequences, $u_{t,I}$ and $u_{t,V}$ (I and V abbreviate infrared and visible image at the time t) in the basis ψ, separately. Consequently, the measurement vectors $y_{t,I}$ and $y_{t,V}$ can be obtained from the selected measurement matrix ϕ.

7.2.2.3 Spatial-Temporal Fusion

Based on the assumptions and notations above, the framework of this fusion method is illustrated in figure 3. The final fused state vector is expressed as follows,

$$\hat{u}_{t,Fusion} = C_{t,I}\hat{u}_{t|t,I} + C_{t,V}\hat{u}_{t|t,V}, \tag{10}$$

where $\hat{u}_{t|t,I}, \hat{u}_{t|t,V}$ stand for the estimate vectors of infrared and visible images, respectively, and $\hat{u}_{t,Fusion}$ stands for the vector of fused image. The weights $C_{t,I}$ and $C_{t,V}$ based on the KFCS remain unknown. Next, we provide a mechanism to determine the unknown weights.

To obtain these weights, as shown in figure 2, the standard Kalman update procedure, based on the estimated support set $T_{t,V}$ and $T_{t,I}$, can be rewritten as

$$\hat{u}_{t|t,M} = \hat{u}_{t|t-1,M} + K_{t,M}\,\tilde{y}_{residual,M}$$
$$\downarrow \qquad \downarrow \qquad \downarrow \qquad , \qquad M \in \{I,V\}, \tag{11}$$
$$\hat{u}_{t|t,M} = \hat{v}_{t,S,M} + \hat{v}_{t,T,M}$$

where $\hat{v}_{t,S,M} = \hat{u}_{t|t-1,M}$ and $\hat{v}_{t,T,M} = K_{t,M}\,\tilde{y}_{residual,M}$ can reflect the spatial and temporal information, respectively. $\hat{v}_{t,S,M}$ can be calculated by the prediction step. And $\hat{v}_{t,T,M}$ is obtained based on the current measurement vector. It is natural to divide the state space into two parts because of its physical meaning. This scheme aims at extracting temporal changes and providing a mechanism to recognize the fusion pattern between these two state vectors. In this chapter, we take advantage of the Tanimoto metric [30] to measure the similarity between visible vector $\hat{v}_{t,O,I}$ and infrared vector $\hat{v}_{t,O,V}$. According to the definition of Tanimoto metric, the similarity between $\hat{v}_{t,O,I}$ and $\hat{v}_{t,O,V}$ is

$$m_O(\hat{v}_{t,I,O}, \hat{v}_{t,V,O}) = \frac{\hat{v}^T_{t,I,O}\,\hat{v}_{t,V,O}}{\hat{v}^T_{t,I,O}\,\hat{v}_{t,I,O} + \hat{v}^T_{t,V,O}\,\hat{v}_{t,V,O} - \hat{v}^T_{t,I,O}\,\hat{v}_{t,V,O}}, O \in \{S,T\}. \tag{12}$$

Then, based on literature [38], the discriminative coefficients in Eq.(12), can be computed as

$$C_{t,O}(m_S,m_T,\varphi_O) = \beta_O^2 \exp(-\eta_{s,O}\|m_S\|)\exp(-\eta_{t,O}\|m_T\|), \tag{13}$$

where $\varphi_O = \{\beta_O, \eta_{s,O}, \eta_{T,O} \mid \eta_{s,O} > 0, \eta_{T,O} > 0\}$, β_O^2 is scale, $\eta_{s,O}$ and $\eta_{t,O}$ stand for the degradation coefficients. It is easy to prove that $C_{t,O}(m_S,m_T,\varphi_O)$ is completely separable and symmetry. Meanwhile, $C_{t,I}$ and $C_{t,V}$ are supposed to be subject to $C_{t,I} + C_{t,V} \leq 1$. The reconstruction image of the fused results can be completed by multiplying the transformation trix ψ' with $\hat{u}_{t,Fusion}$.

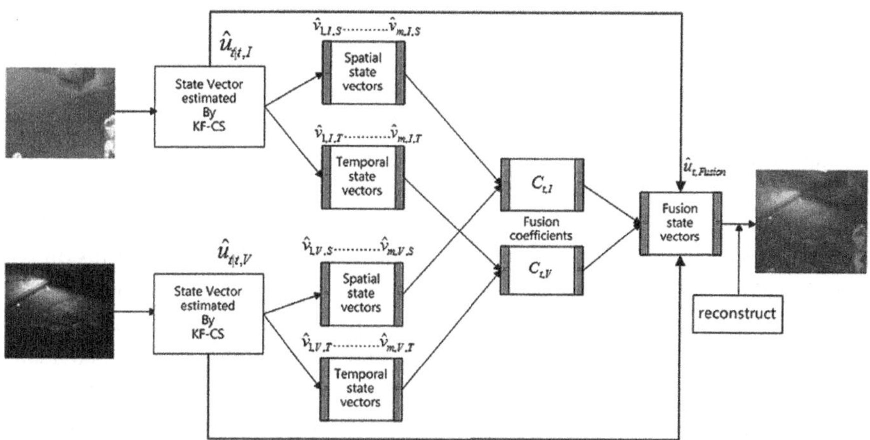

Fig. 3 The schematic representation of the spatial-temporal fusion method

7.2.3   Experiments and Evaluation

7.2.3.1   The Experiment Settings

In this subsection, some preparations for evaluating fusion performance of the proposed fusion method are presented, which include datasets, simulating environments and evaluation criteria. All experiments are conducted in real-world image sequences. In this chapter, some appropriate evaluation criteria are utilized to measure the dynamic fusion performance, including entropy (EN) [1], standard deviation (STD) [1], overall cross entropy(OCE) [7], mutual information (MI) [33][34], edge preservation metric ($Q^{AB/F}$) [35] and the feature similarity metric (FSIM) [36]. Among these criteria, larger values of MI, EN, STD, $Q^{AB/F}$ and FISM indicate better fusion results. And smaller OCE value suggests better image fusion quality. Meanwhile, the minimum of the two FSIM values, denoted by FSIM$_{min}$, is used to connote the actual quality score of the evaluated algorithms. The score of FSIM indicate the features contained in the input image.

Four MRA based fusion methods are compared, which include discrete wavelet transform (DWT), Dual-tree complex wavelet transform (DT-CWT) [4], Low redundant and nearly shift-invariant discrete wavelet frame (LRDWF) [5] and Discrete wavelet frame (DWF) [37]. The number of the decomposition levels are chosen to be 4. The maximum value of the decomposition coefficients is selected for the fused pixel or the significant coefficient. The basis of these fusion methods is chosen to be "db4" uniformly. All algorithms are tested and performed in Matlab 2010a and a PC with a Core i5 2.67GHz CPU shipped with 4G RAM.

7.2.3.2 Results on the First Image Sequences

In this subsection, the visual quality and fusion performance of the fused results are demonstrated and studied. Some samples of the first two image sequences are displayed in figure 4. The fused results of all fusion methods are presented in figure 5. The results of the compared methods are given in figure 5 (a)-5(d). They are similar to each other in term of edges and textures visually. The visual perception of KFCS based fusion method is demonstrated in figure 5(e). It can be noted that its visual quality is better than the compared fusion methods, which indicate that the high frequency portion or temporal information of the KFCS based fusion results is more than the compared methods. The moving objects generated by the KFCS based fusion, marked with rectangles, are more prominent. This may be contributed to the introduction of temporal components in state space. The adaptive weights, generated by spatial-temporal fusion method, are determined by the temporal and spatial components separately. These considerations bring out the improvement on the fusion of the underlying natural scenes. Meanwhile, the regions, edges and contours are sharper than the other fusion methods.

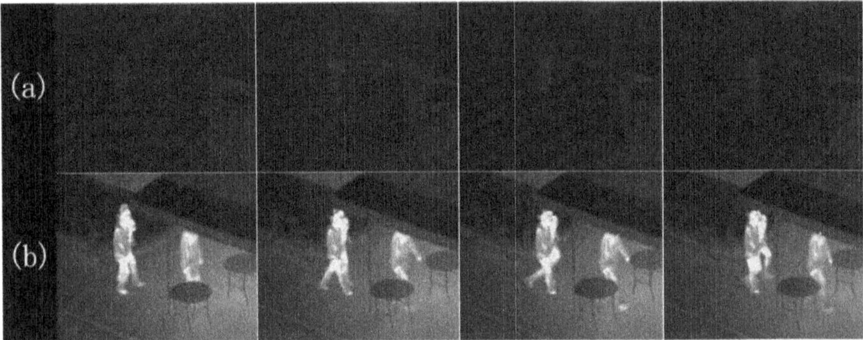

Fig. 4 The first image sequences: (a) visual sequence and (b) infrared sequence (8, 14, 17, and 20 frames, 320×240)

The details of the fusion performance about the fused sequences are demonstrated in figure 6. The graphs from figure 6(a) to figure 6(f) are the dynamic process measured by the six criteria. Compared to the other MRA based fusion methods, it can be observed that our method could obtain promising features locally and globally. First, it is interesting to see that a local increment appears in the plot of the KFCS based fusion method, which is marked by the arrow in figure 6(a). This raise on the MI may be due to the advance of the estimation fusion.

This phenomenon indicates more temporal information is integrated into the resulting image. Second, in a global view on these curves, the outcomes of the EN, STD, OCE and $Q^{AB/F}$ are similar in the all fusion methods. Among these methods, it can be noted that the behaviors of KFCS based on fusion are better than the other methods. Third, for the graph of the OCE of KFCS based on fusion, displayed in figure 6(d), it behaves a more steady way than other fusion algorithms and achieves better fusion performance improvement. Similarly, it is clear that the KFCS's performance in OCE given in figure 6(d) is superior to the other fusion methods. However, the numerical results in figure 6(e) indicate that the performance of the DWF in $Q^{AB/F}$ is better than the other methods. This is somewhat surprising, since the multi-resolution based fusion methods can preserve more edge information of the natural scene than the KFCS to some extent.

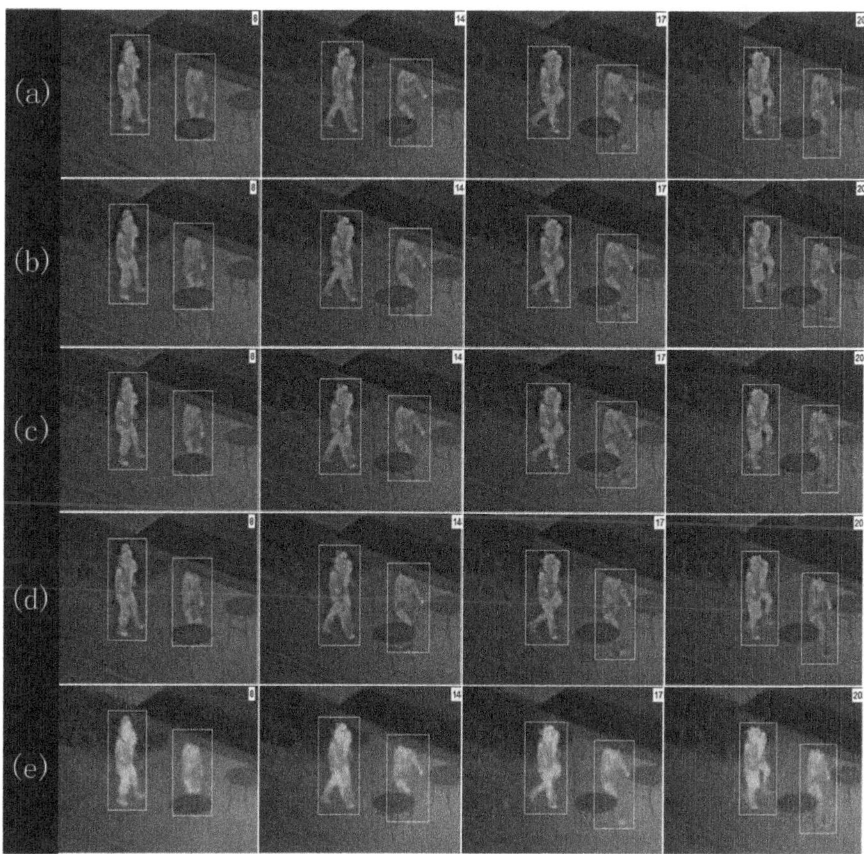

Fig. 5 The fused results of the first image sequences: (a) by the DWT, (b) by the DWF, (c) by the LRDWF, (d) by the DT-CWT, and (e) by the KF-CS

    Z. Jing, H. Pan, and G. Xiao

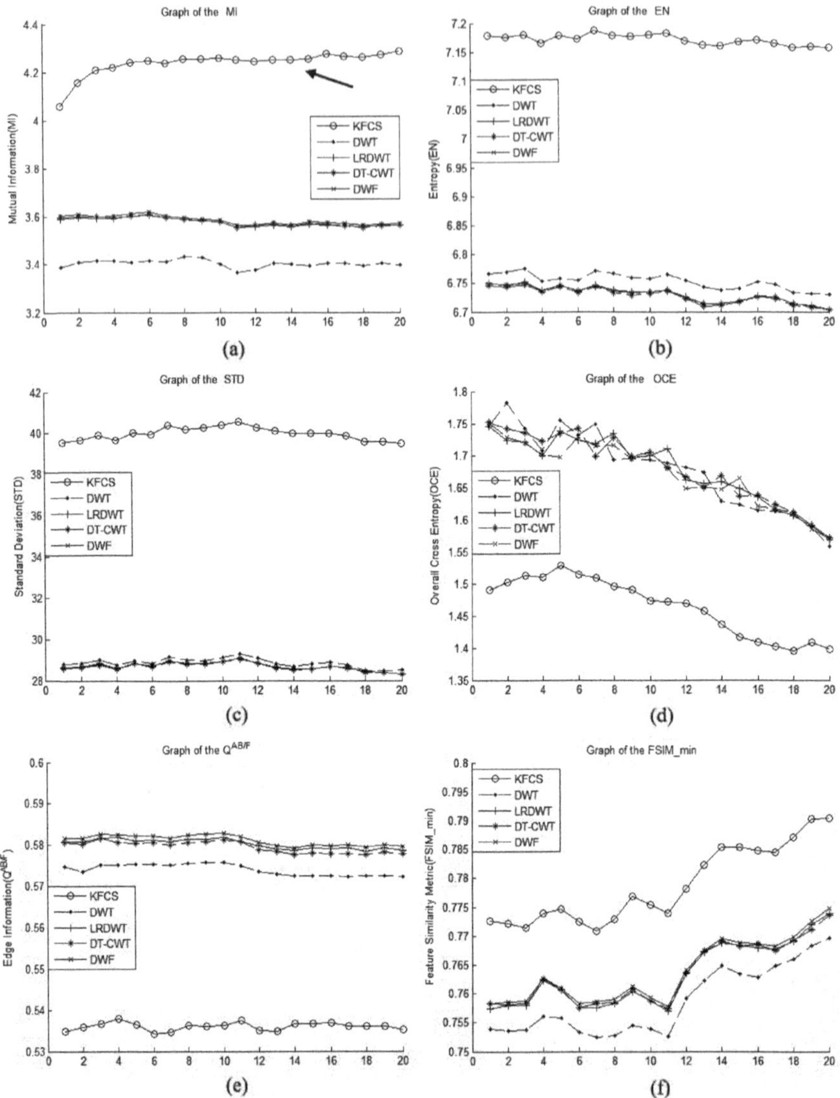

Fig. 6 The dynamic fusion performance compassion on the first image sequence: (a) MI, (b) EN, (c) STD, (d) OCE, (e) $Q^{AB/F}$, and (f) FSIM$_{\min}$.

The average objective numerical results in term of the six evaluation indexes are summarized in Table 1. Based on the definition of the MI, EN, STD, OCE and $FSIM_{min}$, KFCS based fusion method is better than other methods which are shown in bold fonts. The considerable increases on $FSIM_{min}$ value indicate that the KFCS can integrate more low-levels of features, i.e., the phase congruency (PC) and the gradient magnitude (GM), than other fusion methods. It also can observed that the KFCS is superior to other methods because of the consideration of the temporal information. Although the $Q^{AB/F}$ score of KFCS based fusion is not better than the compared methods, the perceptual edge information is extremely approximate to the compared algorithms.

Table 1 The average objective criteria for the first dataset (20 frames)

Algorithm	MI	EN	STD	OCE	$Q^{AB/F}$	$FSIM_{min}$
DWT	3.4043	6.7533	28.8606	1.6780	0.5740	0.7587
DWF	3.5844	6.7299	28.6962	1.6719	**0.5812**	0.7638
DT-CWT	3.5777	6.7275	28.6716	1.6796	0.5795	0.7633
LRDWF	3.5757	6.7306	28.6982	1.6771	0.5803	0.7631
KF-CS	**4.2369**	**7.1718**	**39.9557**	**1.4644**	0.5370	**0.7787**

7.2.3.3 Results on the Second Image Sequences

In this subsection, the pictorial results and objective evaluation of the fused images are presented. Some samples of the input image sequence are illustrated in figure 7, where three persons are moving quickly. This experiment is performed for the evaluation of the local fusion pattern.

The average numerical results based on the fused videos are shown in Table 2. The outcomes of the KFCS are better that the other fusion methods again in term of the criteria of MI, EN, OCE and $FSIM_{min}$.

Table 2 The average objective criteria for the second dataset (20 frames)

Algorithm	MI	EN	STD	OCE	$Q^{AB/F}$	$FSIM_{min}$
DWT	2.2938	6.1489	21.4363	4.5844	0.5840	0.7442
DWF	2.4379	6.1138	21.1872	4.6495	**0.5922**	0.7477
DT-CWT	2.4355	6.1113	21.1438	4.6620	0.5908	0.7472
LRDWF	2.4291	6.1149	21.1939	4.6403	0.5910	0.7466
KF-CS	**3.0233**	**7.3215**	**24.1778**	**0.5103**	0.5397	**0.7652**

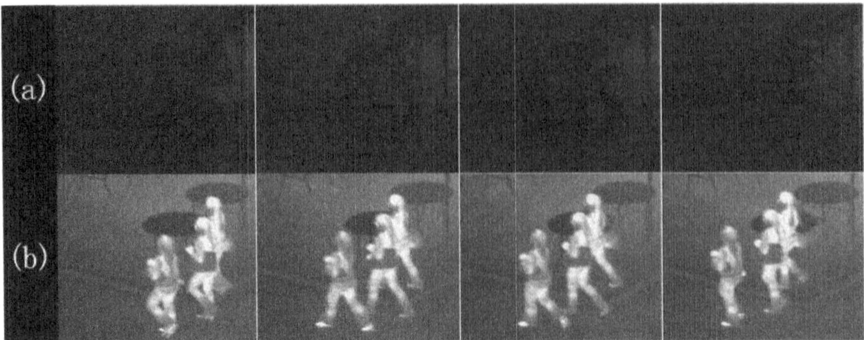

Fig. 7 The second image sequence: (a) visual images and (b) infrared images (6, 14, 16, and 19 frames, 256×256)

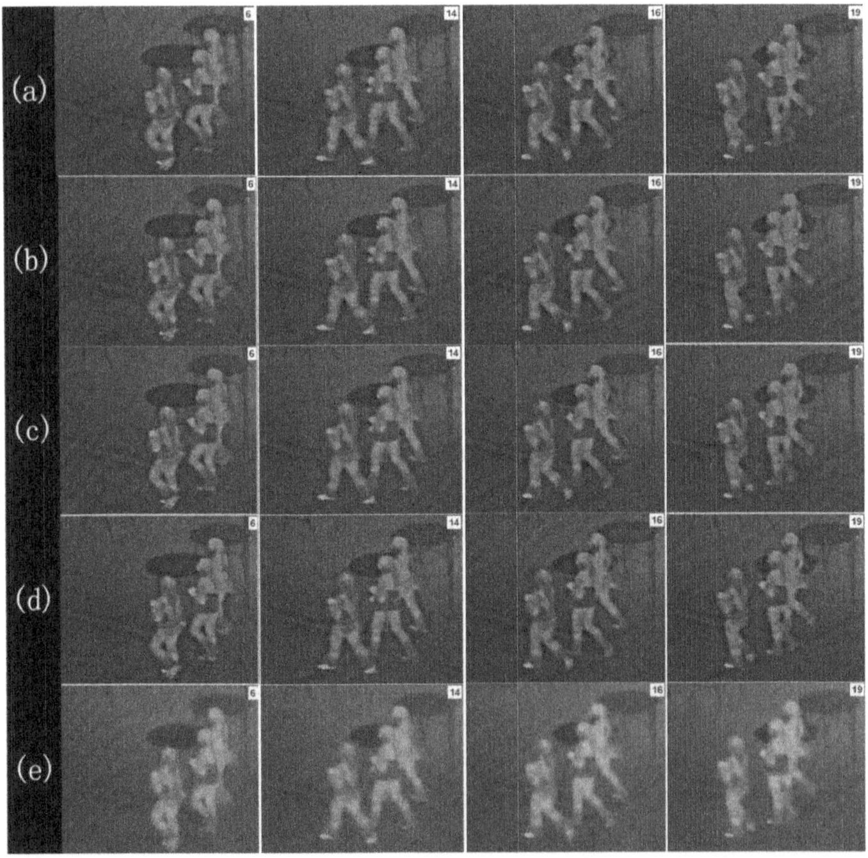

Fig. 8 The fused results of the second image sequence: (a) by the DWT, (b) by the DWT, (c) by the LRDWF, (d) by the DT-CWT, and (e) by the KF-CS

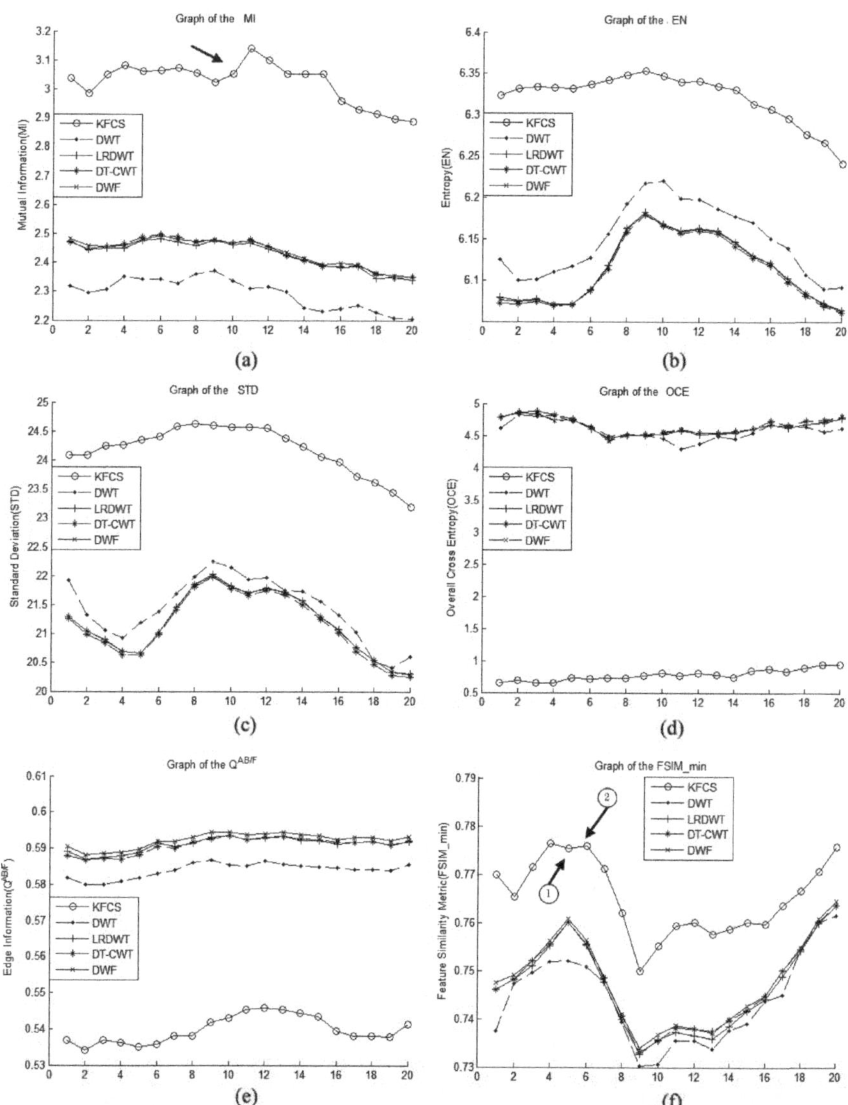

Fig. 9 The dynamic fusion performance compassion on the second image sequence: (a) MI, (b) EN, (c) STD, (d) OCE, (e) $Q^{AB/F}$, and (f) FSIM$_{min}$.

A visual compassion of the fusion results are illustrated in figure 8. It is evident that the KFCS based fusion method can generate a brighter look than the compared methods and combine more salient features. In the meantime, it can enhance the moving object and preserve the details of the individual sensor. The average numerical results of the fusion performance are presented in Table 2. It is clear that the KFCS based fusion method is superior to the other methods quantitatively since the usage of the spatial-temporal fusion method allows us to combine the visual information adaptively. The visual results in figure 8(e) verified the efficiency of the KFCS. Similarly, for the metric $Q^{AB/F}$, the DWF based fusion method can yield a larger score. Although the KFCS based fusion does not perform better in $Q^{AB/F}$, we argue that it can provide an enhancement results on the image features in term of the $FSIM_{min}$. The success of the KFCS based fusion method contributes to the estimation procedure of the time-varying fusion pattern.

Extensive evaluation of dynamic fusion performance of all fusion algorithms are demonstrated in figure 9. Compared with other MRA based fusion algorithms, the quality of the KFCS based fusion method measured by MI is characterized by the local increasing trend in figure 9(a), which is marked by an arrow. This phenomenon is due to the utilization of separable spatial-temporal fusion procedure, which results in integrating local structures of the multi-source images. It can be seen from figure 9(b) to figure 9(d) that the curves of the KFCS based fusion method behave in a more stable way than the other fusion algorithms. Figure 9(f) shows the variation of the criteria $FSIM_{min}$ caused by the estimation error of KFCS fusion, which is marked by arrow ①. Fortunately, the mechanism for updating the support set in the KF-CS's step overcomes this problem. Then, a raise completes in $FSIM_{min}$, which is marked by arrow ②. As the description above, the advancement of the KFCS lies in the integration of the spatial and temporal information.

7.2.3.4 Results on the Third Image Sequences

In this section, a challenging experiment is performed. The main difference to the previous videos is that the background of the third one is static and its cameras are moving. The goal of this experiment is to evaluate the fusion performance under the full dynamic environment. Some samples of the third image sequences are displayed in figure 10.

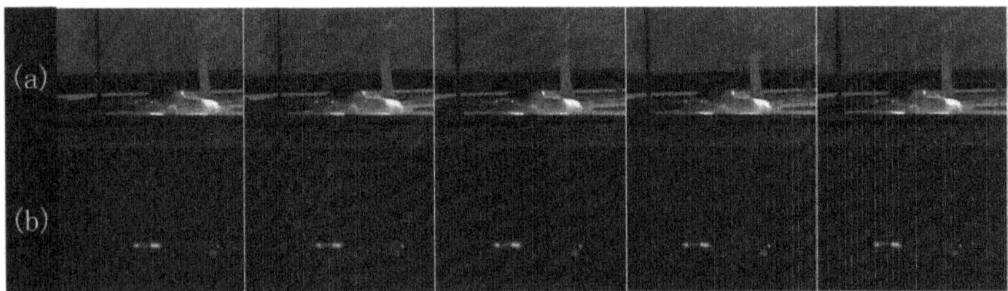

Fig. 10 The third image sequence: (a) visual images and (b) infrared images (3, 8, 14, 17 and 21 frames, 320×240 pixels)

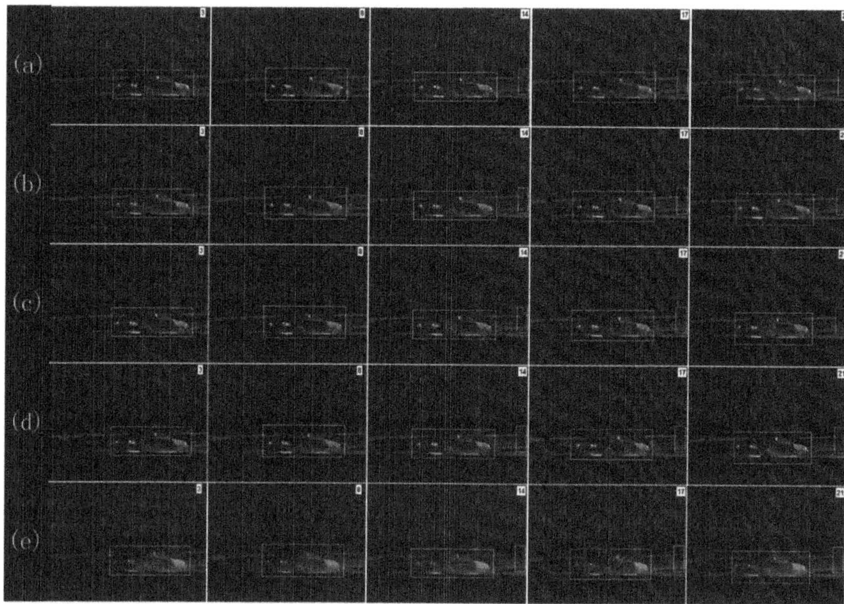

Fig. 11 The fused results of the third image sequence: (a) by the DWT, (b) by the DWT, (c) by the LRDWF, (d) by the DT-CWT, and (e) by the KF-CS.

The sequential spatial details of the five methods' fused results are presented in figure 11. The multi-resolution-like fusion methods are illustrated in figure 11(a)-(d) successively. The detail information of the KFCS based fusion is presented in figure 11(e). It can be noted that the image content of the small objects of the KFCS's fused results, marked by rectangles, is more than the other methods. The proposed separable fusion method achieves better visual quality of the fused results and enhances certain remarkable objects.

The quantitative evaluation of the five fusion methods are presented in Table 3. The outcomes of the KFCS fusion exhibit a stable aspects of the fusion performance in term of the five fusion quality assessment metric and increase the confidence on its effectiveness to fuse temporal information. It can be concluded that the KFCS fusion method is appropriate for integrating the natural characteristics of the spatial-temporal information into a universal formation. Examined the last row of Table 3 and the figure 11(e), the KFCS based fusion can achieve a better visual perception repeatedly. Meanwhile, the local features of the behavior of dynamic fusion performance, displayed in figure 12, also indicate that the KFCS fusion is superior to the other methods. These local behavior of all fusion methods, varied in frame by frame, can be summarized in some aspects. First, it can evident that the local growth trend can be observed in term of the criteria MI and FSIM_{min}, which are marked by arrows in figure 12(a) and 12(f) separately. Second, the STD of the KFCS fusion is creasing globally. This appearance may indicate more salient features are transferred. The graphs of the other fusion methods behave in a more stable way than the ones of the KFCS fusion. The behavior of the KFCS in the figure 12(e), may owe to the consideration of the temporal component.

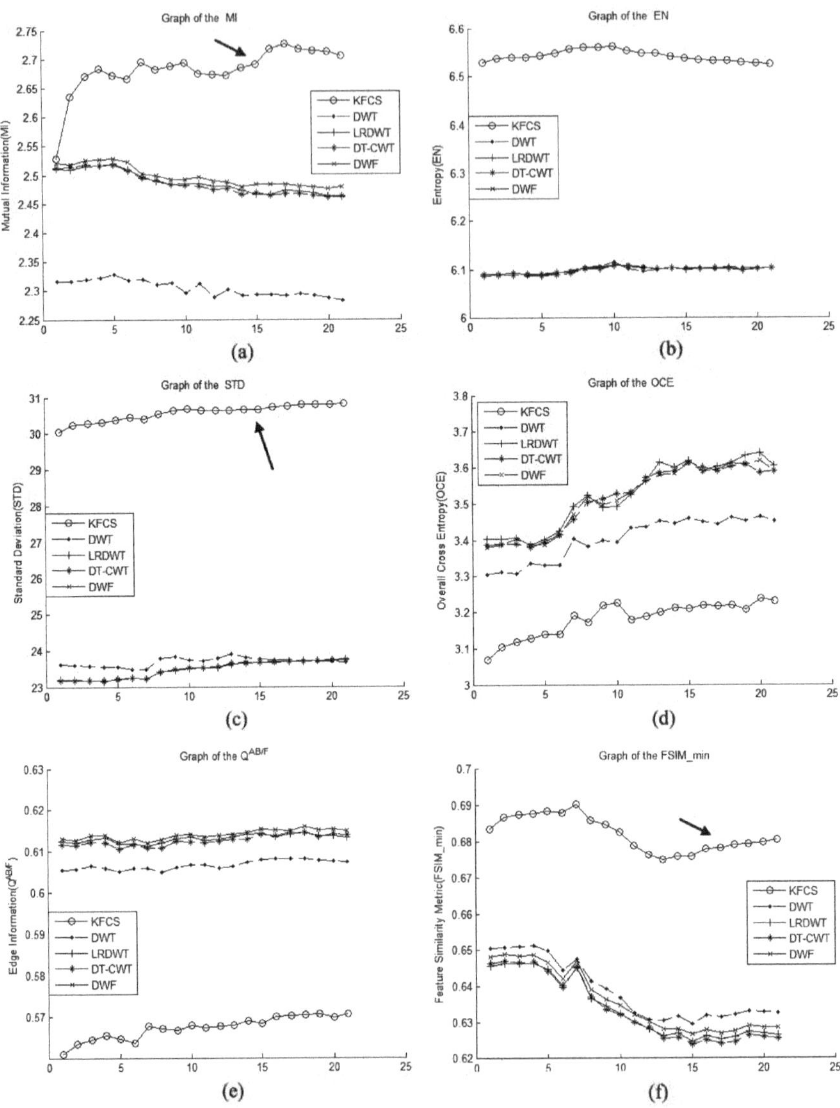

Fig. 12 The dynamic fusion performance compassion on the second image sequence: (a) MI, (b) EN, (c) STD, (d) OCE, (e) $Q^{AB/F}$, and (f) $FSIM_{min}$.

Table 3 The average objective criteria for the third dataset (20 frames)

Algorithm	MI	EN	STD	OCE	$Q^{AB/F}$	$FSIM_{min}$
DWT	2.3044	6.0987	23.6868	3.4027	0.6067	0.6386
DWF	2.4973	6.0990	23.4779	3.5160	**0.6141**	0.6359
DT-CWT	2.4852	6.0977	23.4819	3.5155	0.6127	0.6335
LRDWF	2.4869	6.0988	23.4786	3.5252	0.6131	0.6339
KF-CS	**2.6527**	**6.3695**	**25.5812**	**3.3120**	0.5673	**0.6820**

7.2.4   Discussion on the Fusion Results

We have presented that the combination of the KFCS and the spatial-temporal fusion method is suitable for dynamic image fusion, characterized by image estimation fusion adaptively. Compared to the MRA based methods in term of the objective fusion performance and visual effect frame-to-frame, the difference in the manner may related to the spatial-temporal fusion method and the algorithm used for dealing with the temporal information. It is quite probable that the process of estimation and separation of spatial and temporal information in state space and the discriminative fusion process generate the promising fusion results globally or locally. A remarkable feature of the proposed fusion method is that it has less computational complexity and has simple formation. The quantitative analysis, illustrated in figure 6(e), 9(e), 12(e) and Table 1-3, indicate that the involvement of temporal pattern makes a prominent improvement on the objective fusion performance and verify the effectiveness of the proposed method. Moreover, it can be noted that the presented method can recognize, selectively fuse the time-varying temporal information. This particular fusion procedure can highlight the moving objects locally and lead into a significant improvement on spatial adaptability. Summed these up, visual and objective fusion performance evaluation indicate the separable fusion method is favorable for spatial-temporal image fusion. Significantly, the presented fusion algorithm potentially may pave a new way for high-level application.

Making a visual examination on the details of the fusion results, we can see that the proposed method have some limitations in some aspects. First, it can be observed that smooth effect exists in the fused videos. The reason is that the recovery procedure of compressed sensing may produce some reconstruction error on the fused video. Second, one shortcoming of our method is that some local features may be lost. Third, the proposed method suffers from the prediction error, introduced by Kalman filter. This error may lead to the distortion of local structures of the scene.

7.3 Optimal Remote Sensing Images Method Using Fuzzy Integral

In this subsection, an optimal color image fusion method for remote sensing is presented. The fuzzy integral, is employed to detect and decide optimal weighting coefficients. The means of this section are based on the references [41][42]. These methods can adjust color distortion caused by the intensity, hue and saturation (IHS) transformation. In general, the proposed method may offer an additional insight about the behavior of optimal multi-spectral image fusion.

7.3.1 An Overview on the Fusion Methods and Rules for Remote Sensing

There are various approaches and fusion rules developed for the multi-spectral remote sensing image fusion. First, A common method for this topic is the wavelet-like schemes because of its excellent representation on local features of image such as line, edge, contour, directional information and textures. Second, another obvious approach is IHS transformation [40]. Third, the Brovey transform also is a popular method for this problem. Moreover, the combination of shearlet transform and the pulsed coupled neural network [47] (PCNN) also is an alternative approach for this problem. Beyond these work, the fusion with temporal information [39] also is a good choice for this problem. And the fusion result can detect the changes of the obverted scene.

Beyond these fusion methods, some fusion rules are proposed to handle the resulting problems. The combination of DWT and IHS transform is examined to be an efficient fusion approach for multi-spectral image fusion. However, beyond these issues, it should be noted that the application of IHS transform would lead to color distortion. Moreover, the effect of IHS transform sometimes depends on the choice of the image data. Some researchers have studied the cause of the color distortion [41][42]. In this subsection, some aspects of this problem are discussed. It composes of two key procedures, i.e. the optimal IHS image fusion and adaptive fusion by fuzzy integral. These procedures provide an alternative and common framework to accomplish fusion performance with desired properties. In other words, it is more likely to achieve a desired fusion result within constraint iterative numbers.

7.3.2 Optimal Image Fusion Method with Fuzzy Integral

7.3.2.1 Optimal IHS Image Fusion

In order to tackle previous problem, an optimal image fusion method [41][42] is proposed through a combination of wavelet transform and IHS transform. Its key idea is to utilize the local frequency features of DWT and the excellent ability of IHS transform to extract the low frequency features. This subsection would complete the presentation of the method without recalling the details of IHS transform. The basic steps of optimal IHS image fusion method is demonstrated as follows,

Step 1: IHS transform is applied to the multi-spectral image B, which generated the components of I, H and S. And then the wavelet transform is used to decompose the component I into 2^j levels.

Step 2: The gray histogram of image A is normalized based on the intensity component of image B. Meanwhile, the normalized image A also is decomposed into 2^j levels.

Step 3: The resulted image $I^{'}$ is reconstructed by the inverse wavelet transform after applying the fusion rule to the images generated by **Step 2**.

Step 4: The fused image C is generated by the inverse IHS transform using the image $I^{'}$, H and S.

Step 5: The weight coefficient k_{opt} is recalculated in term of the referred evaluation metric. Correspondingly, the coefficients k_1 and k_2, according to the sub-band, is obtained. It should be pointed out that the **Step 3** to **Step 5** is calculated iteratively for a suitable solution k_{opt}.

Step 6: Some objective functions are taken to assess the efficiency of the weight coefficient k_{opt}.

Remark: To exploit the statistical properties of the decomposition coefficients of the wavelet transform, a window based fusion rule is employed according to different details of DWT's sub-bands. For the high frequency coefficients, the corresponding weight can be obtained by

$$W^k(2^j,x,y) = \begin{cases} W_A^k(2^j,x,y) & D_A^k > D_B^k \\ W_B^k(2^j,x,y) & D_A^k < D_B^k \end{cases}, \tag{14}$$

where $D(2^j)$ denotes the variance of the coefficients at the decomposition level 2^j and $W^k(2^j,x,y)$ for the fusion results. D_A^k signifies the variance coming from the image A within a window mask with a size of $n \times n$, and D_B^k for the image B. And the central point of this window mask is denoted by (x,y).

For the low frequency coefficient, the weight can be determined by

$$A(2^j,x,y) = k_{opt}A_A(2^j,x,y) + (1-k_{opt})A_B(2^j,x,y), \tag{15}$$

where $A_A(2^j,x,y)$ denotes the low frequency coefficient of image A, and $A_B(2^j,x,y)$ for image B. Moreover, k_{opt} is the optimal coefficient.

$$F(k_{opt}) = Max\{E_{SP}(k_{opt}), E_{HF}(k_{opt})\},$$
$$s.t. \quad g(k_{opt}) : 0 \le k_{opt} \le 1, \quad k_{opt} \in D \in R \tag{16}$$

where $g(\bullet)$ denotes the fuzzy function. The symbol E_{SP}, denoted for the evaluation on the spectral information, is defined as follows

$$E_{SP} = Corr(f, f_M) = \frac{\sum_{j=1}^{npix} (f - \overline{f})(f_{M,j} - \overline{f}_M)}{\sqrt{\sum_{j=1}^{npix} (f - \overline{f})^2 (f_{M,j} - \overline{f}_M)^2}}, \tag{17}$$

where $npix$ denotes the numbers of the pixels. The symbol f denotes the fused image, and f_M for the multi-spectral image. Similarly, $\overline{f}$ and $\overline{f}_M$ correspond to the mean of the images respectively.

For the evaluation index E_{HF} on high frequency information, its definition can be expressed as follows

$$E_{HF} = \frac{Corr(f^h, f_H^h) + Corr(f^v, f_H^v) + Corr(f^d, f_H^d)}{3}, \tag{18}$$

where f_H denotes the high resolution panchromatic image, and f^a, f^h, f^v and f^d for the low frequency component, horizontal high frequency component, vertical high frequency component and diagonal high frequency component after the wavelet transform, respectively. Similarly, $f_H{}^a$, $f_H{}^h$, $f_H{}^v$ and $f_H{}^d$ correspond to the components of high-resolution image.

The pictorial procedure is displayed in figure 13. And the fusion rule is described in the previous subsection.

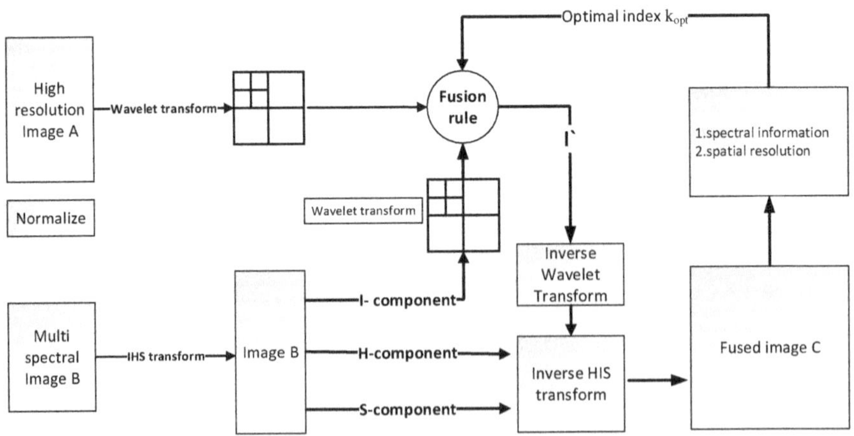

Fig. 13 The workflow of the optimal IHS fusion method

7.3.2.2 Optimal Image Fusion with Fuzzy Integral

The proposed method for improving the optimal IHS fusion method is named as fuzzy optimal fusion (FOF). The key idea of this method is the introduction of the fuzzy integral for the integration of the spectral information and spatial information. The main procedure is the selection of the optimal fusion weights through evaluating the multi-evaluation indexes for the optimal fusion coefficient.

Definition 1: let x is a domain. And e is the measurable fusion range from [0, 1] in the domain x, and $A \in P(X)$. Based on the measure g defined in the set A, the fuzzy integral F of the e function can be mathematically expressed as follows [43-46]

$$F = \int_A e(x) \circ g(x) = \sup_{\alpha \in [0,1]} \min[\alpha, g(A \cap E_\alpha)]$$
$$= \max_{\alpha \in [0,1]}[\min(\alpha, g(A \cap E_\alpha))], \tag{19}$$

where $E_\alpha = \{x \mid e(x) \ge \alpha\}$ and $P(x)$ denotes a power set based on the x. Similarly, function $g(\cdot)$ denotes the fuzzy measure, which can assign an importance scalar factor to x_i when synthetically evaluation applies to this factor based on the fuzzy integral.

The fuzzy function $g(\cdot)$ is an important element of the definition of the fuzzy integral. In this chapter, $X = \{x_1, x_2, x_3, \cdots x_n\}$ is assumed to be set-limited. Consequently, the arbitrary measure can be obtained in the set $A \subset X$ after the fuzzy measure $g_\lambda(x_i)$ for the scale factor set $\{x_i\}$ is assigned.

For the application on combining the multi-spectral and panchromatic images, the domain is chosen to be $X = \{x_1, x_2\}$. More exactly, x_1 is taken for the evaluation index on the spectral information, and x_2 for the spatial information. Meanwhile, the corresponding measures are $g_\lambda(x_1)$ and $g_\lambda(x_2)$ respectively. They can be referred briefly as g_1 and g_2. The overall procedure can be expressed as follows

$$g(\{x_1, x_2\}) = g(\{x_1\}) + g(\{x_2\}) = 1, \quad s.t.\ g(\{x_1\}) = g_1, g(\{x_2\}) = g_2. \tag{20}$$

Closely, $e(x_1) = E_{SP}$ and $e(x_2) = E_{HF}$ suggest the spectral and spatial information evaluation indexes. Then, the fuzzy integral, based on the definition of Eq.(19), can be expressed as follows

$$\int_X e(x) \circ g(x) = \sup_{\alpha \in [0,1]} \min[\alpha, g(X \cap E_\alpha)]$$
$$= \max_{\alpha \in [0,1]}[\min(\alpha, g(E_\alpha))]. \tag{21}$$

For simplicity, the previous Eq.(20) can be rewritten as follows

$$F = \sup\{\min[e(u_1), g(E_1)], \min[e(u_2), g(E_2)]\}, \tag{22}$$

where u_1 and u_2 denote the corresponding position.

7.3.3 Experiments and Evaluation

7.3.3.1 The Experiment Setting

In this subsection, some settings for the experiments are presented. It is assumed that the input multi-spectral images A and the Panchromatic image B are registered simultaneously. Two datasets are taken for the assessment of the referred methods. Some samples, captured by Quick Bird satellite at Jun. 2002, are illustrated in figure 14(a) for the panchromatic image and 14(b) for multi-spectral image. Meanwhile, the spatial resolution of figure 14(a) is 0.61-m. This image has been resampled with 4 times, then the size is 512×512. It can be seen that the image content of the sampled image is about the highway and buildings of the unban city, Shanghai, China.

7.3.3.2 The Experiment Results of First Image

As shown the configurations, the visual results and numerical results are presented in this subsection. The fused result of the FOF involved with fuzzy integral is

(a) The panchromatic image

(b) The multi-spectral image

(c) The result by FOF by fuzzy integral

(d) The IHS fused image

Fig. 14 The original Quickbird satellite image and the fused results

displayed in figure 14(c). For a visual comparative assessment, the visual effect of IHS transform is displayed in figure 14(d). It is evident that the proposed method can overcome the shortcomings of IHS transform apparently in term of the color and the local features of the observed scene. In a contrast, the visual features, produced by IHS transform, is profound and not match the nature and scale of the input images.

7.3.3.3. The Experiment Results of Second Image

In this subsection, visual fusion results for the second images are demonstrated. Similarly, the spatial resolution of the original panchromatic image is displayed in figure 15(a), which is 0.42-m. and the size of this image is 512×512 pixels. These pictures were captured at Oct, 2003. The spatial resolution of the original multi-spectral image is 2.52-meter. It has been re-sampled with 6 time ratios. These two cameras were shipped with a same PAV30 platform. Similarly, visual quality of the FOF method is better than the outcome of IHS transform in terms of color and local textures again. It should be noted that the improvement on local features, such as lines, contour and buildings, is obvious. These fusion results can help the user classifying and recognizing the special objects by considering special structure and shape features.

(a) The panchromatic image

(b) The multi-spectral image

(c) The result of FOF by fuzzy integral

(d) The IHS fused image

Fig. 15 The original Quickbird satellite image and the fused results

Table 4 The compare evaluation indexes for the fused images

Method	Channel	Fig.14		Fig.15	
		CORR	AG	CORR	AG
IHS	R	0.5624	10.6146	0.4214	17.5486
	G	0.4644	11.3663	0.5112	16.4781
	B	0.5060	9.5875	0.5124	14.8623
FOF	R	**0.8488**	8.9448	**0.7878**	16.9718
	G	**0.7691**	9.4102	**0.8040**	15.3258
	B	**0.8247**	7.9353	**0.7986**	14.0322

To assess these fusion results quantitatively, objective evaluation is performed. The numerical results, in terms of two fusion performance metric: correlation coefficient (CORR) and average gradient (AG), are presented in Table 4. We perform the process of performance evaluation in each channel. The outcomes of the FOF method reveal that the effectiveness of the proposed lies in the improvement on the visual quality in terms of removal of color distortion and the availability to scene understanding easily. In a summary, we can conclude that the application of fuzzy integral in FOF method is superior to original fusion method.

7.3.4 *Discussion on the Fusion Results*

In this section, the proposed method is sufficient for fusing the images in the field of remote sensing. The visual quality, in terms of the local geometric features of the fused results, indicates the feasibility of this method and the robustness to color distortion. Moreover, the objective evaluation on the proposed method reveals the efficiency of the combination of the FOF. Specially, the visual effect, displayed in figure 14(c) and 15(c), and the scores in Table 4 verified the effectiveness of the proposed method. Or more exactly, the success of FOF based method is the introduction of fuzzy integral and the objective evaluation guided based fusion process, which defined in Eq.(21).

7.4 Conclusions and Future Research

In this chapter, a spatial-temporal image fusion method is proposed. A computational feasible and efficient solution for improving temporal adaptability is provided. The proposed method exploited temporal information by dividing the state space of KFCS into two components. The fusion process is completed by separable spatial-temporal fusion method. The fusion results, in terms of dynamic fusion performance and visual quality, verified that the effectiveness of the proposed method. This method can tackle some drawbacks of classic estimation fusion schemes [29] and improve temporal consistency. The future research direction may lie in modelling spatial-temporal dependency by more than two images successively.

Beyond the discussion of the fusion method of image sequences, an image optimal fusion method for remote sensing, based on fuzzy integral, is proposed. The main advantage of the proposed method lies in exploiting the advantages of IHS fusion method and DWT fusion method jointly. This method can fulfill the desired requirement in term of object classification and aggression. Some attention may be paid to the decision process for object-based fusion rule in the future.

Acknowledgments. We would like to thank our collaborators, i.e. Bo Jin and Rongli Liu, for many helpful comments on these studies. And this work is supported by the national natural science foundation of China (Grant Nos. 61175028, 60775022) and the Ph.D. Programs foundation of ministry of education of China (Grant Nos. 20090073110045).

References

1. Jing, Z.L., Xiao, G., Li, Z.H.: Image Fusion: Theory and Applications. Higher Education Press, Beijing (2007)
2. Moira, I.S., Jamie, P.H.: Review of image fusion technology in 2005. In: Proceedings of the SPIE - The International Society for Optical Engineering, vol. 5782(1), pp. 29–45 (March 2005)
3. Rockinger, O.: Image sequence fusion using a shift invariant wavelet transform. In: Proceedings of International Conference on Image Processing, vol. 3, pp. 288–291 (1997)
4. Hill, P.R., Canagarajah, C.N., David, R.B.: Image Fusion Using Complex Wavelets. In: British Machine Vision Conference, pp. 1–10 (2002)
5. Yang, B., Jing, Z.L.: Image fusion using a low-redundant and nearly shift-invariant discrete wavelet frame. Optical Engineering 46, 107002 (2007)
6. Chen, T., Zhang, J.P., Zhang, Y.: Remote sensing image fusion based on ridgelet transform. In: Proceedings of International Conference on Geoscience and Remote Sensing Symposium, pp. 1150–1153 (2005)
7. Miao, Q.G., Shi, C., Xu, P.F., Yang, M., Shi, Y.B.: A novel algorithm of image fusion using shearlets. Optics Communications 284(6), 1540–1547 (2011)
8. Wang, H.J., Yang, Q.K., Li, R.: Tunable-Q contourlet-based multi-sensor image fusion. Signal Processing 93(7), 1879–1891 (2013)
9. Tang, L., Feng, Z., ZongGui, Z.: The nonsubsampled contourlet transform for image fusion. In: International Conference on Wavelet Analysis and Pattern Recognition, November 2-4, vol. 1, pp. 305–310 (2007)
10. Kong, W., Lei, Y., Ni, X.: Fusion technique for grey-scale visible light and infrared images based on non-subsampled contourlet transform and intensity-hue-saturation transform. IET Signal Processing 5(1), 75–80 (2011)
11. Kun, L., Guo, L., Song, C.J.: Contourlet transform for image fusion using cycle spinning. Journal of Systems Engineering and Electronics 22(2), 353–357 (2011)
12. Zhang, Q., Wang, L., Ma, Z., Li, H.: A novel video fusion framework using surfacelet transform. Optics Communications 285(13), 3032–3041 (2012)
13. Yin, H.T., Li, S.T.: Multimodal image fusion with joint sparsity model. Optical Engineering 50(6), 067007–067007 (2011)
14. Yang, B., Li, S.T.: Pixel-level image fusion with simultaneous orthogonal matching pursuit. Information Fusion 13(1), 10–19 (2012)

15. Donoho, D.L.: Compressed sensing. IEEE Transactions on Information Theory 52(4), 1289–1306 (2006)
16. Candès, E.J., Wakin, M.B.: An introduction to compressive sampling. IEEE Signal Process. Mag. 25(2), 21–30 (2008)
17. Candès, E.J., Romberg, J., Tao, T.: Robust uncertainty principles: Exact signal reconstruction from highly incomplete frequency information. IEEE Transactions on Information Theory 56(2), 489–509 (2006)
18. Candès, E.J., Tao, T.: The dantzig selector: statistical estimation when p is much larger than n. Annals of Statistics (2006)
19. Luo, X.Y., Zhang, J., Yang, J., Dai, Q.: Image fusion in compressed sensing. In: IEEE 16th International Conference on Image Processing (ICIP), pp. 2205–2208 (2009)
20. Wan, T., Canagarajah, N., Achim, A.: Compressive image fusion. In: IEEE 15th International Conference on Image Processing (ICIP), San Diego, CA, pp. 1308–1311 (October 2008)
21. Divekar, A., Ersoy, O.: Image fusion by compressive sensing. In: IEEE 17th International Conference on Geoinformatics, pp. 1–6 (2009)
22. Pan, H., Jing, Z.L., Liu, R.L., Jin, B.: Simultaneous spatial-temporal image fusion using Kalman filtered compressed sensing. Optical Engineering 51(5), 057005 (2012)
23. Xiao, G., Jing, Z.L.: An Optimal Color Image Fusion Approach Based on Fuzzy Integral (2006)
24. Baraniuk, R., Davenport, M., DeVore, R., Wakin, M.: The Johnson-Lindenstrauss lemma meets compressed sensing. Submitted manuscript (June 2006)
25. Yin, W., Morgan, S., Yang, J., Zhang, Y.: Practical compressive sensing with Toeplitz and circulant matrices. In: Proc. SPIE, vol. 7744 (2010)
26. Liu, H.X., Song, B., Qin, H., Qiu, Z.L.: An Adaptive-ADMM Algorithm With Support and Signal Value Detection for Compressed Sensing. IEEE Signal Processing Letters 20(4), 315–318 (2013)
27. Vaswani, N.: Kalman filtered compressed sensing. In: IEEE 15th IEEE International Conference on Image Processing (ICIP), San Diego, CA (2008)
28. Qiu, C.L., Lu, W., Vaswani, N.: Real-time Dynamic MR Image Reconstruction using Kalman Filtered Compressed Sensing. In: IEEE International Conference on Acoustics, Speech and Signal Processing (ICASSP), pp. 393–396 (2009)
29. Li, X.R., Zhu, Y., Wang, J., Han, C.: Optimal linear estimation fusion I. Unified fusion rules. IEEE Transactions on Information Theory 49(9), 2192–2208 (2003)
30. Xiao, G., Yang, B., Jing, Z.L.: Infrared and visible dynamic image sequence fusion based on region target detection. In: The 10th International Conference on Information Fusion, Quebec, Canada, July 9-12, pp. 80–85 (2007)
31. Xiao, G., Wei, K., Jing, Z.L.: Improved dynamic image fusion scheme for infrared and visible sequence based on image fusion system. In: IEEE 11th International Conference on Information Fusion, pp. 1–6 (2008)
32. Tou, J.T., Gonzalez, R.C.: Pattern Recognition Principles. Addision Wesley Publishing Comp., London (1974)
33. Qu, G.H., Zhang, D.L., Yan, P.F.: Information measure for performance of image fusion. Electronics Letters 38(7), 313–315 (2002)
34. Cvejic, N., Loza, A., Bull, D., Canagarajah, N.: A novel metric for performance evaluation of image fusion algorithms. In: IEC 2005, Prague, pp. 80–85 (2005)
35. Xydeas, C.S., Petrovic, V.: Objective image fusion performance measure. Electronics Letters 36(4), 308–309 (2000)

36. Zhang, L., Zhang, L., Mou, X.Q., Zhang, D.: FSIM: A Feature Similarity Index for Image Quality Assessment. IEEE Transactions on Image Processing 20(8), 2378–2386 (2011)
37. Li, S.T., Kwok, J.T., Wang, Y.: Using the discrete wavelet frame transform to merge Landsat TM and SPOT panchromatic images. Information Fusion 3(1), 17–23 (2002)
38. Michael, S.: Spatial Statistics and Spatio-Temporal Data: Covariance Functions and Directional Properties. Wiley Series in Probability and Statistics (2011)
39. Song, H., Huang, B.: Spatiotemporal Satellite Image Fusion Through One-Pair Image Learning. IEEE Transactions on Geoscience and Remote Sensing 51(4), 1883–1896 (2013)
40. Haydn, R., Dalke, G.W., et al.: Application of IHS color transform to the processing of multisensors data and image enhancement. In: Proceedings of the International Symposium on RS of Arid and Semi-Arid Lands, pp. 559–607 (1982)
41. Xiao, G., Jing, Z.L., Li, J.X., Henry, L.: Analysis of Color Distortion and Improvement for IHS Image Fusion. In: Proceedings of 2003 IEEE International Conference on Intelligent Transportation Systems, vol. 1, pp. 80–85 (October 2003)
42. Xiao, G., Jing, Z.L., Li, J.X., Henry, L.: A united optimum images fusion based on analysis of color distortion. Chinese Optics Letters 2(3), 144–147 (2004)
43. Kumar, A.S., Basu, S.K., Majumdar, K.L.: Robust classification of multispectral data using multiple neural networks and fuzzy integral. IEEE Transactions on Geosciences and Remote Sensing 35, 787–790 (1997)
44. Wang, D.Y., Keller, J.M., Carson, C.A., McAdo-Edwards, K.K., Bailey, C.W.: Use of fuzzy-logic-inspired features to improve bacterial recognition through classifier fusion. IEEE Transactions Systems, Man and Cybernetics, Part B 28, 583–591 (1998)
45. Li, J., Chen, G., Chi, Z., Lu, C.: Image Coding Quality Assessment Using Fuzzy Integrals with a Three-Component Image Model. IEEE Transactions on Fuzzy Systems 12, 99–106 (2004)
46. Li, J.L., Chen, G., Chi, Z.R.: A Fuzzy Image Metric with Application to Fractal Coding. IEEE Transactions on Image Processing 11, 636–643 (2002)
47. Miao, Q.G., Shi, C., Xu, P.F.: A novel algorithm of image fusion based on shearlets and PCNN. Neurocomputing 117(10), 47–53 (2013)

Chapter 8
Precision Cultivation System for Greenhouse Production

I.-Chang Yang and Suming Chen

Abstract. In view of increasing demands of high quality and safe agricultural products, it is important to effectively and precisely manage crop production. Precision agriculture is a site-specific cultivation method, and has been developed by using satellite remote sensing for field crops. However, more and more crops such as vegetables and ornamental plants are grown in greenhouses, and growth status monitoring of these crops is equally important as that for field-grown crops. Consequently, a ground-based remote sensing system for greenhouse precision cultivation needs to be developed. This work aimed to develop a RFID-integrated precision cultivation system based on remote spectral imaging of crops and environmental sensing for greenhouse production. This work illustrated the precision cultivation approach using vegetable seedlings as an example. Ground-based multi-spectral imaging system with plant-oriented sensing algorithm was developed in this study. A mobile environmental sensing system, including temperature, relative humidity and light intensity, was also developed to measure and analyze the spatial distribution of these environmental parameters in the greenhouse. The data and information communication was conducted through a wireless network. Irrigation policies corresponding to PLAI (Projected Leaf Area Index), NDVI (Normalized Difference Vegetation Index) and environmental conditions were established to provide a basis for precision irrigation. A variable rate spraying system was developed to implement the precision irrigation practices based on the analyzed status. This research successfully integrated RFID (Radio Frequency Identification), environment sensing, and remote spectral imaging for developing precision cultivation and traceability systems for the greenhouse production.

I.-Chang Yang · Suming Chen
Department of Bio-Industrial Mechtronics Engineering,
National Taiwan University,
Taipei, Taiwan

© Springer International Publishing Switzerland 2015
H. Leung and S.C. Mukhopadhyay (eds.), *Intelligent Environmental Sensing,*
Smart Sensors, Measurement and Instrumentation 13, DOI: 10.1007/978-3-319-12892-4_8

8.1 Introduction

In view of increasing demands of high quality and safe agricultural products, it is important to effectively and precisely manage crop production. In the last two decades, more and more crops have been cultivated in greenhouses under controlled environment [1, 2, 3, 4]. Although the macro-environment in a greenhouse remains relatively similar, the micro-environment around local confined area could differ drastically. Not only the outside factors such as the cloudy and rainy weather are known to affect the micro environment for plant growth, but also inside factors such as the shade of the greenhouse structure, air distribution and temperature gradients play important roles [5]. Therefore, the development of a reliable and traceable system capable of sensing the whole environment and monitoring factors critical to plant growth is crucial to greenhouse operations. Precision agriculture (PA) approaches have been shown to increase crop quality and yield, as well as decrease adverse environmental impacts by improving cultivation and management control on the farm [6]. Spectral imaging, an essential technology to remote sensing, as well as remote sensing by satellites has led the development of precision agriculture for field crops [7]. PA is a site-specific cultivation, and the information of the position acquired from a global positioning system (GPS) needs to be combined with the plant sensing data [8].

Several vegetation indices were developed to analyze the spectra or spectral imaging information of crops, among which the projected leaf area index (PLAI) and normalized difference vegetation index (NDVI) were the most commonly used [9, 10, 11]. While remote sensing data could be acquired from aircrafts or satellites, such techniques are less feasible in areas of small-sized and segmented farms such as in Asian countries or in the greenhouses. Nowadays, more and more crops such as vegetables and ornamental plants are grown in greenhouses, and growth status monitoring of these crops is equally important as that for field-grown crops [12, 13]. The PA technology hence needs to be modified to be suitable for such applications. Consequently, a ground-based remote sensing system for greenhouse precision cultivation needs to be developed.

8.2 Motivation

In order to develop precision cultivation system for greenhouse production, systems such as ground-based multi-spectral remote imaging [14, 15] and plant-oriented remote-sensing algorithms [16] based on monitoring of plant physiological status need to be studied. Additional to remote sensing information of crops, environmental controls of temperature, relative humidity and lighting are also vital to the crop cultivation in greenhouses [17]. Furthermore, the radio frequency identification (RFID) technology is ideal for indoor greenhouse applications to link the information including sensing information of crops, environmental factors and status of the plants [18]. Therefore, the objective of the work was to integrate RFID (Radio Frequency Identification), environment sensing, and remote spectral imaging for developing precision cultivation and traceability systems in the greenhouse.

8.3   Sensing and System Development

The entire system consists of several subsystems, and they are linked by the RFID system as shown in Figure 1. An operating platform is based on the wireless technology, and the terminal devices (such as PC, smart phone, mobile devices, and *etc.*) can link to the platform to monitor and control the operations in the greenhouses. The kernel subsystems include the multi-spectral imaging system (MSIS), the environmental factor measurement system (EFMS), the crop production traceability system (CPTS) and the variable-rate spraying system (VRSS), which are all operated under the precision local positioning system (PLPS) and linked by the radio frequency identification (RFID) system.

The web image monitoring system (WIMS) was developed in early 2000s, and it was a convenient tool for the greenhouse owner or operators to observe the real-time conditions in the greenhouse, and is still very useful today. The remote environment control system (RECS) is the combination of the basic environment control and the remote control technology. The environment control strategy and the final operation commands can be sent to the RECS to control the hardware such as fans, shade curtain, fan and pad, mist facilities and *etc.* to keep an appropriate and stable inner environment for the growth of crops in the greenhouse.

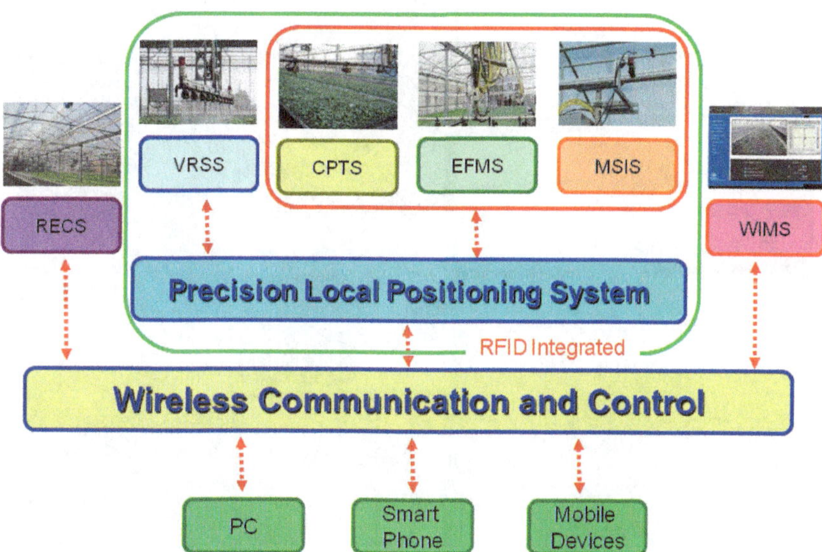

Fig. 1 Remote IT platform for precision cultivation consists of RECS (remote environment control system), VRSS (variable rate spraying system), CPTS (crop production traceability system), EFMS (environmental factors measurement system), MSIS (multi-spectral imaging system) and WIMS (web image monitoring system), based on the wireless technology for the communication and control.

8.3.1 *Remote Sensing and Monitoring*

The multi-functional remote sensing system (Figure 2) is composed of multi-spectral imaging system (MSIS), environmental factors measurement system (EFMS), RFID information system and a carrier [3, 4, 18]. The spray boom used in the greenhouse could be adopted as a carrier. As the carrier cruises along the length of the benches in the greenhouse, multi-spectral images of the plants on benches, environmental measurements of temperature, relative humidity and light intensity throughout the greenhouse can be obtained. The images and environmental measurements are linked using the RFID information system. In contrast to current greenhouse practices nowadays, in which environmental measurements are made only at a few specific locations, the environmental sensing subsystem developed in this study provides environmental data over a spatial map of the greenhouse area. Thus, the contour profiles of temperature, relative humidity, and lighting conditions can be used to improve the degree of

Fig. 2 (A): Multi-functional remote sensing system, (B): the dual band CCD cameras for the multi-spectral imaging, (C): the quantum sensor, (D): the temperature and the relative humidity sensors, (E): thermocouples, (F): lens with filter and filter mount, (G): color checker, (H): RFID information system with RFID reader, antenna, and RFID tags.

uniformity of environmental conditions. The multi-spectral images, when correlated to physiological status of the plants, provide information essential for the development of plant-oriented remote sensing algorithms by which site-specific needs of plants, such as irrigation and nutrient management, could be managed. The cabbage seedlings grown in the plug trays in the greenhouse will be used as an example for the purpose of discussions in this work.

8.3.1.1 Multi-spectral Imaging System

The software needed for the multi-spectral imaging system was developed based on Matlab 6.5 and LabVIEW 7.1 to enable the control of measurements, the data transfer, wireless communication and image processing. The following image processing steps were applied to the crop images with RFID information before conducting further analyses.

> **A. Spatial Calibration:** Spatial calibration was conducted based on a physical calibration frame of known dimensions and the following transform equations [19],

$$x' = c1x + c2y + c3xy + c4 \tag{1}$$
$$y' = c5x + c6y + c7xy + c8 \tag{2}$$

> where (x, y) are the true coordinates at the corners of the calibration frame; (x', y') are the coordinates of the corners in the raw images.

> **B. Image Stitch:** After the multi-spectral imaging system completed its acquisition of the 154 fixed-location crop images along the carrier track in the greenhouse, the images were calibrated and then stitched to form a complete image for each channel of multi-spectral imaging system. The image stitch program was developed based on the minimum value of D in the following equation:

$$D = \sum_{x=1}^{i} \sum_{y=1}^{j} [A(x, y) - B(x', y')]^2 \tag{3}$$

> where $A(x, y)$ represents the gray level of each pixel in the selected region of the target image; $B(x', y')$ is the gray level of each pixel in the corresponding region of next adjacent image. The minimum of D value means the most likeness of the images (Figure 3). A ColorChecker (GretagMacbeth Co., New Windsor, NY, USA) was used as a reference for the gray level calibration required in the image stitch process.

$$A = \alpha + B\beta \tag{4}$$

$$\alpha = \bar{A} - \beta \bar{B} \tag{5}$$

$$\beta = \frac{\sum_{i=1}^{k} (A - \bar{A})(B - \bar{B})}{\sum_{i=1}^{k} (B - \bar{B})^2} \tag{6}$$

where A represents the gray level of each pixel in the selected region of target image; B is the gray level of each pixel in the corresponding region of next adjacent image. $\bar{A}$ and $\bar{B}$ are averaged gray levels.

C. Image Segmentation: Image segmentation using a transformation formula of RGB colors was performed to separate the background from the canopy in the color image [20] :

$$Y = a \cdot R + b \cdot G + c \cdot B \tag{7}$$

where R (Red), G (Green), B (Blue) are the components of the color image; and a, b, c are weighted parameters. It was found that the best set of weighted parameters for canopy segmentation were a = -1.25, b = 1.9, c = -1.25. The green pixels were most important for distinguishing the canopy and thus were more heavily weighted [3].

Fig. 3 A completed image of the seedling crop canopy was obtained by stitching the images acquired at multiple locations along the traveling direction of the multispectral imaging system in the greenhouse

The multi-spectral remote imaging system was constructed using both color and B/W IEEE1394 CCD cameras (AVT Marlin F145-C2, F145-B2; Allied Vision Technologies GMBH, Germany) and a Barebone SG31G2 computer (Shuttle Co., Taiwan) with a P1394V frame grabber. The electronic shutter speed and gain value were controlled for automatic and correct exposures using software developed in this study with LabVIEW (National Instrument Co., USA) and Matlab (Mathworks Co., USA). A band filter with peak wavelength at 780 nm (NIR) was used for the F145-B2 B/W CCD camera to obtain near infrared images. Red, green and blue band images (R, G and B) were obtained from the F145-C2 color CCD camera. The projected leaf area index (PLAI) for the plant canopy was calculated by the following definition:

$$PLAI = \frac{P_{Canopy}}{P_{Tray}} \tag{8}$$

where P_{Canopy} represents the number of pixels of the canopy on each tray, and P_{Tray} is the total pixels of entire tray area. The regression relationship between the day of crop growth (with the representative images of seedlings) and the projected leaf area index (PLAI) was shown in Figure 4. The index of PLAI could provide the information of the crop growth status for the system to calculate the scores to evaluate the irrigation level.

The red band and NIR band images thus obtained was used to calculate the normalized difference vegetation index (NDVI) [10]. The NDVI index is an indication of crop growth and vegetation; it is defined as following:

$$NDVI = \frac{NIR-R}{NIR+R} \tag{9}$$

Where NIR represents the spectral reflectance at NIR wavelength; R is the spectral reflectance at red waveband.

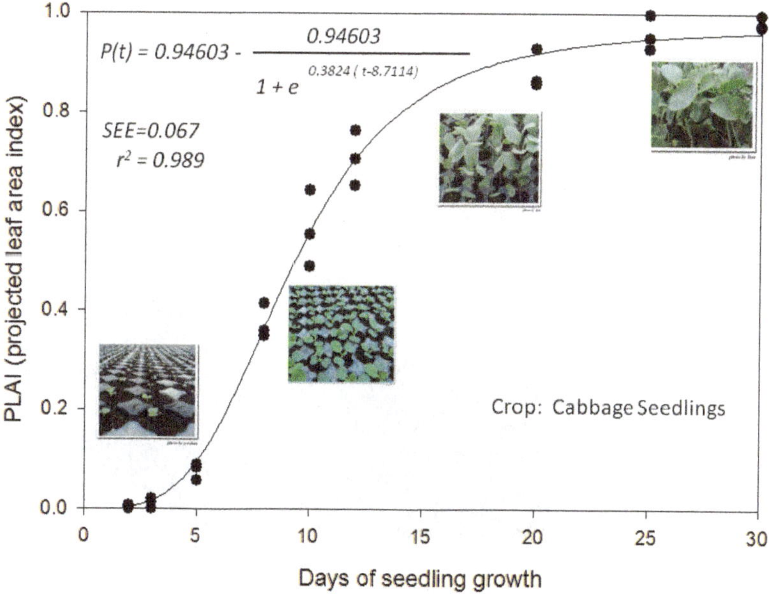

Fig. 4 The regression relationship between the day of cabbage seedling growth and the projected leaf area index (PLAI). The representative images of seedlings in different stages are shown in the figure. [18]

8.3.1.2 Environmental Factors Measurement System

The environmental factors measurement system (EFMS) consisted of four thermal couples, a humidity transmitter and a quantum sensor. The measurement signals of these environmental sensors were transmitted to the Barebone computer through the PCI-6023 data acquisition interface card (National Instruments Co. USA). Datasocket (a module of LabVIEW) and a wireless web device (WL167G, ASUS) were incorporated to enable remote data transfer for this system. Different contour layers of environmental factors could be obtained separately by EFMS. Temperature, relative humidity, and light intensity (Figure 5) are the major factors that the system is able to obtain via the carrier cruising along the rail in the greenhouse. The data would be recorded step by step with 60-cm interval. The hot spot in the greenhouse or the shading of the facility structure could be identified from the mapping of the environmental factor information layers (Figure 6).

The traditional environmental factor measurement is usually single or several measurements in a big greenhouse. More sensors will increase the cost when the sensors were fixed at the specific locations in the greenhouse. In this study, EFMS provide the possibility to measure the environmental factors at different locations all over the greenhouse. The EFMS via boom cruising could provide 2-dimensional contour profiles of Temperature, relative humidity, and light intensity in the entire greenhouse which could not be provided by the traditional fixed single point (or a few locations) measurement. The traditional measurement could

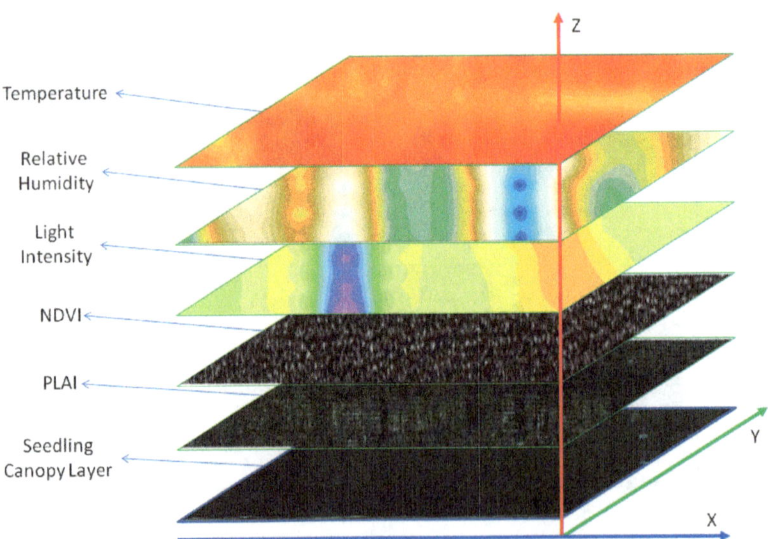

Fig. 5 Mapped contour profiles for site-specific greenhouse managements including the temperature, relative humidity, light intensity, NDVI, and PLAI layers over the crop canopy layer. [18]

not be applied to the precision cultivation due to the lack of site environmental factors information. The two dimensional data could provide the GIS (Geographic Information System) mapping as shown in Figures 5 and 6. It is similar to the GIS analysis for precision agriculture in the field via the satellite remote sensing. The ground-based or the indoor site-specific treatments could be attempted by using the environmental factors measurement system and precision local positioning system (PLPS).

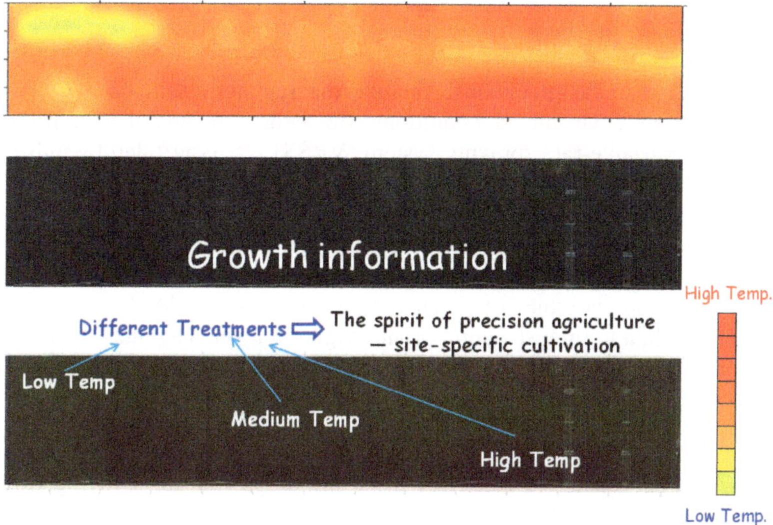

Fig. 6 The GIS mapping of growth information, environmental factors and vegetation indices provides the decisions for site specific precision cultivation in the greenhouse

8.3.1.3   Web Image Monitoring System

The web image monitoring system (WIMS) was built with several dome cameras which can remotely control via internet by users. The camera could be controlled to zoom in/out to focus on the object. The motion detection could be accomplished using WIMS as well. WIMS is a network video surveillance system with different functions with it. The network video camera set up at the location in the greenhouse can be used to observe every corner of the field area, and the image data was transmitted via the wireless platform to the information center in the control room at remote location from greenhouses.

The WIMS consisted of a network video server WebEye B101 (WebGate, Inc., GunpoSi, Korea) which can provide high quality image resolution of 720 x 486, network dome camera which can be rotated 360 degrees with the machine, and wireless base station.

The motion detection and the setting point cycle monitoring via the system were the major functions which were adopted for WIMS. The dome cameras were set to focus on the main gate, control panel, fan and pad, and several vital points in the greenhouses. When the motion detector was triggered by human actions, irrigation operations, or *etc.*, the motion would be recorded by the imaging system. Therefore, operators via the wireless platform could easily observe the triggered images to know the events. The system provides not only the warning signs but also record different incidents.

8.3.2 *Precision Irrigation Control*

The operations in the greenhouse include the remote sensing of the crop and measurement of environment conditions, the irrigation, and the environmental control. The variable-rate spraying system (VRSS) was controlled by pulse width modulation (PWM). The irrigation would be operated after the multi-spectral images were obtained by MSIS. Figure 7 shows the step by step variable rate spraying system with multi-spectral imaging system and RFID information on the boom.

The irrigation volume of each boom cruising could be estimated by experimental data to infer the appropriate irrigation level. The theoretical spraying irrigation volume was calculated by the following formula [21, 22].

$$Lt = Ls \times (W/v) \times r \tag{10}$$

Lt is the theoretical volume, and Ls is the single nozzle spraying flow rate (L/sec). W is the interval between two nozzles, and v is the boom cruising speed (m/s). The parameter "r" is the ratio of "Duty Ratio/Cycle." The value of Ls is 0.0225 L/sec under the pressure of 3.0×10^4 kg/m^2.

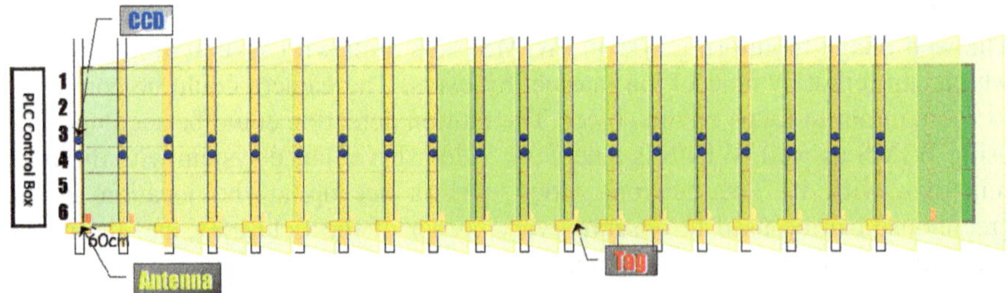

Fig. 7 The step by step variable rate spraying system with multi-spectral imaging system and RFID information on the boom

Fig. 8 The stitched scene of the seedlings in the greenhouse, and the variable-rate spraying system

The irrigation level can be calculated based on the nozzle flow rate, and in practice, modified as needed according to the boom cruising speed, the PWM parameters, and the water pressure in the hose. The stitched scene of the seedlings in the greenhouse is shown in Figure 8. The operation of variable-rate spraying system is illustrated in Figure 8 (C).

8.3.3   Crop Production Traceabiity System

A crop production traceability system was established, and this study explored the applications of the passive radio frequency identification (RFID) system in the greenhouse. Regarding the performance, the best relative distance and angle were 0.225 m and 56°, respectively. RFID-integrated multi-functional remote sensing system developed in this study included crop spectral image analysis, spatial variation analysis of environmental factors, and wireless information transmission and the scheme of this system is shown in Figure 9. The RFID information was recorded with the boom location to analyze images and environmental factors of the crops. In addition, the related database fed by the RFID system could also provide additional information such as greenhouse operators, crop species, and cultivation management information (Figure 10). The system can reduce the cost of the human resources needed for the operations in the greenhouse. For example, the NDVI-based look up table (LUT) was used by the water management module to enable automated irrigation precisely suitable for individual plants. Similar precision cultivation control could be implemented for other greenhouse factors based on information obtained from the RFID database.

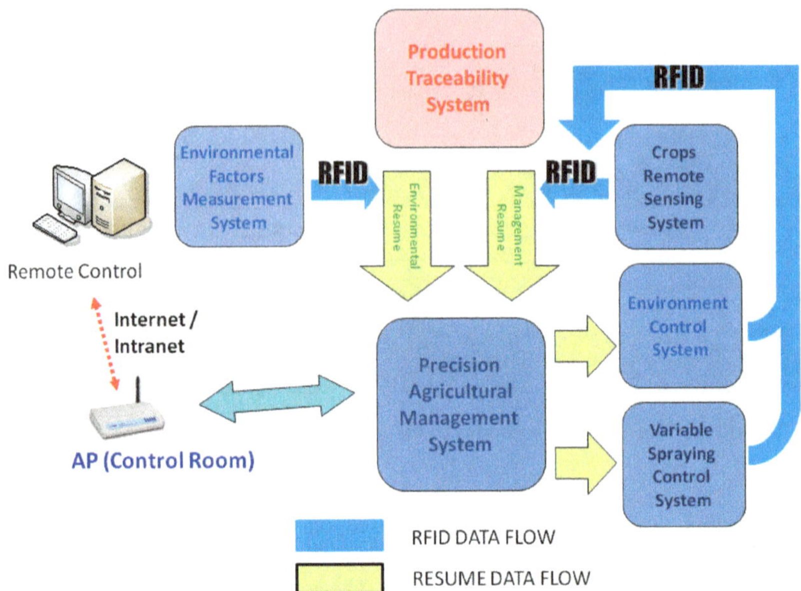

Fig. 9 Scheme of RFID-integrated multi-functional remote sensing system. The production traceability system includes environmental resume and management resume for precision greenhouse management.

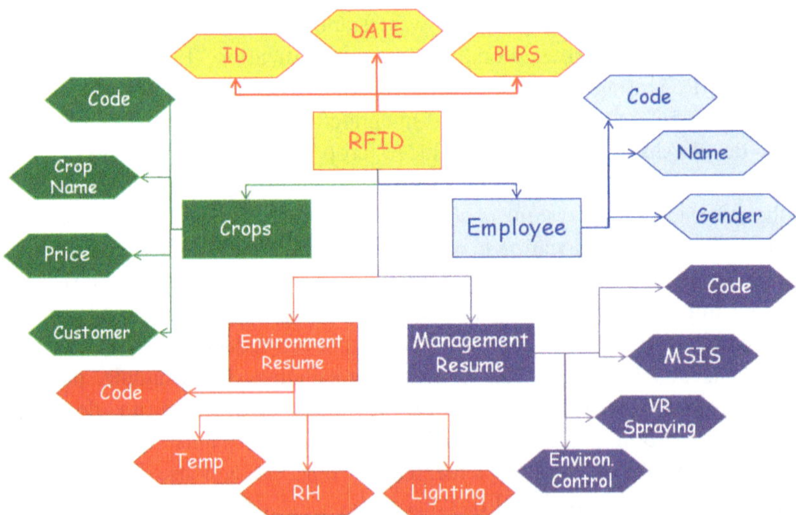

Fig. 10 Data scheme of RFID information system. The human resource, seedling, environment, and management information can all be linked by using the RFID system. [18]

Fig. 11 The interface of the production traceability system (I) (The production resume system)

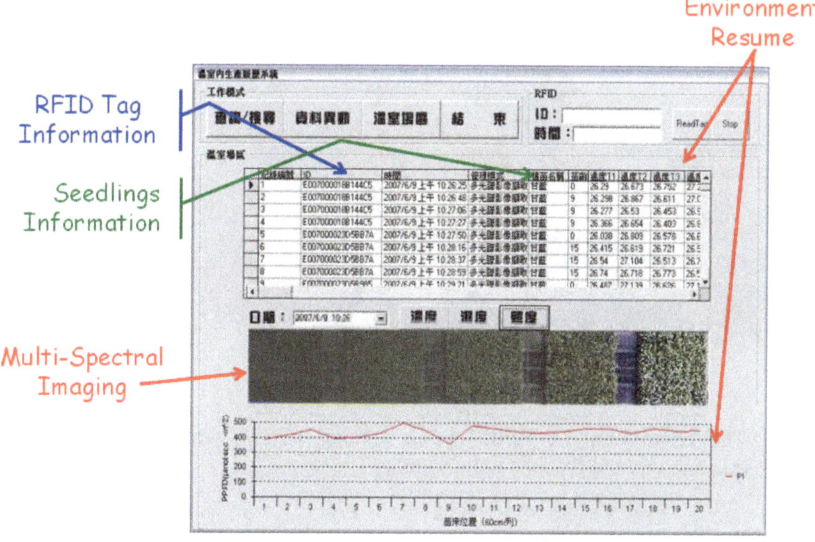

Fig. 12 The interface of the production traceability system (II)

The user interfaces of the RFID system were shown in Figures 11 and 12. The basic information of seedlings, multi-spectral imaging, the environment resume including temperature, relative humidity, and light intensity were all linked by RFID. Therefore, when the RFID tag was read in the greenhouse by the cruising boom, all information and the treatments or operations of crops would be recorded.

8.4   System Integration and Applications

The scheme of RFID-integrated multi-functional remote sensing system was previously shown in Figure 1. The subsystems mentioned in previous sections now need to be integrated, and the relationship of RFID system among remote sensing and monitoring, greenhouse operations and production traceability can further be illustrated in Figure 13. The key element for the success of system integration is the development of local positioning system in the greenhouse since the GPS (Global Positioning System) signals from satellite is not available. We will elaborate the development of local positioning system in Section 4.1.   To demonstrate the applications, we will give a case study as an example in Section 4.2: an application of the integrated system developed in this study for the precision irrigation in a cabbage seedling nursery in Yun-Lin County, Taiwan.

Fig. 13 The relationship of RFID system among remote sensing and monitoring, greenhouse operations and production traceability

8.4.1   *Local Positioning System*

Due to satellite GPS signal is not available in the greenhouse, the precision local positioning system (PLPS) was developed in this work using a photo-encoder and the proximity switch (as shown in Figure 14) to provide the precision locations during sensing and measurement in the greenhouse. The sensing blocks were mounted on the boom rail every 0.6 m. The movement distance of the boom could be calculated by the photo-encoder, and reset by the proximity switch every 0.6 m to avoid error accumulation. There were 205 ± 1 pulse signals generated in each 0.6-m distance cycle, and the error of the distance measurement was ± 2.92 mm. The design gives the local positioning system extremely high accuracy. Therefore, accurate distance information and precise locations could be provided by the precision local positioning system without using the satellite signal.

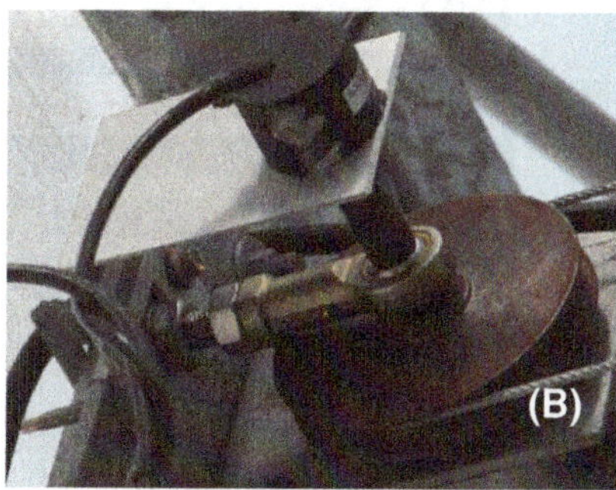

Fig. 14 The composition of the precision local positioning system (PLPS). (A): The proximity switch on the boom set, which could be triggered by the indicators along the boom rail. (B): The distance of the boom travelled along the rail was calculated by the rotary optical encoder.

8.4.2   Precision Irrigation – An Example

The precision irrigation in a cabbage seedling nursery in Yun-Lin County, Taiwan is used as an example to discuss the application of integrated system on remote IT platform for precision cultivation developed in this study.

The criteria for determining irrigation to be given to seedlings, the PLAI and NDVI were employed to calculate the first step watering level for the cabbage seedlings as shown in Table 1. The criteria were separated into five levels including level A, B, C, D and E. For level A, the PLAI value was lower than 0.1. For level B, the value was in the range of 0.1 and 0.4, and for level C was between 0.4 and 0.85. The PLAI criterion values of levels D and E were both higher than 0.85; and the NDVI value (0.75) was added into the criterion for evaluation of the watering level. The relationship between amount of irrigation versus NDVI can be investigated (Figure 15). At level E, the most water is given for irrigation, while level A irrigation is the minimum.

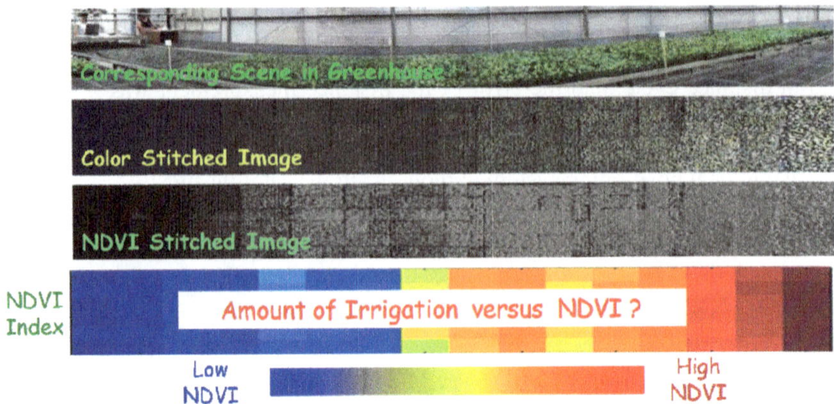

Fig. 15 The relationship between amount of irrigation versus NDVI can be investigated

However, because the micro-climate environments of seedlings were always different and changing in the greenhouse, the environmental factors such as the temperature, the relative humidity, and the lighting intensity were also used to determine the irrigation final decision. High temperature, low relative humidity, and high light intensity can make a microclimate environment lose water easily and increase the evaporation rate of the plant. The environmental factors give different weighted values to calculate the extra watering score to adjust the final irrigation decision, and the calculation look-up table is shown in Table 2. The simple calculation formula is shown as follows:

$$\text{Score} = 0.5\text{T}\,(0,1) + 0.3\,\text{RH}(1,0) + 0.2\text{LI}\,(0,1) \tag{11}$$

where T(0, 1), RH(1, 0), and LI(0, 1) represents the temperature, the relative humidity, and the light intensity. The output values of the T (temperature) and LI (light intensity) will be 1 when the measurement values are higher than the setting threshold. The output of the RH will be 1 when the RH measurement is lower than the setting threshold. The threshold of the score was 0.6 to increase the watering level in the previous calculation. Considering the objective situation, the irrigation volume of each boom cruising could be estimated by the experiment data to infer the concrete irrigation level. Therefore, level D, C, B, A (Table 1) are 75%, 50%, 25%, and 0% of Level E, respectively. Figure 16 shows the adjustment of irrigation level due to environmental factors, and Figure 17 illustrates the graphical user interface of irrigation decisions based on RFID-integrated precision cultivation system.

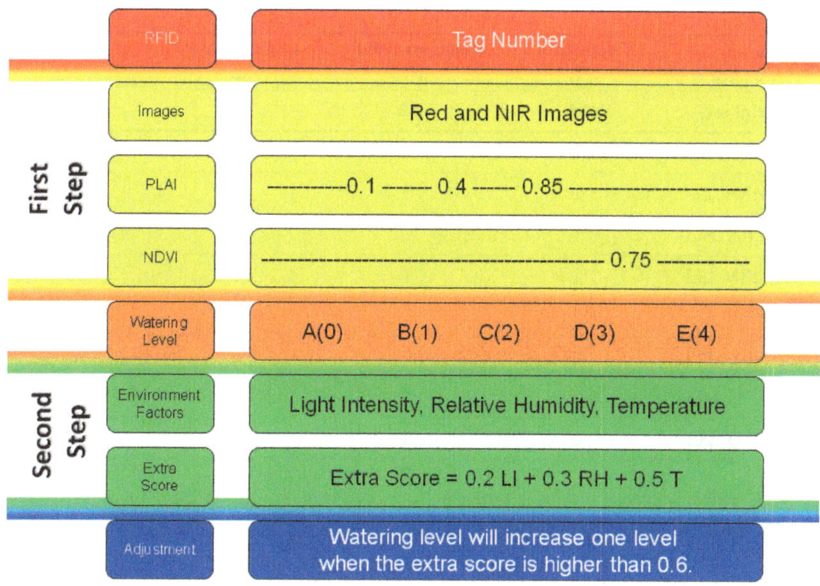

Fig. 16 The final irrigation level could be calculated by the multi-spectral imaging information and adjusted by the environmental factors

Table 1 Table of irrigation levels. [18]

Indices	Criterion Values				
PLAI	< 0.1	$\geq$0.1 & < 0.4	$\geq$0.4 & < 0.85	$\geq$ 0.85	$\geq$ 0.85
NDVI	--	--	--	< 0.75	$\geq$ 0.75
Irrigation Level	A	B	C	D	E
Environment Factors	Consideration from different environment conditions				
Irrigation Final Decision	Irrigation level pluses the extra score using environment factors				

Table 2 Extra score for irrigation adjustment. [18]

Environment Factors	Temperature (T)	Relative Humidity (HR)	Light Intensity (LI)	Score
Weighted	0.5	0.3	0.2	
HT, HRH, HLI	1	0	1	0.7
HT, HRH, LLI	1	0	0	0.5
HT, LRH, HLI	1	1	1	1
HT, LRH, LLI	1	1	0	0.8
LT, HRH, HLI	0	0	1	0.2
LT, HRH, LLI	0	0	0	0
LT, LRH, HLI	0	1	1	0.5
LT, LRH, LLI	0	1	0	0.3

Note: The extra score calculation table for the different weighted environmental factors. HT, LT: represent the condition which is higher or lower than the temperature threshold that set by the user. LRH, HRH: represent the condition which is lower or higher than the relative humidity threshold. HLI, LLI: represent the condition which is higher or lower than the light intensity.

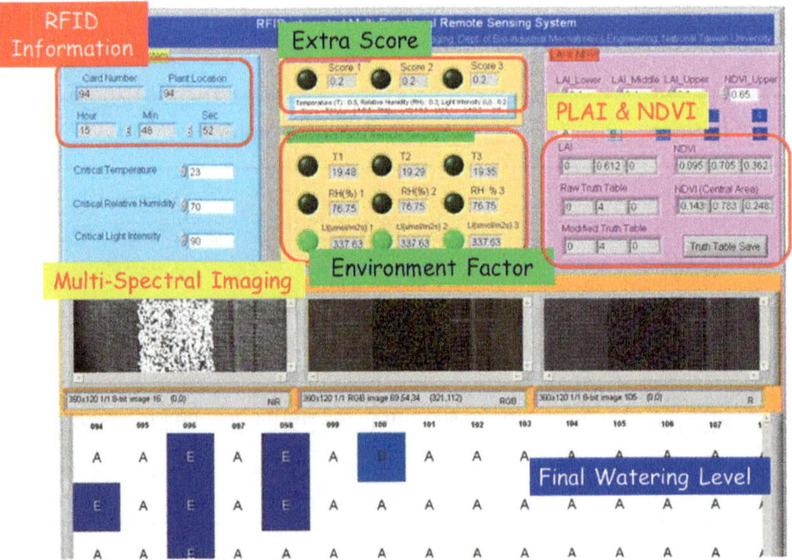

Fig. 17 The implementation of the RFID-integrated precision cultivation system includes (1) RFID information, upper left; (2) extra score, environment factor measurements, upper center; and (3) criteria of the vegetation indices, upper right. The multi-spectral images and the irrigation level tables are shown in the center and lower sections, respectively.

8.5   Conclusions

The quality of agricultural products is affected by the weather and environment. A RFID-integrated system on remote IT platform for precision cultivation was successfully developed in this study. The concept of the crop production traceability system was conducted in this work to build up a quality monitoring and control system to allow the producer's full control over the environment and management conditions in the greenhouse. The remote sensing information of crop and environmental factors were acquired by the remote control multi-functional sensing system. The criteria of the crop irrigation were determined by vegetation indices (PLAI and NDVI) and environmental factors including temperature, relative humidity, and light intensity. The production database of the crop was linked by RFID on each site of crop spectral images and environmental sensing. The system developed in this study will improve the management and production, and also serve as a demonstration for precision cultivation in greenhouses. Precision cultivation including irrigation, fertilizing, crop disease detection are only some of the examples for future applications using the integrated system developed. Many other crops grown in greenhouses are also suitable for precision cultivation based on the same concept and similar technologies discussed in this paper.

Acknowledgments. The authors thank Wu-Pin Seedling Supplier for providing their greenhouse facilities for the experiments, and Miss Diane E. Chan in the USDA Environment Microbial and Food Safety Laboratory for paper editing assistance. We appreciate Prof. Kuang-Wen Hsieh, Prof. Yu-I Huang, Dr. Chia-Tseng Chen, Dr. Yu-Liang Chen, Dr. Chao-Yin Tsai, Mr. Hong-Chi Lu, Mr. Chin-Lun Chang, Miss Hui-Mei Lin, Miss Jou-Ching Li, Mr. Wei-Ting Tu, and Mr. Chun-Chi Chen for their support and help in these projects. We also thank the grants of the projects (COA: '93AS-6.1.1-FD-Z1', '94AS-6.1.2-FD-Z1', '95AS-7.1.2-FD-Z1', '96AS-9.1.1-FD-Z1') from Council of Agriculture, Taiwan.

References

[1] Chen, S., Li, M.T., Chen, C.T., Lin, Y.C., Huang, C.W., Wu, T.H.: Remote sensing of crop growth characteristics in greenhouses. In: Chen, S., Lin, T.T. (eds.) Proceedings of International Symposium on Design and Environmental Control of Tropical and Subtropical Greenhouses. Acta Horticulturae, vol. 578, pp. 295–301 (2002)

[2] Chen, S., Tsai, C.Y., Hsieh, J.F., Hung, C.H., Chiu, Y.C., Hsieh, K.W.: Growth status monitoring and quality evaluation for bio-production and products using spectral sensing techniques. In: Proceedings of the Second International Symposium on Machinery and Mechatronics for Agriculture and Bio-Systems Engineering, Keynote Speech, KN-9-23. Kobe University, Kobe (2004)

[3] Chen, S., Lu, H.C., Hsieh, K.W., Huang, Y.I., Chen, C.T., Yang, I.C.: Development of multifunctional remote sensing system for greenhouse production. In: Proceedings of SPIE International Symposium on Optics East e Sensors and Photonics for Applications in Industry, Life Sciences, and Communications, vol. 6381, p. 638103-1. The International Society for Optical Engineering, Boston (2006)

[4] Yang, I.C., Chen, S., Huang, Y.I., Hsieh, K.W., Chen, C.T., Lu, H.C.: RFID-integrated multifunctional remote sensing system for seedling production management. In: Food Processing Automation Conference Proceedings. The American Society of Agricultural and Biological Engineers, Providence (2008)

[5] Kittas, C., Bartzanas, T., Jaffrin, A.: Temperature gradients in a partially shaded large greenhouse equipped with evaporative cooling pads. Biosystem Engineering 85(1), 87–94 (2003)

[6] Marino, B.D.V., Geissler, P., O'Connell, B., Dieter, N., Burgess, T., Roberts, C.: Multispectral imaging of vegetation at biosphere 2. Ecol. Eng. 13, 321–331 (1999)

[7] Han, S., Hendrickson, L., Ni, B.: Comparison of satellite remote sensing and aerial photography for ability to detect in-season nitrogen stress in corn. ASAE Paper No. 011142. ASAE, St. Joseph, Mich. (2001)

[8] Hache, C.: Site-specific crop response to soil variability in an upland field (M. Sc. thesis). Tokyo University of Agriculture and Technology, Division of Environmental and Agricultural Engineering, Tokyo (2003)

[9] Plant, R., Munk, D., Roberts, B., Vargas, R., Rains, D., Travis, R.: Relationships between remotely sensed reflectance data and cotton growth and yield. Trans. ASAE 43(3), 535–546 (2000)

[10] Lizaso, J.I., Batchelor, W.D., Westgare, M.E.: Using the normalized difference vegetation index and a crop simulation model to predict soil spatial variability. Trans. ASAE 45(4), 1217–1222 (2002)

[11] Kacira, M., Ling, P., Short, T.: Machine vision extracted plant movement for early detection of plant water stress. Trans. ASAE 45(4), 1147–1153 (2002)

[12] Chen, S., Li, M.T., Chen, C.T., Lin, Y.C., Huang, C.W., Wu, T.H., Hsieh, K.W.: Remote sensing of crop growth characteristics in greenhouses. In: Chen, S., Lin, T.T. (eds.) Proceedings of International Symposium on Design and Environmental Control of Tropical and Subtropical Greenhouses. Acta Horticulturae, vol. 578, pp. 295–301 (2002)

[13] Chen, S., Tsai, C.Y., Hsieh, J.F., Hung, C.H., Chiu, Y.C., Hsieh, K.W., Jiang, J.A., Chen, R.L.C., Yang, H.C., Chen, C.T., Yang, I.C., Yang, C.W., Wu, T.H., Li, M.T., Huang, C.W., Huang, C.C., Tsai, C.C., Yang, C.K., Brimmer, P.: Growth status monitoring and quality evaluation for bio-production and products using spectral sensing techniques. In: Proceedings of the Second International Symposium on Machinery and Mechatronics for Agriculture and Bio-Systems Engineering, Keynote Speech, KN-9 ~ 23, Kobe University, Kobe (2004)

[14] Yao, H., Tian, L., Noguchi, N.: Hyperspectral system optimization and image processing. ASAE Paper, No. 011105. ASAE, St. Joseph, Mich. (2001)

[15] Kostrzewski, M., Waller, P., Guertin, P., Haberland, J., Colaizzi, P., Barnes, E.: Groundbased remote sensing of water and nitrogen stress. Trans. ASAE 47(1), 291–299 (2002)

[16] Chen, S., Li, M.T.: Monitoring. In: Proceedings of the 6th International Symposium on Fruit, Nut, and Vegetable Production Engineering Held in Potsdam 2001, pp. 603–608. Institute of Agricultural Engineering Bornim e.V. (ATB), Potsdam (2001)

[17] Chen, S., Chen, C.T., Yang, I.C., Li, M.T., Huang, C.W., Huang, C.C., Hsieh, K.W.: Precision greenhouse cultivation using multi-spectral imaging techniques. In: Proceedings of APO Multi-Country Study Mission on Precision Farming – Resource Papers, Taoyuan, Taiwan. Sponsored by Asian Productivity Organization and Council of Agriculture, Taiwan (2006)

[18] Yang, I.C., Hsieh, K.W., Tsai, C.Y., Huang, Y.I., Chen, Y.L., Chen, S.: Development of an automation system for greenhouse seedling production management using radio-frequency-identification and local remote sensing techniques. Engineering in Agriculture, Environment and Food 7(1), 52–58 (2014)

[19] Gonzalez, R.C., Woods, R.E.: Digital image processing. Addison-Wesley Publishing Company Inc., New York (1992)

[20] Bulanon, D.M., Kataoka, T., Ota, Y., Hiroma, T.: A segmentation algorithm for the automatic recognition of Fuji apples at harvest. Biosyst. Eng. 83(4), 405–412 (2002)

[21] Hsieh, K.W., Chen, S., Shiu, W.H., Chang, S.H., Lai, P.H., Chen, C.Y.: Study on precision variable rate spraying system in the greenhouses. In: Proceedings of the Third International Symposium on Machinery and Mechatronics for Agricultural and Bio-Systems Engineering (ISMAB 2006), pp. 317–322. Korean Society for Agricultural Machinery, Korea (2006)

[22] Hsu, W.H.: Study of precision variable-rate spraying control system in greenhouses. MS Thesis. National Chung-Hsin University, Department of Bio-Industrial Mechatronics Engineering, Taichung (2006)

Chapter 9
Environment Monitoring System Based on IEEE 1451 Standard

A. Kumar and G.P. Hancke

Abstract. This chapter describes the work that has been done to design a environment monitoring system for smart building. The proposed system can also be used for the monitoring the indoor air quality and indoor environmental parameters. The sensor array is implemented using electrochemical sensors. In order to realize the target design goals of the sensor node or wireless standard transducer interface module and wireless network capable application processor were developed based on IEEE1451 standard. The graphical user interface (GUI) is implemented in LabVIEW 9.0. NCAP is connected to the STIM through a zigbee communication. The level of indoor environment parameters and information regarding STIM can be seen on the graphical user interface of the NCAP. After integration of the sensors are recalibrated using the static chamber and potentiometer adjustment technique. The EM system is low cost, energy efficient, and portable.

9.1 Introduction

Global greenhouse gas emissions rise every year due to human activities. In South Africa, there is various outdoor air quality monitoring stations. Studies have shown that the indoor air quality of buildings is sometimes more polluted than the outdoor air quality [1]. The problem investigated in this chapter was to develop a wireless sensor system which monitors indoor gas emissions in smart buildings. Smart buildings use technology to make the operation of buildings more efficient [2], [3]. This can include monitoring and control systems. An environmental

A. Kumar · G.P. Hancke
Electrical, Electronics and Computers Engineering Department
University of Pretoria, South Africa
e-mail: anuj.kumar@up.ac.za, gp.hancke@city.edu.hk

© Springer International Publishing Switzerland 2015
H. Leung and S.C. Mukhopadhyay (eds.), *Intelligent Environmental Sensing,*
Smart Sensors, Measurement and Instrumentation 13, DOI: 10.1007/978-3-319-12892-4_9

monitoring system allows building management to study pollutant levels in the air, to identify polluting sources and to ensure proper ventilation is implemented in the building. This system can also be used for indoor air quality (IAQ) audits and it can be implemented as the monitoring part of a heating, ventilation, and air conditioning (HVAC) control system.

Common greenhouse pollutants include CO_2, SO_2, NO_2 and CO. These gasses negatively affect the health of those who come in contact with it and can cause effects such as sick building syndrome [4], [5] and are presented in Fig. 1. The purpose of this product was to offer a relatively low cost dedicated indoor air quality measurement system. Along with air quality, conditions such as temperature and humidity were also monitored. The engineering challenge was to design a relatively low cost multiple node wireless sensor network. Different wireless protocols and sensor types were investigated. The sensor nodes had to be battery powered. This would allow the user to easily move the sensor from one room to another when monitoring a building's indoor air quality. Limitations that were considered are power consumption and the range of the wireless data transmission. Because the device is battery operated, the device has to use available power very efficiently. The wireless range of the device is also limited because of the low power specification. The sensors used, were bought off-the-shelf, therefore the accuracy depends on the manufacturers. Much research has been done on various air qualities and environmental parameter monitoring system [5]-[14]. The intended contribution of this chapter was to design a sensor network which can measure multiple environmental parameters and transmit them wirelessly to be displayed and stored on a user interface application. The system also incorporates an buzzer system which alerts occupants of a room if a dangerous level of a parameter is measured. Low frequency transceivers were

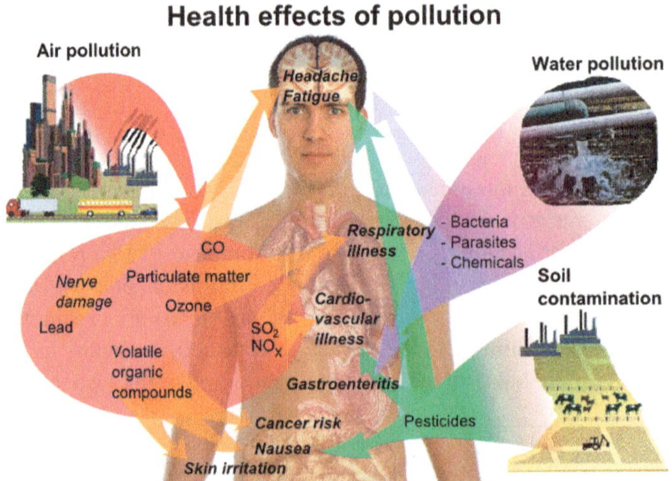

Fig. 1 Health effects of pollution [5]

used to improve the indoor transmission range and to reduce power consumption. The sensors used in the sensor nodes were chosen to minimize power consumption whilst remaining accurate. To reduce the cost of the wireless sensor network, the sensor nodes, the sink node and the user interface were designed from IEEE1451.1, IEEE1451.2, and IEEE802.15.4 standards.

9.2 Developed Environment Monitoring Systems

Environment monitoring system (EMS) is a complete real time monitoring and data recording system. It automatically measures and records the air quality and the environmental parameters. It is essential to monitor indoor environment to provide thermally comfortable and toxicant free environment for occupant's better health [5]-[6].

The detailed block diagram of developed EM system is shown in Fig. 2. The wireless transducer interface module (WSTIM) is linked to wireless network capable application processor (WNCAP) through zigbee communication. The developed EM system can be used of detecting the concentration level of greenhouse gases such as SO_2, NO_2, CO, CO_2, O_2 with environmental parameter (temp. and humidity). The sensors output are processed through signal conditioning circuit and the integrated signals are connected to the inbuilt ADCs channel of the processor. After processing and integration, the sensors results are sent to a network capable application processor (NCAP) PC through zigbee module and are also saved in micro memory card (MMC) according to transducer electronic data sheet. The results are displayed on the graphical user interface (GUI) running on a PC. The mentioned design of EM system is a self-contained unit which makes it relatively easy to add extra sensor nodes.

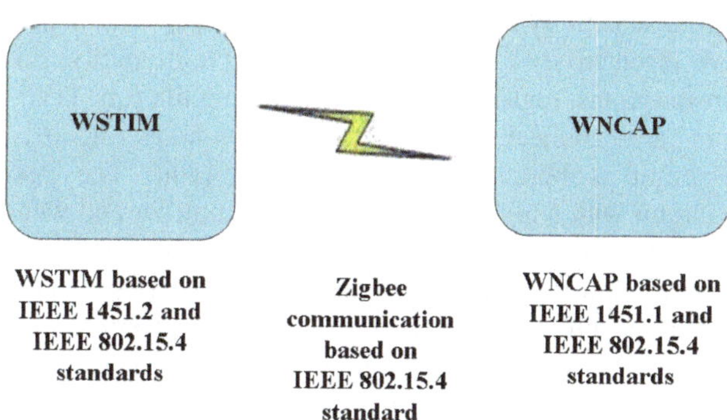

Fig. 2 Block diagram of developed EM system

9.3   Wireless Standard Transducer Interface Module

A wireless smart transducer interface modules (WSTIMs) or wireless sensor nodes consist of several components such as sensor with signal processing circuit, a power supply unit, memory, and a transceiver unit [14]. The wireless sensor node is designed based on the IEEE802.15.4 and IEEE1451.2 standards and an array of electrochemical sensors are used [15], [16]. The developed wireless sensor module is capable of communicating with NCAP through zigbee, handling the actuators interface, and supporting transducer interface electronic data sheet (TEDS). The developed seven sensing modules are connected to the seven channels of the proposed design node and used the ADC bus. Also, one channel is free for the used in the near future. A PIC 18F4550 microcontroller is chosen to support all above functions and to support the development of the node. The developed standard transducer interface module has eight channels out of which seven are used to interface the developed sensing modules while the other one channel is free for further processing. The block diagram of the developed WSTIM or wireless sensor node is shown in Fig. 3.

9.3.1   Sensor Array

The selected sensor has several advantages such as low power consumption, low cost, high accuracy, and capable of detecting different gases. The detailed explanation of the selected electrochemical sensors including their usage, advantages and disadvantages are given in [17]-[21].

The low power consumption and low cost electrochemical sensors with additional temperature and humidity sensors are suitable to use as an array for cost effective and energy efficient environment monitoring system to measure the air pollutant gases with oxygen and environmental parameter. The electrochemical sensors have high sensitivity and selectivity to detect gas with improved property such as relative insensitivity to fluctuations in relative humidity, EMF/RF noise, low power consumptions, high response time, and long life time [19].

The sensors are connected to signal conditioning circuits. The signal conditioning circuit is based on potentiostatic circuit. The designing of a potentiostatic circuit with a high input bias current amplifier and without precision will impact the sensor sensitivity resulting in increased sensor to sensor variation. Hence, the precise, ultra low input bias current amplifier (less than 5nA), such as LMP7721, OP90 and OP296 are used to design the potentiostatic circuits. The inbuilt amplifiers, OP90 and LMP7721 improve the circuit performance and allow the electrochemical sensor to detect low gas concentration with high accuracy. The input bias current of these amplifiers LMP7721 and OP90 is 3fA and 4nA, respectively, at room temperature (25°C) [6]. The power consumption, response time, sensing range, and operating voltage of the implemented sensor module are given in Table 1.

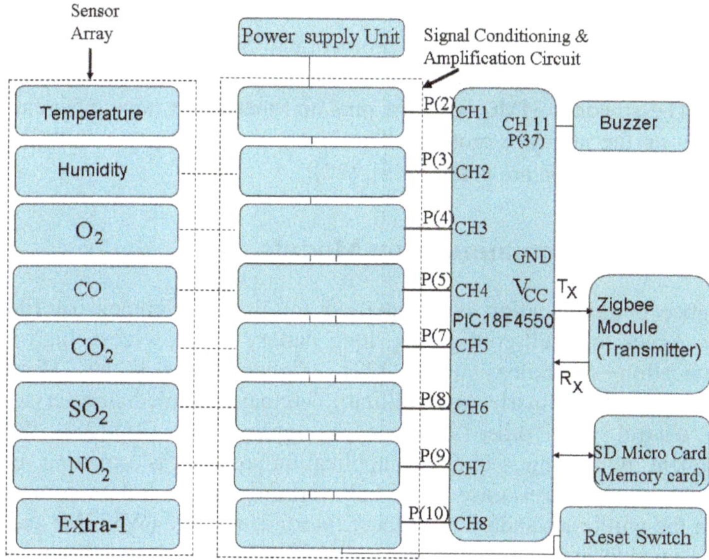

Fig. 3 Block diagram of wireless standard transducer interface module

Table 1 Implemented Sensor Module

Sensor Module	Range	Power Consumption	Response Time	Operating Voltage
CO	0.5ppm – 20ppm	9.093mW	60sec	±7.0V - ±9.0V
SO_2-D_4	0.04ppm – 2ppm	0.72mW	40sec	±2.0V - ±3.3V
CO_2	50ppm – 800ppm	1.1434mW	5minute	±3.0V - ±3.3V
O_2	15% - 65%	3.6292mW	5minute	±5.0V - ±9.0V
NO_2	0.01ppm – 0.5ppm	11.2712mW	1minute	±7.0V - ±9.0V
LM35CZ	$15^0C – 70^0C$	1.7673mW	2sec	±5.0V
HIH4000	0% - 100%	1mW	15sec	5.0V
SO_2-BF	0.04ppm – 5ppm	12.722mW	60sec	±7.0V - ±9.0V

9.3.2 MMC Interface Module

The PIC 18F4550 microcontroller and MMC interface module has been developed. The MMC is a flash memory storage device designed to provide high capacity, non-volatile, and rewritable storage in a small size. These devices are being frequently used in many electronic consumer goods such as cameras, computers, GPS systems, mobile phones, and PDAs etc. The capacity of the MMC can be increased at any time for further use. Presently, the available capacities of these memories lie in the range of 128MB to 32GB.

The MMC can be interfaced to microcontroller using two different protocols: the SPI (Serial Peripheral Interface) protocol and SD (Serial Digital) protocol. As the SPI protocol is being widely used, it is preferred over SD protocol in this module. The standard MMC has nine pins and these pins have different function depending on the interface protocol. The function of each pin in both the SD and SPI modes of operation are given in [6], [22].

9.4   Wireless Communication Module

The zigbee communication is being used for the development of EM system because zigbee is a self-configuring, long battery life, low cost, high reliability communication technology [13], [22]. Zigbee network has distinguished applications such as smart farms, military (viginet), telemedicine services, home device control and other commercial applications [23]-[28]. Real-world environment monitoring is one such application area that is attracting researchers around the world in response to global warming. The wireless communication between the wireless standard transducer interface module (WSTIM) and wireless network capable application processor (WNCAP) is achieved through a zigbee communication and XBee-PRO S2 module is selected. The XBee-PRO S2 modules are capable for transferring the data for both indoor and outdoor line of sights [22]. To begin with, we have created a network between the coordinator (WNCAP) and node (WSTIM) through X-CTU. In our system networks, there is one coordinator and several WSTI modules. However, all developed modules (coordinator and end Module) are on same PAN (private area network) ID. If the develop system is working properly, then it normally establishes a connection between the WNCAP-PC and the WSTI modules. And the communication between zigbee coordinator and NCAP is through a serial communication. Serial communication depends on the UART (universal asynchronous receiver transmitter). The RF module can work in four modes of operation namely, idle, transmit, receive, and sleep modes [10]. The wireless connection of WSTIM and zigbee coordinator with PC system is shows in Fig. 4.

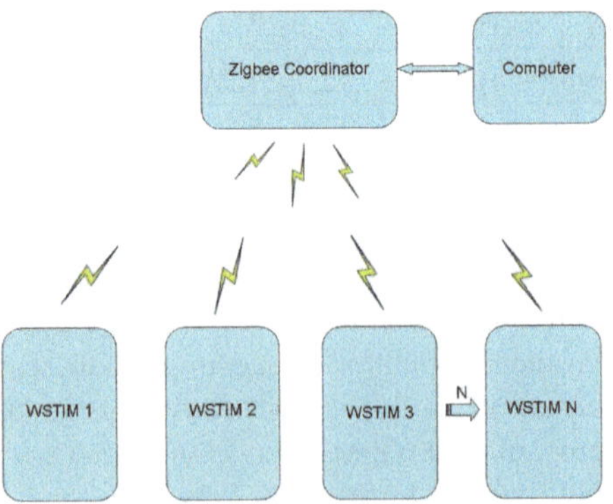

Fig. 4 Wireless communication of WSTIM and zigbee coordinator with PC system

9.5 WNCAP Module

The wireless network capable application processor (WNCAP) is developed based on the IEEE802.15.4 standards [16]. The developed WNCAP module is shown in Fig. 5. The WNCAP module is divided into two parts, the zigbee coordinator and graphical user interface (GUI). The GUI is designed in LabVIEW 9.0. It performs the all functionalities such as signal reception, visualization, and (micro memory card) MMC data storage. The data saved correspond to the fixed of the sample value and we have designed the sample ranges from 1sec to 100sec. The on-line WNCAP module front panel is shown in Fig. 6.

The developed front panel has the start button, knob for setting sample interval, digital and graphical output, data save, and stop button. The front panel is designed for eight sensing modules but we have used only seven sensing modules and the last one can be used in the future. The block diagram represents the controlling and conditional structures of operations for CO_2 sensor module is given in Fig. 7. i.e. internal oblong reprents the controlling and conditional structures of operations, and outer oblong represents the while loop. In partcular, the CO_2 block diagram the flag 101 is set on the conditional structure i.e. flag 96 of controling structure and add of 5 (ADC5-connected to the sensor module). The other gases and environmental parameters sensor modules block diagram are the same as in Fig. 7, only difference the conditional and controlling structure parameters.

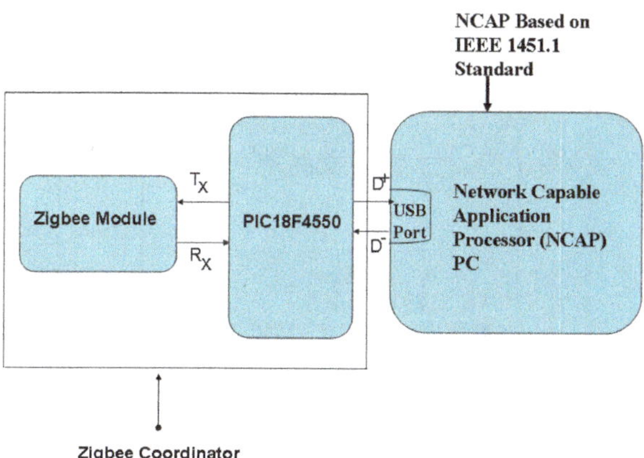

Fig. 5 Block diagram of the WNCAP module

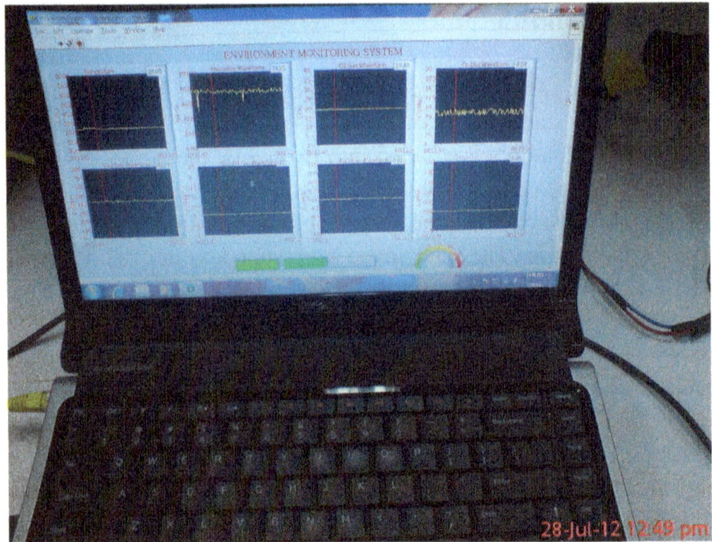

Fig. 6 Front panel of WNCAP

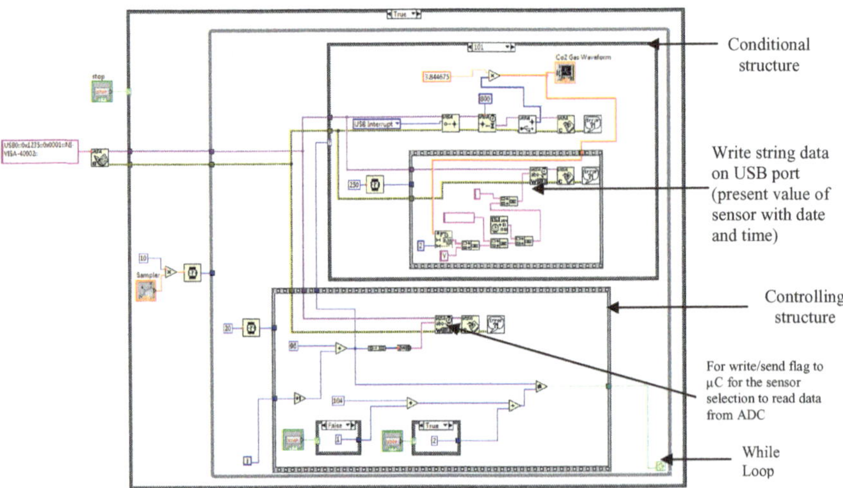

Fig. 7 CO2 Block diagram

9.5.1 Interface between Zigbee Coordinator and NCAP PC

An interface between zigbee RF module (receiving mode) and GUI PC based on the IEEE1451.2 and IEEE802.15.4 standards is developed [15], [16]. The IEEE1451.7 standard based USB interface between zigbee coordinator and GUI PC is developed in the EM system. There are various advantages of the USB based interface such as reliability, hot pluggable, low power consumption, low cost, etc. Another most important advantage of the USB based interface module is that USB PC port provides the 100mA at 5V power for external use. As a result, no additional power source is needed for the zigbee coordinator. USB supports the serial data transfers in four ways, namely, interrupt transfers, control transfers, isochronous transfer, and bulk transfer [6]. In this chapter, the interrupt driven transfer protocol was chosen. The LabVIEW based GUI requires driver software for the I/O communication with zigbee coordinator. In these days, many types of drivers are used for communicating to instruments. In this chapter, we make use of VISA drivers. VISA is a support for serial instruments or standard application programming interface (API) for instruments I/O communications. The detailed information on the different communication drivers including VISA sub palette in LabVIEW, advantages, disadvantages can be found in [30].

9.6   Re-calibration of the System

Calibration is the important issue in the designing and development of the instruments for the system reliability. Measurement instrument is required to be calibrated at a time intervals with the intent that the desired accuracy and quality level is maintained. The static and dynamic chamber methods for the gas sensor module calibration are available. We used the static chamber method to calibrate the sensor array. They define the zero value and span of the sensor, also, they used the synthetic air and unpolluted nitrogen to calibrate the zero value of the sensor [6], [8]. We also used the factory calibrated sensors in the development of sensor module. And after integration we choose the static chamber method for the gas sensor calibration and procedure was reported in [22], [29].

To convert the sensor output voltage to the temperature measured, the sensor has to be calibrated. A MTD82 multimeter with a thermocouple was used to measure the actual temperature. The values are compared to the real LM35 temperature values. The values were chosen when it was observed that the temperature on both sensors has settled. It can be seen that the LM35 has an accurate temperature reading. The sensor measurements do not have to be calibrated. The humidity sensor was calibrated by using the EM5510 multimeter with a built-in relative humidity sensor.

9.7   Results and Discussion

The main aim of the research work completed is to develop an environment monitoring system (EMS) which is proficient in measuring common indoor air pollutant concentrations and environmental parameters using a sensor array of electrochemical sensors. The system is based on the IEEE 1451, ASHRAE55-2004, and ISO7730 standards. Having developed the electrochemical sensor array, the standard transducer interface module (STIM), and the network capable application processor (NCAP) program have also been successfully developed. The STIM and NCAP modules were developed using the guidelines provided by the IEEE 1451.2, IEEE 1451.1, and IEEE 802.15.4 standards.

The developed EM system has three main parts: (i) STIM, (ii) zigbee communication, and (iii) NCAP. Although the NCAP is dependent on self powered PC or Laptop it does not consume power. Thus total power consumption of the developed EM system is the sum of power consumed by STIM module only.

The sensors were recalibrated through the static chamber method with potentiometer adjustment technique. This was done to verify the accuracy of the developed system. The current indoor air pollutant and environmental parameter levels can be directly read from the NCAP graphical user interface. Online data is saved on a memory card (MMC) to be used for further processing.

These sensors are highly sensitive to EMF/RFI, and due care was taken to shield the system to prevent the degradation of the sensor performance by using appropriate PVC holders and multicore PCB. The recommended temperature and humidity specified by the manufacturer for proper functioning of these sensors is within temperature range of 21°C to 27°C and relative humidity of less than 50% at 21°C -27°C.

A set of real time field measurements of the indoor gases, namely, CO, CO_2, NO_2, SO_2, O_2, sensors were recorded in a normal laboratory environment. Fig. 8 show their concentration levels together with temperature and humidity levels. Table 2 shows the minimum and maximum levels, as well as, one hour mean of the indoor air quality with thermal parameter data in-situ.

Table 2 Indoor Air Quality with Thermal Parameter Measurement Data In-Situ

Indoor air Quality with Thermal parameter	Minimum Level	Maximum Level	Exposure based on an hourly mean
CO	7.7ppm	9.02ppm	8.60125ppm
CO_2	411ppm	434ppm	428.68ppm
SO_2-BF	0.07ppm	0.15ppm	0.11ppm
NO_2	0.01ppm	0.03ppm	0.02125ppm
O_2	19.68%	20.57%	20.12%
SO_2-D_4	0.07ppm	0.15ppm	0.11ppm
Temperature	26°C	28.5°C	27.3°C
Humidity	56%	67%	58%

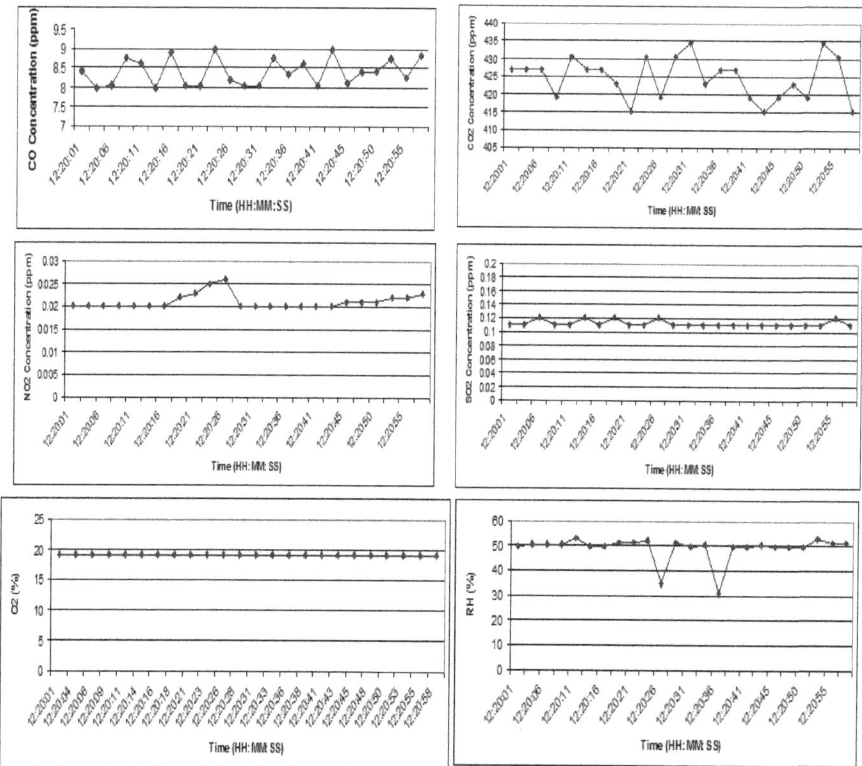

Fig. 8 Real time in-situ monitoring of the environmental parameter

9.8 Conclusions

The environment monitoring system has been successfully developed in compliance with the IEEE 1451, ASHRAE55–2004, and ISO7730 standards. It consists of the following functional blocks: the STIM and the NCAP. The STIM driver, a part of NCAP model, uses the network communication capabilities of LabVIEW 9.0. Eight "smart" transducers have plug and play capability, i.e., STIM can easily be moved from one NCAP to another. The electrochemical sensors can be successfully used to monitor target gas concentrations and environmental parameters in real time.

The usage of these types of sensors adds several advantages to a system, such as, low power consumption, low cost, fast response, ability to produce online measurement, etc. The calibration of the sensors with the appropriate accuracies is of prime importance in monitors and maintains of inbuilt environment and buildings automation.

References

[1] National environmental management: air quality act: amendments to the 2007 national framework for air quality management in South Africa: proposal, Government gazette, no. 36161, pp. 3–101 (2013)

[2] USEPA: basic information and indicators in the United States, United States Environmental Protection Agency, 2nd edn. (2005)

[3] IEA climate change and the energy outlook, Tech. Rep., pp. 173–184 (2009)

[4] Brook, R.D., Franklin, B., Cascio, W., Hong, Y., Howard, G., Lipsett, M., Luepker, R., Mittleman, M., Samet, J., Smith, S.C., Tager, I.: A Statement for Healthcare Professionals From the Expert Panel on Population and Prevention Science of the American Heart Association. A. H. A. J., 2655–2671 (2004)

[5] Kumar, A., Kim, H., Hancke, G.P.: Environmental monitoring system: a review. IEEE Sensors Journal 13(4), 1329–1339 (2013)

[6] Kumar, A., Singh, I.P., Sud, S.K.: Energy efficient and low cost indoor environment monitoring system based on the IEEE 1451 standard. IEEE Sensors Journal 11(10), 2598–2610 (2011)

[7] Rodriguez-Sanchez, M.C., Borromeo, S., Hernandez-Tamames, J.A.: Wireless sensor networks for conservation and monitoring cultural assets. IEEE Sensors Journal 11(6), 1382–1389 (2011)

[8] Kularatna, N., Sudantha, B.H.: An environmental air pollution monitoring system based on the IEEE 1451 standard for low cost requirements. IEEE Sensors Journal 8(4), 415–422 (2008)

[9] Postolache, O.A., Pereira, J.M.D., Girao, P.M.B.S.: Smart sensors network for air quality monitoring applications. IEEE Transaction on Instrumentation Measurement 58(9), 3253–3262 (2009)

[10] Jelicic, V., Magno, M., Brunelli, D., Paci, G., Benini, L.: Context-adaptive multimodal wireless sensor network for energy-efficient gas monitoring. IEEE Sensors Journal 13(1), 328–338 (2013)

[11] Yang, H., Qin, Y., Feng, G., Ci, H.: Online monitoring of geological CO_2 storage and leakage based on wireless sensor networks. IEEE Sensors Journal 13(2), 556–562 (2013)

[12] Lin, H.C., Kan, Y.C., Hong, Y.M.: The comprehensive gateway model for diverse environmental monitoring upon wireless sensor network. IEEE Sensors Journal 11(5), 1293–1303 (2011)

[13] Akyildiz, I.F., Vuran, M.C.: Factors influencing WSN design. In: Wireless Sensor Networks, 1st edn., pp. 37–51. Wiley (2010)

[14] Nieves, A.R., Madrid, N.M., Seepold, R., Larrauri, J.M., Larrinaga, B.A.: A UPnP service to control and manage IEEE 1451 transducers in control networks. IEEE Transaction on Instrumentation Measurement 61(3), 791–800 (2012)

[15] IEEE Standard for a Smart Transducer Interface for Sensors and Actuators - Transducer to Microp. Commun. Protocols and Transducer Electr. Data Sheet (TEDS) Formats, IEEE standard 1451.2-1997 (1997)

[16] Institute of Electrical and Electronics Engineering, Inc., IEEE Std. 802.15.4-2003, IEEE standard for information Technology-Telecommunications and Information Exchange Between Systems-Local and Metropolitan Area Networks, New York (2003)

[17] Chen, C., Tsow, F., Campbell, K.D., Iglesias, R., Forzani, E., Tao, N.: A wireless hybrid chemical sensor for detection of environmental volatile organic compounds. IEEE Sensors Journal 13(5), 1748–1755 (2013)

[18] Wilson, D.M., Hoyt, S., Janata, J., Booksh, K., Obando, L.: Chemical sensors for portable, handheld field instruments. IEEE Sensors Journal 1(4) (December 2001)

[19] Alphasense Sensors Data Sheet, SO2-D4, CO2-D1, CO-CF, NO2-A1, O2-A1: Alphasense Ltd., Sensor Technology House, Great Notley, U. K. (June 2009)

[20] NSC Data Sheet, LM35CZ: Precision Centigrade Temperature Sensor, National Semiconductor Corporation, USA (November 2000)

[21] Honeywell Data Sheet, HIH4000: Humidity sensors, Honeywell Company, Illinois, USA (March 2005)

[22] Kumar, A., Hancke, G.P.: Energy efficient environment monitoring system based on the IEEE 802.15.4 standard for low cost requirements. IEEE Sensor Journal, doi:10.1109/JSEN.2014.2313348

[23] Kim, Y., Evans, R.G., Iversen, W.M.: Remote sensing and control of an irrigation system using a distributed wireless sensor network. IEEE Transactions on Instrumentation and Measurement 57(7), 1379–1387 (2008)

[24] Ekuakille, A.L., Trotta, A.: Predicting VOC concentration measurements: cognitive approach for sensor networks. IEEE Sensors Journal 11(11), 3023–3030 (2011)

[25] Yan, R., Sun, H., Qian, Y.: Energy-aware sensor node design with its application in wireless sensor networks. IEEE Transaction on Instrumentation Measurement 62(5), 1183–1191 (2013)

[26] Ali, A.R., Zualkernan, I., Aloul, F.: A mobile GPRS-sensors array for air pollution monitoring. IEEE Sensors Journal 10(10), 1666–11671 (2010)

[27] Tsow, F., Forzani, E., Rai, A., Wang, R., Tsui, R., Mastroianni, S., Knobbe, C., Gandolfi, A.J., Tao, N.J.: A wearable and wireless sensor system for real-time monitoring of toxic environmental volatile organic compounds. IEEE Sensor Journal 9(12) (December 2009)

[28] Mukaro, R., Carelse, X.F.: A microcontroller based data acquisition system for solar radiation and environmental monitoring. IEEE Transaction on Instrumentation Measurement 48(6), 1232–1238 (1999)

[29] Baekelmans, J., Witte, W.D., Mario, V., Vos, T.: US Patent: EP2099199A1 (2008)

[30] NIC LabVIEW Manual, National Instruments Corporate, Austin, USA (August 2009)

Chapter 10
Application of Wireless Sensor Networks Technology for Induction Motor Monitoring in Industrial Environments

Ruan D. Gomes, Marcéu O. Adissi, Tássio A.B. da Silva,
Abel C. Lima Filho, Marco A. Spohn, and Francisco A. Belo

10.1 Introduction

In an industrial environment, mechanical systems driven by electric motors are used in most production processes, accounting for more than two thirds of industry electricity consumption. Regarding the type of motors usually employed, about 90% are three-phase AC induction based [1], mainly due to its cost effectiveness and mechanical robustness [2].

Torque is one of the main parameters for production machines. In several industry sectors, torque measurements can identify equipment failure, making their monitoring essential in order to avoid disasters in critical production processes (e.g., oil and gas, mining, and sugar and alcohol industries) [3].

For decades, researchers have studied methods and systems for determining the torque in rotating shafts. There are basically two lines of study: direct shaft torque measurement [4, 5, 6, 7], and estimated torque measurement based on motor electrical signals [8, 9, 10].

In most cases, the methods for direct shaft torque measurement are the more accurate. However, they are highly invasive, considering the coupling of the measurement instrument between the motor and the load [11]. Moreover, some of these techniques have serious operational challenges [12].

Ruan D. Gomes · Marcéu O. Adissi
Federal Institute of Education, Science, and Technology of Paraíba, Paraíba, Brazil
e-mail: `ruan.gomes@ifpb.edu.br`, `marceueterra@gmail.com`

Marcéu O. Adissi · Tássio A.B. da Silva · Abel C. Lima Filho · Francisco A. Belo
Federal University of Paraíba, Paraíba, Brazil
e-mail: `borgestassio@gmail.com`, `{abel,belo}@les.ufpb.br`

Marco A. Spohn
Federal University of Fronteira Sul, Chapecó, SC, Brazil
e-mail: `marco.spohn@uffs.edu.br`

© Springer International Publishing Switzerland 2015
H. Leung and S.C. Mukhopadhyay (eds.), *Intelligent Environmental Sensing*,
Smart Sensors, Measurement and Instrumentation 13, DOI: 10.1007/978-3-319-12892-4_10

The estimated torque from the motor's electrical signals (i.e., current and voltage) makes the system less invasive, but it is less accurate when compared to direct measurement systems. There are problems, such as noise in signal acquisition, those related to numerical integration, and low levels of voltage signals at low frequencies. However, in many cases, high precision is not critical, and low invasiveness is required.

There are different methods to measure efficiency in induction motors, which are based on dynamometer, duplicate machines, and equivalent circuit approaches [13]. However, their application for in-service motors is impractical, because it requires interrupting the machine's operation for instrumentation.

There are some simple methods for in-service efficiency estimation: nameplate method, slip method, and current method [14]. These methods present as the main limiting factors the low accuracy, estimative based on nominal motor data, and the need of typical efficiency-versus-load curves. In the ORMEL96 method [32], the efficiency is obtained from an equivalent circuit that is generated from the motor nameplate and the rotor speed measurement. In the OHME method [33], the efficiency estimation is performed from the input power measurement and data from the motor nameplate.

Hsu et al. [34] presented the air-gap torque (AGT) for energy efficiency estimation. In [35], the air-gap torque is also used to measure efficiency in a much less invasive manner. The AGT method can be employed without interrupting the motor operation, and it is not based on the motor nameplate. This method is usually more accurate than the other methods described previously [14].

The analysis of induction motors' failures consists on digital sampling from sensors strategically integrated to these machines, and the processing of these signals as well. Among the sensing methods used for induction motor fault diagnosis, there are two major lines of study: methods which verify variations in voltage and current signals [15, 16, 17], and methods based on vibration analysis using accelerometers, which are based on parameters such as displacement, velocity, and acceleration for detecting faults in motors [18, 19].

Among the methods based on sampling current and voltage signals, the most commonly used is the motor current signature analisys (MCSA) [20, 21, 22], which consists in detecting faults from the spectral analysis of the armature current, generally used in stationary operation regimes. We can also cite the Gabor analysis [23], the analysis via Fractional Fourier Domain [24], and the finite element method [25]. Recent studies also showed the feasibility of using air-gap torque values for fault detection in induction motors [25, 26, 27]. Artificial intelligence has also been used for both the analysis of the armature current [28] and vibration [29].

Among the methods based on vibration, we can cite the wavelet processing technique [30] for transient signals, and the classical methods based on fourier transform for steady state signals [31].

Traditionally, energy monitoring and fault detection in industrial systems are performed in an off-line manner or through wired networks. The installation of cables and sensors usually have a higher cost than the cost of the sensors themselves [36]. Besides the high cost, the wired approach offers little flexibility, making the network deployment and maintenance a harder process. In this context, wireless networks present a number of advantages compared to wired networks as, for example, the ease and speed of deployment and maintenance, and low cost [37]. In addition to that, Wireless Sensor Networks (WSNs) provide self-organization and local processing capability. Therefore, these networks appear as a flexible and inexpensive solution for building industrial monitoring and control systems.

Nevertheless, the use of WSNs, when developing automation systems for industrial environments, presents a number of challenges that should be faced. Wireless networks have unreliable communication links, what can be aggravated with noise and interference in the communication spectrum range. The unreliability of the transmission medium in wireless networks makes it difficult to define quality of service guarantees. Studies on the application of WSNs in industrial environments, aiming at replacing wired systems, have been extensively explored in recent years [10, 36, 38, 39, 40, 41, 42, 43, 44, 45].

Most WSN applications employ the IEEE 802.15.4 standard for wireless communication. This standard allows the deployment of a large network of sensors in various industrial segments [39]. This standard has been employed also in the mechatronics field [46, 47]. In comparison with other standards, such as IEEE 802.11 (*WiFi*) and IEEE 802.15.1 (*Bluetooth*), the IEEE 802.15.4 standard has advantages related to energy consumption, scalability, reduced time for node inclusion, and low cost [48].

This chapter discusses about all aspects of the implementation of motor monitoring systems employing the WSN technology, including methods for torque, efficiency, and speed estimation, and a discussion about the challenges of using WSN in industrial environments. In the last section we present an industrial WSN for monitoring torque, efficiency, and speed, developed in a previous work [10]. The described system does all the data processing locally, transmitting to the base unit only the targeted parameters previously computed. Thus, there is a large reduction in the amount of transmitted data, enabling real-time and dynamic monitoring of multiple motors, even with a high data rate acquisition in the analog to digital converters.

We also present studies on the relation between the WSN performance and the quality of the communication medium in the network operating environment. As a result, we observed the correlation between Packet Error Rate and Spectral Occupancy in the band used for communication. We use the term Spectral Occupancy to denote the induced power on the channel used by radios for communication.

The impact resulting from the addition of new interference sources in the environment is also analyzed. Through these studies and a theoretical analysis, it was demonstrated that employing nodes with local processing capabilities is essential for this type of application, reducing the amount of data transmitted over the network and allowing monitoring even in high interference scenarios. In addition to that, our work provides insights for guiding the development of new technologies and protocols for industrial WSNs.

10.2 Motor Monitoring

Induction motors are widely used and essential for many industrial processes. These motors usually operate in adverse environmental conditions, such as high temperature, high humidity, high vibration, and dust, which can lead to motor malfunction. Motor failures raise overall cost due to repairment and unexpected process shutdown, which can lead to high losses. Therefore, reliable monitoring of induction motors is important for controling the process and to avoid unscheduled shutdowns [38]. In control systems and motor drives, which operate based on measured or estimated values of the current motor condition, input data must be reliable and on time.

Methods for motor monitoring that use only measures of currents and voltages are preferred due to their low cost and noninvasiveness. Allied to these methods, advances in wireless communications technologies, embedded systems, and electronic sensors, allowed the development of low cost monitoring systems using the Wireless Sensor Network (WSN) technology [36].

The implementation of motor monitoring systems using the WSN technology has significant advantages over traditional monitoring methods such as off-line monitoring or wired based networks. Combining noninvasive estimation techniques, embedded processing, and data transmission through a WSN one can build flexible low cost systems.

10.2.1 Efficiency of Motor Energy Conversion

The electrical machine is the main practical way of converting electrical energy into mechanical or mechanical energy into electrical. This is achieved through an electromagnetic coupling, as shown in Figure 1.

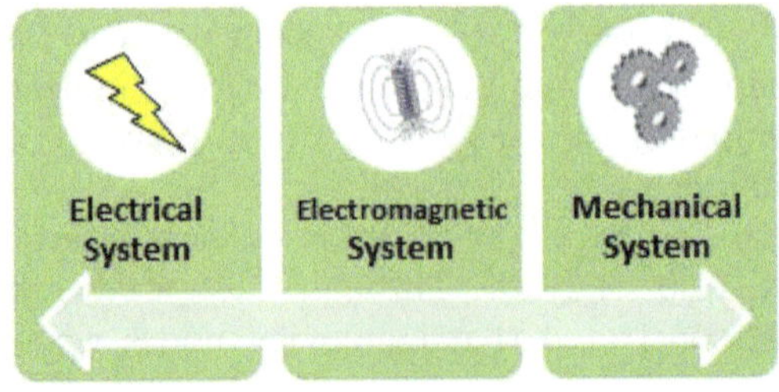

Fig. 1 Energy conversion in the electric motor

In the process of converting electrical energy into magnetic, and converting magnetic energy into mechanical, losses inevitably occur. The electrical input power, (P_{in}), is transmitted to the air gap region after accumulating losses in the stator and core. From the air gap region, the power is transmitted to the rotor, accumulating the mechanical and additional losses. The developed mechanical power, (P_{out}), depends on the slip and on the power in the air gap region.

Assessing the motor's energy efficiency is equivalent to computing its losses. Thus, it is very important to define them. Before describing the main methods for efficiency estimation, we will describe the expected losses, (K), in a three phase induction motor.

In the IEEE 112 standard Part 5 [13] four types of losses in squirrel-cage induction motors are defined, as in equation 10.2. For induction motors with wound-rotor, brushes contact losses should also be taken into account. The losses are graphically represented in Figure 2.

$$k = P_{in} - P_{out} = K_S + K_{CORE} + K_{MEC} + K_{SL} \qquad (10.1)$$

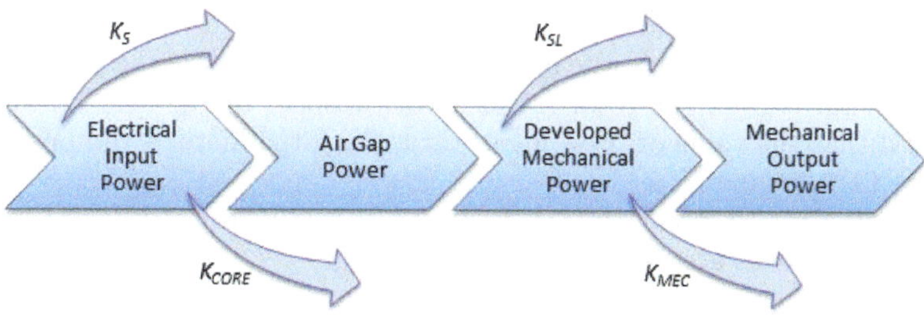

Fig. 2 Simplified motor losses diagram (adapted from [35])

Stator copper loss, (K_S), is the Joule loss in the stator windings. This loss can be computed with the equation $K_S = 1.5I^2R$, assuming the three phases of the motor, where I is the rms line current, and R is the average resistance between two phases. For a well balanced motor, $R = 2R_S$, where R_S is the stator resistance. The copper losses also vary with the temperature and frequency of the supply voltage.

Core loss, (K_{CORE}), results from Foucault currents and hysteresis in the magnetization of the core. This loss varies little remaining almost the same when the motor is operating with full load or empty. The hysteresis loss depends on the core material and the frequency of the magnetic field. The Foucault currents losses are divided into classical losses and excess losses. These losses vary with the magnetic material and its geometry, which are often laminated to reduce losses. The loss in the core can be determined by subtracting the mechanical loss and the copper loss in the stator out of the total losses. The total loss can be obtained in a test with empty load.

Mechanical loss, (K_{MEC}), is composed of losses associated to the friction between moving parts (the bearings, between the rotor and the air, and in the rings) and by the motor ventilation (external and internal). This loss depends on the motor size, geometry, rotor speed, and the lubrication of bearings and fan blades. Thus, it varies with the motor and the process where it is inserted. Its value is practically constant and independent of the load imposed. Mechanical losses may be determined through a linear regression on the power versus voltage square curve. At least three points with different input voltages must be used. The desired power of each point is the total loss obtained from the empty test subtracted from the loss in the stator copper. The curve must be extended linearly to zero voltage, where the power at this point is the engine mechanical loss.

Stray load losses, (K_{SL}), are the losses that have not been anticipated in the losses described in the last paragraphs. These losses are a consequence of non-ideal characteristics of a real motor, taking place in the stator or rotor. They are the result of nonlinear phenomena of different nature which are difficult to quantify [49]. Stray load losses increase the dissipation of energy and vibration, reducing the efficiency and useful torque developed by the motor. The main causes of additional losses are imperfections in the geometry of the rotor and the stator, maldistribution of the stator windings along the slits, the effect of the magnetic field at the ends of the machine, as well as metal bars of the rotor inclined with respect to the axis . It should be noted that the value of these losses varies with the lifetime and operation regime of the engine.

In the IEEE 112 standard Part 5 [13] three ways to determine the value of additional losses are described:

- Indirect measurement: Subtract the sum of known losses from the total losses. The remaining value represents the additional losses.
- Direct measurement: The components of the fundamental frequency and high frequency components of the additional losses are determined from different tests and equations. The sum of these two components comprise the total additional loss.
- Assumed measurement : When acceptable to the standards or the specifications of the machine, the additional losses can be assumed according to Table 1.

Table 1 Additional losses as defined by the IEEE std-112

Nominal Power (kW)	Additional Loss (% of the load)
≥ 1 and ≤ 90	1.8 %
≥ 91 and ≤ 375	1.5 %
≥ 376 and ≤ 1850	1.2 %
≥ 1851	0.9 %

The choice of the method for losses's estimation in a given machine depends on several factors. Often it is economically or technically infeasible to employ more accurate methods, forcing the choice of simpler methods.

10.2.2   Estimation Methods

Direct measurement methods usually employ electromechanical devices, providing high accuracy. However, these devices present limitations on their application, such as the high cost, low mechanical strength, lack of space for instalation, and low noise immunity. Besides that, such devices can modify the machine's inertia. In some cases, it might even be impossible to install direct measurement devices.

Several estimation methods have been proposed for indirectly determining torque, speed, and efficiency with minimal invasion to the monitored system. Even though indirect measurement leads to less accurate measures, sometimes it is the only possible option due to the advantages it offers in comparison to direct measurement devices [10]. Many advances have been made in the development of methodologies for estimating parameters of induction motors. However, there is a long way to go to the development of an ideal methodology. The challenge is on finding a method to estimate the parameters of the transient and steady state, with a low error, requiring low computational effort, simplicity and robustness, and complying to the needs of industry.

Shaft Torque Estimation

In any system in which there is energy conversion or transfer of energy through mechanical devices, the torque is an important data to evaluate the process [50]. Therefore, there is a constant search for methods for shaft torque measurement in rotating mechanical systems, which are essential in several industrial production systems. However, currently most of the torque measurements are performed with high uncertainty or extreme invasivity.

The methods for measuring the force are well known for hundreds of years, but the first methods for torque measurement in rotating shafts were presented later. The first instrument for torque measurement was proposed by the French physicist Gaspar de Prony, after the advent of the steam engine in 1700.

Torque is a rotational force, or a force through a distance. Torque can be divided into two major categories, either static or dynamic. The torque is considered static if it is applied to a non-accelerating object. When the rotational torque has angular acceleration, it is known as dynamic torque [12].

Depending on the application, there are excellent instruments for measuring static torque, but there are still serious challenges to measure dynamic torque [3]. Although there are several techniques for direct measurement of shaft torque, torque transducers used by these methods do not meet the existing needs of the industrial demand. They are highly intrusive, have a high cost, and are difficult to mount on the motor shaft, which make these methods not applicable for most industrial applications.

Over the years, many shaft torque estimation methods have been proposed. Following, the most important methods are discriminated according to the classification commonly used in the literature.

- Nameplate Method

  The simplest method to determine the motor's output torque uses nominal data provided by the manufacturer. The nominal torque, (T_n), can be determined by the ratio between the nominal output power, (P_n), and the nominal rotacional speed, (ω_n), as in Equation 10.2.

  $$T_n = \frac{P_n}{\omega_n} \tag{10.2}$$

  This method is non-invasive, but not necessarily the motor operates at its nominal power. In fact, the torque on the motor shaft depends on the driven load demands. Therefore, this method is inadequate to monitor motors that are coupled to a variable load or with a constant load different of the nominal one.

- Slip Method

  This method directly relates the measured slip, (s), to the shaft torque, (T), also considering the motor nominal values. Thus, the shaft torque, T, can be estimated according to Equation 10.3.

  $$T = \frac{s}{1-s} \cdot \frac{1-s_n}{s_n} \cdot T_n \tag{10.3}$$

  This method drawback is on the assumption that the sliding is equal to the percentage of the load. Furthermore, the real nominal speed may vary around 20% with respect to the speed informed on the nameplate [51]. This method can be improved by including a component that indicates the real voltage variation, following the Ontario Hydro method [52], according to Equation 10.4.

  $$T = \frac{s}{1-s} \cdot \frac{1-s_n}{s_n} \cdot T_n \cdot (\frac{V}{V_n})^2 \tag{10.4}$$

- Current Method

  Similarly to the slip method, the ratio between the measured current, (I), and the nominal current, (I_n), is used to estimate the shaft torque (T). This method provides a better estimate compared to the slip method, since the curve of load versus the current variation is slightly non-linear [52].

Besides, it is also necessary measuring the shaft rotational speed, (ω), as defined in Equation 10.5.

$$T = \frac{I}{I_n} \cdot \frac{\omega_n}{\omega} \cdot T_n \tag{10.5}$$

- **Segregated Loss Method**

Using the sum of losses, (K), and the speed, one can get the torque on the motor shaft by subtracting the losses of input torque, according to Equation 10.6.

$$T = T_e - \frac{K}{\omega} \tag{10.6}$$

Where T_e is the input torque given by the ratio between the electrical input power, P_{in}, and the synchronous speed, ω_s.

Due to the difficulty on measuring each loss, and the extensive invasiveness of the measuring procedures, the usage of the segregated loss method becomes impracticable to estimate torque for motors in operation.

- **Air Gap Torque Method**

In an induction motor, the air gap is the region between the stator and the rotor, where the electromechanical conversion process takes place (Figure 3). The air gap torque (AGT) is the conjugate formed between the rotor and the stator magnetic flux. According to Equation 10.7, the estimation of the air-gap torque can be performed noninvasively taking current and voltage measurements from the electric motor [8].

$$T_{ag} = \frac{p\sqrt{3}}{6}\{(i_a - i_b)\int[v_{ca} + r(2i_a + i_b)]dt \tag{10.7}$$

$$+(2i_a + i_b)\int[(v_{ab} - r(i_a - i_b)]dt\}$$

Where:
p - is the number of motor poles;
i_a, i_b - are the motor line currents, in ampere;
v_{ca}, v_{ab} - are the motor power line voltages, in volt;
r - is the resistance of motor armature, in ohm;

Equation 10.7 can be applied using instantaneous and simultaneous acquisitions of i_a, i_b, v_{ca}, v_{ab}, and a measured value of r. It is valid both for motors connected in Y, with no connection to the neutral, or Δ. The integrals correspond to the stator flux linkages. Air gap torque equations has also been used in many works based on other types of motors [53, 54, 55].

The torque on the shaft can be estimated by subtracting the losses occurring after the air-gap process of electromechanical energy conversion, according to Equation 10.8 [34].

$$T_{shaft} = T_{ag} - \frac{K_{MEC}}{\omega} - \frac{K_{SL}}{\omega}$$

(10.8)

Where:

ω - is the rotor speed;

K_{MEC} - are the mechanical losses;

K_{SL} - is the stray-load loss;

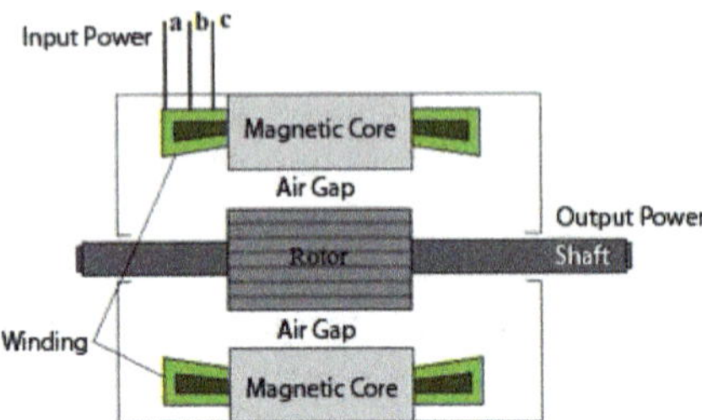

Fig. 3 Inside the motor, indicating the air gap position (Lima Filho, 2009)

Energy Efficiency

The increasing search for machines with high efficiency ratings, implies the development of more reliable and more accurate methods for measuring and monitoring the efficiency of such machines. Moreover, these methodologies must be as simple as possible, to meet technical and economic constraints.

The efficiency determination is complex due to the difficulty in measuring or estimating all variables involved. To measure the mechanical and electrical values, high invasive equipments are usually necessary, presenting high cost and difficult deployment. The difficulty on measuring efficiency increases when it is necessary to perform real-time measurement, with the machine in operation, in transient mode, and with power variation.

Both the scientific community and manufacturers have worked together on the subject related to efficiency measurement of electric machines, particularly for induction motors. Therefore, the standards for efficiency measurement play an important role in this field.

Some standards are listed below:

- The Standard 112 of the Institute of Electrical and Electronic Engineers (IEEE) is composed by five methods that form the basis for NEMA MG-1, and C390, which are used in the U.S. and Canada, respectively.
- Standards 60034-2 and 61972 of the International Electrotechnical Commission (IEC), adopted in Europe;

- Standard JEC37 of the Japanese Electrotechnical Committee (JEC), adopted in some Asian countries.

The methods described in these standards are usually not applicable in the field, but they are useful as a reference model for other possible methods.

The way to perform the estimation of losses or output power defines the fundamental differences among the various methods for estimating efficiency. An estimation method can consist of one single basic method or it can be based on a combination of different basic methods. We have compared the basic methods according to their theoretical bases and sources of error. The basic methods under consideration are:

- Direct Shaft Torque Method;
- Nominal Values Method;
- Slip Method;
- Current Method;
- Equivalent Circuit Method;
- Segregated Losses Method;
- Air Gap Torque (AGT) Method;

The methods are commonly classified in the literature by their level of invasiveness to the process and the estimation accuracy. The level of invasiveness is determined by the type of data that need to be acquired, the physical installation of the devices, and their cost. The accuracy is evaluated by comparing with a reference measurement.

The graph in Figure 4 compares the seven methods in a simple and general form. The x-$axis$ represents the error, and the y-$axis$ the level of invasiveness of the system. The size of the circle represents the dispersion, since some theoretical bases can use multiple and slightly different theoretical methods, covering a larger area.

The ideal method needs to have a minimum error and a low invasiveness. The method that is closest to the ideal is the air gap torque method (AGT).

One or more of the following steps may be involved in the methods for efficiency estimation:

- Speed measurement by using a tachometer on the shaft;
- The use of nominal data;
- Measurement of electric current;
- Measurement of electrical voltage;
- Measurement of the stator winding's resistence;
- Measurement of the winding temperature;
- Direct measurement of shaft torque;
- Tests with different loads and conditions: with no load, at nominal load, with variable load, with reverse rotation, with locked rotor.

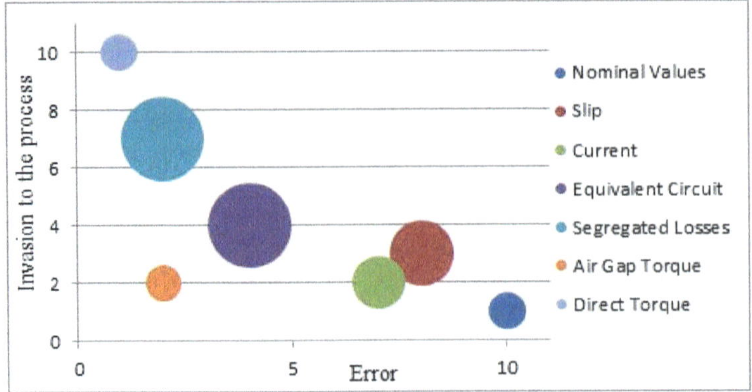

Fig. 4 Invasion and Accuracy in the Estimation of Efficiency (Adapted of [52, 33])

The test with no load is only possible if the engine can be decoupled from the applied load, so it is considered an invasive test for the process. In this case, as the mechanical power output is null, the electrical input power is equal to the sum of the losses with the no load motor [13].

10.3 Wireless Sensor Networks

The development of new technologies in the fields of sensors, integrated circuits, and wireless communication has enabled the development of low-cost devices, with processing and wireless communication capabilities. These low-cost devices has enabled the development of Wireless Sensor Networks (WSN), which are networks formed by low-cost embedded systems equipped with sensors, or actuators, and transceivers for communication via radiofrequency.

These sensors can produce responses to physical changes in the environment where they are deployed, such as temperature, humidity, or magnetic field. Overall, the WSN has as main objectives monitoring the environment where it is employed or performing objects' tracking [56]. Devices equipped with processors can do more than just transmit the collected measurements from the sensors, they can also perform local processing, which allows to generate new information while reducing network traffic [10].

In a WSN the measures collected from the sensors are transmitted to a sink node (sometimes called controller or monitor), possibly through multiple hops via retransmissions by intermediate nodes. The data received by the sink node can be used locally or transmitted to external stations through other networks (usually the Internet). The WSN's nodes can also be mobile, which requires that the network have self-organization capabilities [57]. In addition, other aspects may influence the dynamics of the network topology, such as obstructions in the environment and interference in the spectrum band used for communication.

The WSN have severe resource constraints such as low communication range, low bandwidth, and limited processing and storage capacity. In many cases, there is also low energy availability (i.e., energy supply via batteries). Therefore, such restrictions need to be taken in consideration in the development of technologies and protocols for this type of network.

Specific applications, such as environmental monitoring and industrial monitoring, have specific characteristics and requirements. Therefore, the deployment of sensor networks must necessarily take into account requirements of the target application. In general, there are key features that should be provided by the WSN, such as security, robustness, reliability, throughput, and adequate determinism. Among these characteristics, the lack of reliability is the main reason for many users not deploying wireless devices in general [58].

10.3.1   Industrial Wireless Sensor Networks

In an industrial WSN (IWSN), sensor nodes are deployed in machinery for monitoring critical parameters such as vibration, temperature, pressure, and efficiency. Measurements are transmitted wirelessly to a sink node, which later provides the gathered information for analysis in a central station. Based on this information, it is possible to repair or replace devices before major damages take place [40].

Although WSNs have several advantages, the deployment of this technology presents some challenges. Wireless communication is inherently unreliable and subject to a larger number of transmission errors when compared to wired networks, mainly due to channel failures and interference. Nodes can suffer interference from the co-existence with other nodes in the network, from the co-existence with other networks, and also from other technologies operating in the same frequency range [59].

In industrial environments, there may be other sources of interference such as thermal noise, interference from motors and devices that cause electromagnetic interference in the band used for communication [40]. The reliability of the WSN depends on the propagation environment, the chosen modulation scheme, transmission power, frequency used, as well as several other parameters. In general, industrial wireless systems are prone to high error rates, and often with high variance [60].

Furthermore, in industrial environments there is a large amount of machinery and metallic components, which are highly reflective, increasing path loss [61]. Industrial environments with a large amount of reflective objects have much more multipath components and exhibit high RMS spread delay and high maximum excess delay, which increases intersymbol interference [62] [63].

Typical interference sources present in industrial environments include industrial electric motors, frequency inverters, welding equipment, and wireless communications equipment, such as wireless phones and wireless LANs. However, most of these interferences, especially those related to industrial equipment, such as motors and frequency inverters, may cause interference within a few hundred MHz range [62], which can disrupt communication on proprietary systems operating at the same frequency range, while not interfering on systems based on the 2.4 GHz Industrial Scientific and Medical (ISM) band. On the other hand, wireless networks, such as *WiFi* and Bluetooth networks, and microwave ovens can introduce high destructive interference in WSNs [10, 64, 65, 66, 67].

Industrial monitoring systems need to measure signals that change rapidly, in a dynamic manner [41]. Applications such as efficiency monitoring and fault detection in induction motors fit into this type of application. Due to the limitations of WSNs, mainly regarding the low bandwidth and the lack of reliability in the transmissions, the implementation of such systems becomes even more challenging.

10.3.2 *Coexistence Issues in Unlicensed ISM Bands*

Due to the increasing number of applications using wireless technologies, the amount of information traveling through wireless links tend to increase. Therefore, the radio spectrum available for wireless communication tends to get more polluted, increasing interference and reducing overall communication quality.

Currently, most radios use static spectrum allocation. This approach tends not to work satisfactorily when a large number of devices is present in the environment sharing the same channel [68]. The ZigBee protocol, for example, does not switch channels during periods of high interference. Instead, it uses only a low duty cycle and a medium access control mechanism to minimize the amount of collisions [65].

Many communication devices operate in the unlicensed 2.4 GHz band, including the standards IEEE 802.15.1 (Bluetooth), IEEE 802.11 (WiFi) and IEEE 802.15.4, as well other devices such as cordless phones, wireless computer peripherals, and microwave ovens. As the ISM band is an unlicensed band, no user has priority over others. The only restriction for users using this band is on the signal strength. The power constraint is used to limit interference among coexisting systems. As there is no protection from other concurrent users, it is necessary to develop efficient coexistence mechanisms to allow the operation of systems in unlicensed bands with an appropriate reliability.

The wireless communication standards, such as IEEE 802.11 and IEEE 802.15.4, define a set of channels available on the unlicensed band. As these standards share the same spectrum band, there is an overlap between the channels, which can lead to interference between radio stations operating in the same environment. Figure 5 shows the induced power on the frequency components of the 2.4 GHz band for an experiment including a WiFi network operating on channel 6, one microwave oven, and an IEEE 802.15.4 network operating on channel 13 [10]. In section 10.4.5 we show some results on the performance of a WSN for motor monitoring under the interference of a WiFi network and a microwave oven.

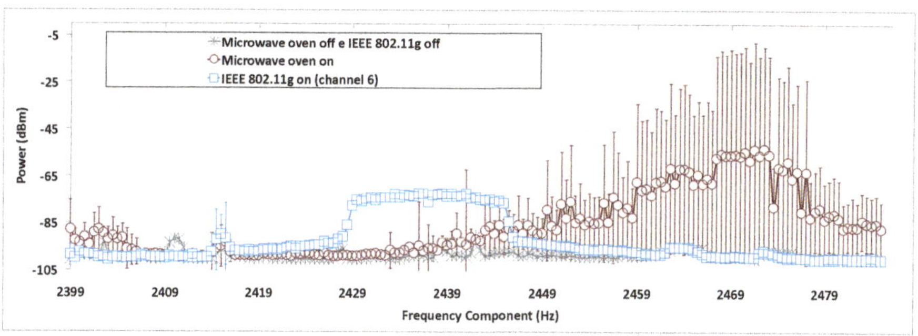

Fig. 5 Induced power on the frequency components [10]

10.3.3   IWSN Standards IEEE 802.15.4

The IEEE 802.15.4 standard provides wireless communication with low power consumption and low cost for monitoring and control applications that do not require large bandwidth. In comparison to other wireless communication standards, such as IEEE 802.11 and IEEE 802.15.1, the IEEE 802.15.4 standard has advantages related to energy consumption, scalability, reduced time for node inclusion, and low cost [48].

The IEEE 802.15.4 standard defines the physical layer and the medium access control layer. It defines three frequency bands for communication: 868 MHz, 915 MHz, and 2.4 GHz. The standard divides the spectrum bands in twenty-seven channels, one for the 868 MHz band, ten for the 915 MHz band, and sixteen for the 2.4 GHz band [69]. These bands are unlicensed. Thus, the radios that comply with this standard share the medium with devices that implement other technologies. However, as the spectrum is divided into channels, it is possible that multiple networks share the band simultaneously without causing much interference.

Nodes in an IEEE 802.15.4 network can be of two types: FFD (Full Function Device) or RFD (Reduced Function Device). FFD nodes can act as the network coordinator or an end node. The network coordinator is responsible for the initialization and synchronization of the network, and generally acts as the sink node of the WSN. In addition, it can serve as a bridge between various IEEE 802.15.4 networks. FFD nodes can also act as routers between nodes, without requiring a coordinator. Through routers, an IEEE 802.15.4 can be expanded and get a large coverage. RFD nodes can act only as end nodes, which are responsible for sensing or actuation, but they can interact with only one FFD node in the network [69].

The topology of an IEEE 802.15.4 network can be organized in three ways: star, mesh, and tree. Figure 6 illustrates the three topologies.

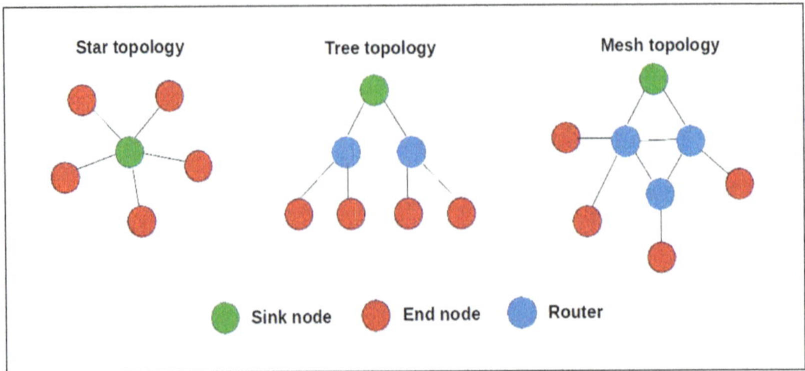

Fig. 6 Topologies supported by the IEEE 802.15.4 standard

A WSN with star topology has a coordinator node and multiple end nodes. The coordinator manages the network and receives the packets directly from end nodes via a single hop.

A WSN with tree topology is hierarchical, in which each set of end nodes communicates with a particular router, which in turn communicates with the coordinator.

A mesh network operates in an ad hoc manner, in which a router can communicate with any other directly without the coordinator intervention. In this topology, the network has self-organization capabilities, and it can automatically adjust itself since its startup or when nodes join or leave the network at any time later. The mesh topology also allows increasing the network scope, since the communication with the coordinator can take place via multiple hops [70].

Even though there are three supported topologies, the network layer protocol (e.g., the routing algorithm) is the user's responsibility whenever employing the IEEE 802.15.4 standard.

The radios complying to the IEEE 802.15.4 standard have a maximum nominal throughput of 250 Kbps when operating in the 2.4 GHz band. However, the channel is underutilized due to the CSMA/CA (Carrier Sense Multiple Access with Collision Avoidance) protocol, where each packet must wait at least a backoff period before any transmission attempt. Experimental studies showed that the real maximum achieved throughput is around 153 Kbps [71].

Even though the achievable throughput is relatively low, it is enough for many WSN applications, because the nodes usually do not transmit a large amout of data. However, some applications require large sample rates, resulting on a large amount of data to be transmitted, which is the case for some motor monitoring applications such as efficiency monitoring, and failure detection. However, by employing local processing capabilities in the end nodes, it is possible to reduce the amount of data transmitted over the network.

The IEEE 802.15.4 standard defines two operation modes. The first one is called *non-beacon mode*, and it is based only on the CSMA/CA mechanism. Before each transmission, a node must wait for a random backoff period of time. After the backoff period, the device checks whether the transmission medium is free or not. If the channel is free, the device starts the transmission; otherwise, the device waits for another random backoff period, and then attempts to access the transmission medium again. The second operation mode is called *beacon mode*.

In the *beacon mode*, a structure called superframe is defined. The superframe is bounded to network beacons sent by the coordinator which are then split into sixteen equal time slots. All transactions must be completed within the time period between two beacons. The superframe has an active portion and an inactive portion. The active portion is split into three parts: a beacon, a contention access period (CAP), and a contention free period (CFP). The beacon is transmitted without using CSMA/CA, and CAP starts immediately after its transmission. The devices communicate during CAP using CSMA/CA based on the time slots (slotted-CSMA/CA). The CFP, if present, starts immediately after the CAP and extends to the end of the active portion. The transmissions in CFP are not based on CSMA/CA.

For low latency applications or applications that require a specific throughput, the coordinator may dedicate portions of the active superframe, called guaranteed time slot (GTS). The coordinator may allocate up to seven GTS slots, and a GTS can use more than one time slot.

Some upper layer protocols have been proposed for IEEE 802.15.4-based WSNs. Following, we present a short description for some of such protocols, as well as implementation issues that impact on the communication reliability.

Zigbee

The most employed protocol in WSNs' applications is the ZigBee protocol [18] [48] [72, 73, 74, 75, 12]. This protocol has a number of desirable characteristics for WSN applications as, for example, low power consumption, and low cost.

ZigBee's network protocol supports the three available topologies (i.e., star, tree, and mesh), allowing the implementation of an ad hoc WSN. When using a mesh topology, the routing process becomes more complex, but the robustness and fault tolerance of the network increases, due to the ability to find and maintain routes. ZigBee does not implement mechanisms to mitigate the co-existence problem.

MiWi

The MiWi protocol [76], developed by Microchip ^{TM}, is an alternative for small networks with at most 1000 nodes. Theoretically, a ZigBee network can contain up to 65536 nodes; although, in practice, it is not recommended having more than 3000 nodes in a single network [75]. Another factor that limits the size of MiWi networks is the maximum number of allowed hops (i.e., four hops). The MiWi Pro protocol supports up to 65 hops.

An interesting feature that sets MiWi apart from ZigBee is MiWi's ability to perform dynamic channel switching. This mechanism, called *Frequency Agility*, is optional and allows moving the network to operate into a different channel once the current operating conditions are not favorable. To set the new channel, a node, called *initiator*, performs an energy scan in all channels to find the least busy one. After that, the *initiator* broadcasts a message to all nodes conveying information regarding the new channel. If a node does not receive the broadcast message from the *initiator* (probably due to a transmission failure), it performs a resynchronization after many recurring failed transmissions. Resynchronization consists on scanning all channels to find out the channel currently in use by the network.

Although this mechanism tends to improve communication quality in general, it incurs an overload on *initiators*. The network may spend much time without providing new data, if the *initiators* perform scans very often. Another important factor is the scanning period. In case it is too long, it is possible to obtain greater accuracy in estimating the best channel; however, the network will be idle for a long period of time. On the other hand, if the scanning period is too short, the network spends little time idle, but then it might present lower accuracy on estimating the best available channel.

Although the *Frequency Agility* mechanism is provided by the MiWi suite, there is a strong dependence on the application layer, since the application determines when a scan and a possible channel switch must happen.

WirelessHart

The WirelessHART standard is considered the first open communication standard designed for wireless industrial monitoring and control applications [77]. The other standards, such as ZigBee and Bluetooth, do not completely meet the requirements of industrial applications.

The WirelessHART is based on the physical layer of IEEE 802.15.4, but it implements its own link layer. It is based on the 2.4 GHz ISM band, but adopting only 15 channels, because channel 26 is not allowed in some countries [78]. Instead of using CSMA/CA as defined by the IEEE 802.15.4 standard, it implements a MAC layer with Time Division Multiple Access (TDMA). By using TDMA, it reduces collisions and power consumption [79].

To improve the co-existence with other networks and other technologies based on the 2.4 GHz band, the standard implements a frequency hopping mechanism. A mechanism called *Blacklisting* is defined, in which channels with high level of interference are avoided. However, the blacklist is not done automatically, but by a network administrator.

In a WirelessHART network all nodes on the network must be able to perform routing based on a mesh topology with redundant routes. This feature allows increasing the reliability and fault tolerance, since redundant routes can replace obstructed paths. The routes are generated by a central entity (Network Manager). The network manager is also responsible for scheduling time among the nodes of the network, ensuring the correct operation of the TDMA mechanism.

WirelessHART networks are centralized, because the entire network operation is managed by a single entity. In MiWi or ZigBee networks, end nodes discover their route to the destination. Moreover, each node can decide when to initiate a transmission independently, using the CSMA/CA mechanism. In WirelessHART, the network manager defines the moment when each node should transmit or receive packets.

Petersen et al. [78] performed studies on the performance of WirelessHART radios. The radios' performance was evaluated when radios were subject to interference from three IEEE 802.11g access points (operating in channels 1, 6, and 11). The results showed that during interference periods, nodes experienced an average packet error rate of 27.2%. However, in the experiment all channels were enabled. With a better management of blacklists, the performance could be improved. Nevertheless, radios do not have the ability to enable and disable channels automatically.

A large latency of around two seconds was noticed for the network operating without interference, and around 2.7 seconds when operating in coexistence with the access points. Thus, we can see that the overhead due to the mechanisms implemented to increase reliability implies on a large latency.

WirelessHART is a recent standard, released in 2007. Until 2009 there was no complying component available on the market [79]. However, more experimental studies should be conducted to verify the performance of WSN that comply with this standard.

ISA100

The Instrumentation, Systems, and Automation Society (ISA) idealized the ISA100 standard [80], which is also designated for industry. As the WirelessHART, the ISA standard is based on the IEEE 802.15.4 physical layer, but defines its own MAC layer. The MAC layer characteristics are very similar to the characteristics presented on WirelessHART. It also applies TDMA and frequency hopping to improve reliability. The network layer is a bit different, since it uses header formats based on the IP protocol [77].

Comparison among the Standards

Table 2 presents a brief comparison among the standards under consideration with respect to some aspects.

Table 2 Comparison among the standards

Standard	*ZigBee*	*MiWi*	*WirelessHART/ISA100*
Supported frequency bands	868 MHz, 915 MHz, and 2.4 GHz	868 MHz, 915 MHz, and 2.4 GHz	2.4 GHz
Physical layer	IEEE 802.15.4	IEEE 802.15.4	IEEE 802.15.4
MAC layer	IEEE 802.15.4	IEEE 802.15.4	Custom
Medium access mechanism	CSMA/CA	CSMA/CA	TDMA
Co-existence mechanisms	-	*Frequency Agility*	Frequency hopping, and Blacklisting
Definition of routes	Distributed	Distributed	Centralized
Redundant routes	No	No	Yes

ZigBee is the only protocol that presents no special mechanism for co-existence. The MiWi protocol provides a mechanism for switching channels, but there is still much dependence on the application layer.

On the other hand, the WirelessHART and ISA100 standards offer more complex mechanisms to improve the coexistence for industrial WSN. The main drawbacks are the heavy network centralization, and the high communication latency, which results in a low information delivery rate [81]. Furthermore, from [78] we can see that, if there is no proper blacklist management, network performance can suffer a significant drop in the presence of interference.

WirelessHART and ISA100 also implement redundant routes, which can increase the reliability, since multiple paths may be defined for data transfer. However, as this mechanism is implemented at the network layer, it can also be implemented in radios that comply with the physical and MAC layers of IEEE 802.15.4.

Although WirelessHART and ISA100 are intended for industrial WSN applications, these are pretty new standards, and they do not have high availability of complying transceivers on the market. On the other hand, there is a wide availability of transceivers that implement the physical and MAC layers of IEEE 802.15.4 and are compatible with ZigBee and MiWi.

10.3.4   Embedded Systems

Embedded systems are computer systems usually dedicated to a particular task. Generally, these systems play a critical role and they are embedded into larger systems. This definition covers almost all embedded systems, although some systems allow the execution of various tasks, such as smartphones, and digital TV [82].

Embedded systems are employed in a myriad of areas, such as industrial applications, medical applications, automotive applications, and telecommunications. Specifically in the industry, embedded systems are usually embedded in systems for process control and industrial automation.

The development of embedded systems must comply with a set of restrictions that are not commonly found in general purpose computing systems, which are usually based on personal computers. The general-purpose systems generally need only to be concerned with the functional requirements of the system, and sometimes a few non-functional requirements, which are sometimes not critical. On the other hand, the design of an embedded system almost always presents a number of constraints that must be addressed, such as: performance, power consumption, reliability, and size. Thus, the major challenge in the development of embedded systems is not only implementing the required functionality, but to implement the functional requirements at the same time that all restrictions (which are often trade-offs) are met.

Many embedded systems must also deal with time constraints. These systems are known as real-time systems. Thus, for a system to run correctly, it needs not only to perform a task correctly, but run it at the correct time.

Basic Embedded System Architecture

There are many ways to develop a digital embedded system. The project can be based on a programmable logic device, an ASIC, or based on a microprocessor or microcontroller.

Once the design of an integrated circuit is a time consuming and costly task, the reuse of hardware components and modifying only the software for a particular application makes developing embedded systems based on microprocessor or microcontroller an interesting alternative. Microprocessors and microcontrollers are usually manufactured in large scale and they are used in a wide range of applications, so its cost is greatly reduced.

Although the execution of a software in a microprocessor suffers with the overhead for fetching and decoding the instructions, the modern microprocessors and microcontrollers implement techniques which makes the software execution quite efficiently. Modern RISC processors can execute one instruction per clock cycle most of the time, and high-performance processors are capable of executing multiple instructions per cycle. Therefore, the overhead can be compensated by internal parallel processing using (e.g., pipelined functional units and multiple processing units).

Figure 7 illustrates a basic architecture of an embedded system containing the following components: sensors (represented by the orange arrow), actuators (represented by the green arrow), input, and output interfaces, processing unit, and memory.

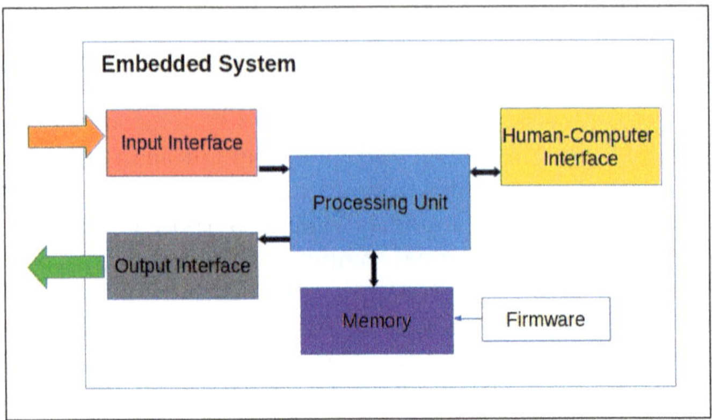

Fig. 7 Basic architecture of an embedded system

The sensors are responsible for retrieving data from the object or environment in which they are inserted, such as voltage and current measures from motors. Such data is often acquired in analog format, then the input interface may have signal conditioners, which make the measures compatible with the hardware system used, and the analog to digital converters. The actuators are responsible for causing changes in the environment or object in which they are inserted. As well as sensors, actuators can be analog, making a signal conditioning circuit and a digital to analog converter a possible requirement.

In addition to the actuators and sensors, the system can communicate with users through a human-computer interface. This interface can connect components such as buttons and displays that allow the user to interact directly with the embedded system.

The input and output interfaces can also be used to communicate the embedded system with other computer systems using a transceptor. For example, in a WSN, embedded systems that form the sensor nodes communicate with each other through a wireless interface.

The core of the embedded system is the processing unit, which receives external data, performs computations on that data and generates results that are available through the output interface. The memory stores the firmware, which is responsible for controlling the system, and also stores data during the firmware execution.

In general, embedded systems have reduced size. Likewise, its processing and storage capabilities are also reduced. Moreover, these systems often have limited power source [83]. Usually, the microprocessors or microcontrollers

used in embedded systems have a much smaller processing capacity than the
processors used in personal computers, but they also have advantages, such as
low cost and low power consumption. The amount of available memory in an
embedded system is also much smaller than the amount of memory found on
a personal computer. These restrictions should be taken into consideration
when developing the embedded software, which should have a small code,
respecting the limit of the program memory size, and also use little data
memory [84].

Besides considerations about the limitations of memory, and processing
power, the design of an embedded system should ensure reliability and respect
its time constraints, since many times these systems are part of larger system
performing a critical task.

10.3.5 *WSN-Based Motor Monitoring Systems*

This section describes some works [10] [36] [41] [38] [42, 43, 44, 45] focusing
on the application of WSN in industrial environments. There is a relatively
small amount of work towards the development of monitoring and control
systems in industry based on WSNs. This is due to the complex requirements
of the system and severe work environment [41]. Some recent works address
the performance evaluation of radios in an industrial environment [66] [65]
[85, 86, 87], while some other works address the challenges of using WSN
technology in industry [40, 60] [39] [88, 89].

Salvadori *et al.* [38] proposed a digital system for the evaluation of power
usage, diagnosis, control, and supervision of electrical systems by employing
WSNs. The system is based on two hardware topologies responsible for signal
acquisition, processing, and transmission: intelligent sensor modules (ISMs),
and remote data acquisition units (RDAUs). However, only wired communication RDAUs are used to perform acquisition of voltage and current of
motors. ISMs were used only for temperature measurements. The work fo-
cuses mainly on the energy consumption of sensor nodes and it does not
provide detailed studies on transmission errors and communication channel
quality.

Hsu and Scoggins [34] presented a method to estimate motor efficiency
from the air gap torque, which is obtained from the motor electrical signals (current and voltage). It is the noninvasive method for determining
torque and efficiency that has less uncertainty [14]. Recent works have also
used this technique to estimate the efficiency and torque of induction motors [36][10][44]. These studies have also employed WSN for data transmission.

Bin Lu *et al.* [36] identify in their work the synergies between WSNs and
analysis of motors based on electrical signals, also following a noninvasive
pattern. They propose a scheme to apply WSNs for online and remote monitoring and fault diagnosis of industrial motors.

The main limitation of the work presented in [36] derives from the low throughput provided by the WSN based on IEEE 802.15.4, since the proposed system does not employ local processing. Thus, it is necessary to transmit a large amount of data to estimate the desired parameters. This limits, among other things, the acquisition rate from the sensors, which consequently limits the accuracy of the estimation. In a WSN with a large number of nodes, the situation becomes even worse, since all nodes share the same physical medium. Moreover, the wireless networks are inherently unreliable, which can result in transmission errors, affecting the estimation process.

Liqun Hou *et al.* [42] developed a motor monitoring system using WSN with local processing. A prototype was implemented and validated in a single-phase induction motor in laboratory. Motor current signature analysis (MCSA) is employed in this application, where motor stator current signal waveforms are given under different working conditions. Using local processing, a reduction of around 90% was obtained in the amount of data transmitted for analyzing the motor. Liqun Hou et al. [41] also developed a system for fault detection in motors using accelerometers and WSN. In this system, it was obtained a reduction of 99% in the amount of data transmitted for performing the failure analysis task when using local processing. However, both [42] and [41] do not performed a detailed analysis of the WSN performance. For example, the information delivery rate was not verified and the experiments were not performed in realistic environments.

Hu [43] [44] presents a DSP-based system for motor monitoring using the air-gap torque method, and WSN for data transmission. It was proposed the estimation of various parameters such as power factor, efficiency, speed, and torque. However, the tests were conducted in laboratory, which does not characterize a realistic experiment. As in [42] and [41], there was no detailed study on the impact of using local processing on the WSN performance.

Esfahani *et al.* [45] developed a multisensor wireless system for condition monitoring of induction motors. They used multiple sensors in a common platform, and current signature analysis and vibration monitoring were employed. Although they do not performed detailed studies about communication performance, they emphasized the need to employ local processing and robust network protocols for attaining good enough quality of service.

In a previous work [10], we developed an embedded system integrated into a WSN for online dynamic torque and efficiency monitoring in induction motors. The air-gap torque method was employed for estimating the shaft torque and motor efficiency. The computations for estimating the targeted metrics are performed locally and then transmitted to a monitoring base unit through an IEEE 802.15.4 WSN.

Experimental tests were performed to analyze the torque values obtained by the system, and then compared with torque values based on the workbench dynamic model. In section 10.4.5, we show an experimental study aiming at identifying the correlation between spectral occupancy and Packet Error Rate

(PER) for the proposed WSN. The experiments were conducted inside a shed, with typical characteristics of industrial environments.

The study demonstrated that the addition of new interference sources can significantly affect the spectral occupancy, by also having a direct impact on the communication performance. Even for harsh condition scenarios, the system was able to provide useful monitoring information, since all processing is done locally (i.e., only the computed metric is transmitted over the network). Without local processing, it might be impossible to use the WSN technology for this particular application, considering an unreliable transmission medium.

10.4   A IWSN for Torque and Efficiency Monitoring of Induction Motors

10.4.1   The Employed Estimation Method

After analyzing the main methods of shaft torque estimation, we have chosen the AGT method to implement our system, due to its low invasiveness and good accuracy. This system was first presented in our journal paper [10].

For speed estimation we have developed a simple and innovative estimation method. This section presents in detail the developed methods and their implementation. The embedded system for monitoring induction motors is fully described, as well as the components employed for its construction.

This study proposes a new specific estimation method to be implemented in an embedded system. The requirements to implement the embedded system are:

- Minimal invasion: it is not acceptable to stop the process for a long period of time to install the devices; it is also undesirable to perform invasive tests and installation of new components in the motor. When operating, the system cannot interfere with the motor or the process itself.
- Maximum accuracy: the estimated values should be reliable and as close as possible to the real values.
- Digital processing into an embedded system: the estimation procedures should be performed by a digital processor, which requires sensors and analog to digital converters with appropriate sample rates. The adopted method should also have low computational complexity requiring low memory.
- Estimation in real time: the time needed for measuring, processing, and transmitting the data must be short enough for making it possible to operate in real time, providing an adequate amount of information per time period.
- Minimum cost: the embedded system must be built out of off-the-shelf and accessible components.

The procedure to perform the estimations consists of a set of routines with certain functions. Figure 8 illustrates a simplified sequence of these routines.

For the device installation, it is necessary to survey the initial data for configuring the internal software, following the form in Figure 9.

Air Gap Torque Estimation

To implement the theoretical AGT method in the embedded system, some steps are required. The integrals in Equation 10.7 represent the magnetic coupling related to the air gap torque. When computing the integrals, the initial phase of the counter-electromotive force is taken into account, because it causes an offset in the signal due to the integration constant. Physically, this offset represents a constant magnetic flux that does not exist during real operation of the machine.

Figure 10 presents, with generic vectors, the real stator flux, (λ_s), the offset of the flux, (λ_{offset}), and the estimated flux, $(\hat{\lambda}_s)$.

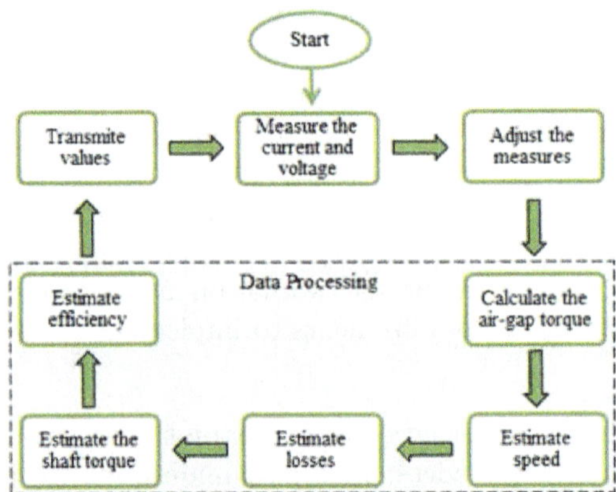

Fig. 8 Sequence for the air gap torque estimation

Form for Installation					
Motor Information			**Test no Load**		
Manufacturer			Air gap torque estimation		N
Model			Electric power measured		W
Number of poles		poles	Speed measured		rpm
Nominal power		kW	**Test with Constant Load**		
Nominal speed		rpm	Air gap torque estimation		N
Stator resistance		Ω	Known torque		N
Supply frequency		Hz	Electric power measured		W
			Speed measured		rpm

Fig. 9 Form to configure the instrument

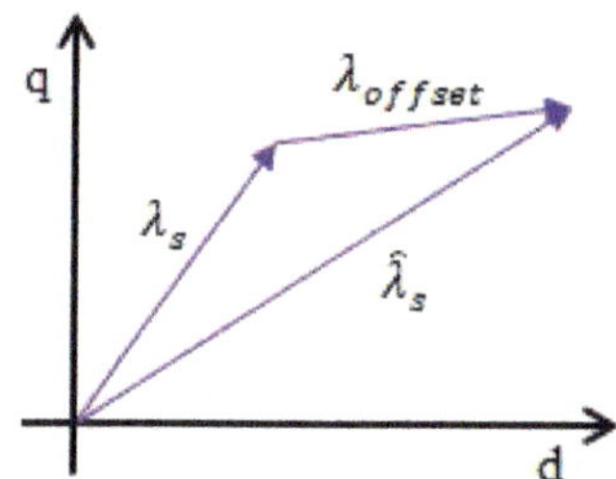

Fig. 10 Spatial representation of the integration offset

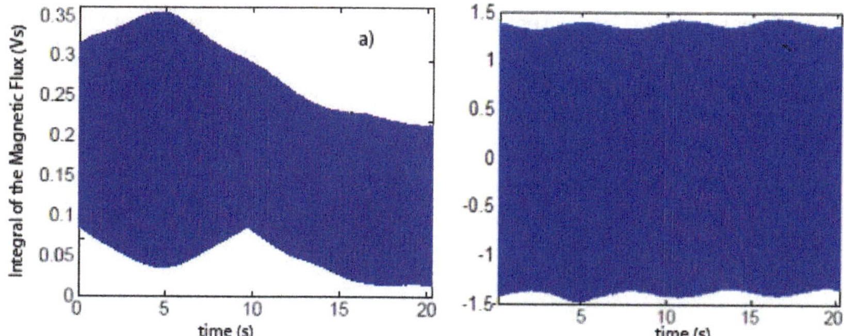

Fig. 11 Integration (a) without compensation, and (b) with offset compensation

Figure 11 shows a sine signal after integration without compensating the offset values, and Figure 11 shows the same signal with offset compensation.

To solve this problem, there are different methods proposed in the literature. The commonly used method is the combination of low pass filter [90] and high-pass filters [91]. However, filters introduce phase distortion.

When the integration is carried out through digital converters and processors, it is easier to identify and subtract the offset. The software implemented subtracts the arithmetic mean for each integration cycle. Thus, the DC component is removed from the sine wave with good accuracy, without phase distortion and with low computational complexity.

The stator resistance can be measured by means of an ohmmeter applied to the motor's terminals while it is off. If it is not possible to measure it directly, the nominal resistance should be used instead.

The stator resistance changes according to the motor conditions, mainly subject to temperature oscillations [92]. Several algorithms have been proposed for estimating stator resistance for different applications [93] [94], including those for determining air gap torque [35]. As the proposed system is a real time embedded system, we did not implement the algorithm for determining the stator resistance, since it would increase the overall processing time. However, the stator resistance has a small impact on the air gap torque computation, as demonstrated in Table 3 (section 10.4.1).

A mean filter was employed to make the air-gap torque values more stable over time. This filter does not require much processing time while being effective.

Speed Estimation

Measuring directly the rotor speed, (ω), can be impractical in some cases. Several methods of sensorless rotor speed estimation have been proposed. These methods follow two categories: one employing an induction motor model, and the other derived from the analysis in the frequency spectrum of voltage and electric current [95]. The method proposed by Ishida et al. [96], based on the electrical voltage, uses techniques of digital signal processing to detect the harmonics generated from the rotor slots. However, it requires high rotor speed and stability [52].

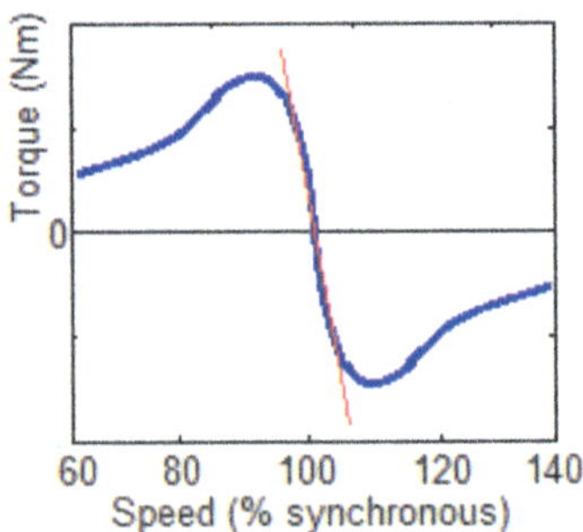

Fig. 12 Relation between Torque and Speed

Ferrah et al. [97] and Hurst et al. [98] used the fast Fourier transform to extract the rotor slots harmonics out from the electric current spectrum. In this case, it requires a large number of samples per cycle and a high processing power. The method also requires information from the motor usually not available from the factory specifications.

The methods mentioned above do not work well when the speed is close to the synchronous speed and in dynamic systems with variable torque and vibration.

A conventional induction motor has a speed variation of less than 10% to the synchronous speed when it is being used from no load to full load. In the normal operation region, close to synchronous speed, the motor presents an almost linear relationship between its torque and its angular velocity (as can be seen in Figure 12). Thus, a procedure for curve linearization can be adopted. To perform this linearization, two points are needed to relate torque and speed.

Because of the problems on implementing the traditional speed estimation methods, a new simple method was developed. The method establishes a direct relation between the angular speed and the air gap torque. This method is well adapted to the constraints of the embedded system. We observed that air gap torque and speed have also an almost linear relationship. In section 10.4.3, we will validate this approach through experiment results.

According to Equation 10.9, one must know the speed for estimating the shaft torque. Therefore, the speed is estimated first, directly from the air gap torque.

The linear relationship parameters shall be determined from two known points in the curve. If it is possible to perform a test with no load, and a test with constant load, the linear relationship parameters can be determined accurately. Otherwise, the chosen points correspond to the situation when the torque is nominal (i.e., nominal speed), and when it is zero (i.e., synchronous speed). This last approach uses nameplate information, without requiring any invasive approach into the system.

The nominal values in the nameplate may present an up to 20% error [51]. Therefore, to configure the system for a given motor, it is desirable measuring the nominal speed for more accuracy in the real time estimation.

Losses Estimation

To determine the motor shart torque by means of the AGT method, it is required to consider the losses resulting from electromechanical conversion (i.e., mechanical losses and additional losses).

Mechanical losses (i.e., K_{mec}) are caused by friction and windage, and depend on each motor. For a more accurate estimation, a test with no load must be performed with the instrument installed at the motor terminals. In this case, as the shaft torque is zero, the estimated air gap torque equals the mechanical losses.

If the test with no load is impossible, the mechanical losses are assumed to be 3.5% of the nominal motor power, as suggested by the Ohme method [35].

For a more accurate estimation of additional losses (K_{ad}), a test with known load is required. Thus, the estimated value for the air gap torque minus the applied torque relate to the losses in place. If such test is not feasible, the additional loss is defined according to the method proposed in the IEEE std E1-112 [13]; i.e., the loss is proportional to the size of the motor, according to Table 1

Shaft Torque Estimation

Computing the shaft torque ($\hat{T}$) from the air-gap torque, ($\hat{T}_{ag}$), the mechanical losses, ($\hat{K}_{mec}$), additional losses, ($\hat{K}_{ad}$), and the estimated rotor speed, ($\hat{\omega}$), should be considered according to Equation 10.9.

$$\hat{T} = \hat{T}_{ag} - \frac{\hat{K}_{mec}}{\hat{\omega}} - \frac{\hat{K}_{ad}}{\hat{\omega}} \tag{10.9}$$

Mechanical losses (i.e., friction and windage, $\hat{K}_{mec}$) vary according to the particular motor and the industrial process in place. If it is not possible to estimate the losses, then a no load test is required. The additional loss (i.e., stray-load loss, $\hat{K}_{ad}$) result from nonlinear phenomena of different natures, difficult to quantify; however, it can be based on a percentage of the motor power [13].

Table 3 quantifies the influence of the values necessary to determine the torque shaft. Even with low accuracy on determining the losses, resistance, and rotor speed, we can see that accuracy limitations incur on low error on the estimated torque value. On the other hand, inaccuracies when measuring voltage, current, and on air-gap estimation leads to high error on torque estimation.

Table 3 Parameters' impact on estimations

Variable	R	P_n	$\hat{\omega}$	V_{ab}	V_{ca}	I_a	I_b	$\hat{K}_{ad}$	$\hat{K}_{mec}$	$\hat{T}_{ag}$	$\hat{T}$
Margin of erros	30%	5%	5%	5%	5%	5%	5%	5%	5%	5%	5%
Variance on Torque ($\hat{T}$)	0.2%	0.3%	0.2%	4.8%	0.1%	4.5%	0.6%	0.1%	0.2%	5%	-
Variance on Efficiency (η)	0.2%	0.3%	5%	0.8%	0.5%	0.7%	0.4%	0.1%	0.2%	5%	5%

Efficiency Estimation

Energy efficiency, (η), is defined as in Equation 10.10, where P_{in} is the input power, and P_{out} is the output power. In induction motors, the output power is the mechanical power delivered to the load.

$$\eta = \frac{P_{out}}{P_{in}} \times 100\% = 100\% - \frac{K}{P_{in}} \times 100\% \qquad (10.10)$$

Despite the simplicity of Equation 10.10, efficiency estimation is often difficult due to the complexity of measuring or estimating all the required variables, which belong to different categories.

The instantaneous electric power input, (P_{in}), of the AC motor is computed from direct measurements of voltage and electric current, according to Equation 10.11.

$$P_{in} = I_a V_a + I_b V_b + I_c V_c = -V_{ca}(I_a + I_b) - V_{ab}I_b \qquad (10.11)$$

The useful power output, ($\hat{P}_{out}$), is the product between the shaft torque, ($\hat{T}$), and angular speed, ($\hat{\omega}$), according to Equation 10.12.

$$\hat{P}_{out} = \hat{T}\hat{\omega} \qquad (10.12)$$

Therefore, efficiency is computed according to Equation 10.13.

$$\eta = \frac{\hat{T}\hat{\omega}}{-V_{ca}(I_a + I_b) - V_{ab}I_b} \times 100\% \tag{10.13}$$

Table 3 quantifies the influence of each parameter required for computing the efficiency on error estimation. We can see that efficiency varies directly with the speed and shaft torque. Thus, to perform an accurate error estimation, speed and torque estimation must be equally accurate.

10.4.2 Embedded System

Figure 13 depicts the WSN proposed in this paper. End nodes are composed by the embedded systems located close to the electric motors. The values of motor voltage and current are obtained from the sensors, and the embedded system performs the processing for determining the values of torque, speed, and efficiency. Information obtained after the processing are transmitted to the base station through the WSN.

Depending on the distance between end nodes and the coordinator, it may not be possible to achieve direct communication, due to the radio's limited range and the interference present on the environment, among other factors. Therefore, the communication among nodes and coordinator can be done with assistance of routers.

Figure 14 shows a simplified block diagram of the proposed embedded system. For current measurement, Hall Effect sensors are employed due to their robustness and non-invasiveness. Transformers with grain-oriented core are used to measure the voltage between phases, which provide the voltages in the secondary and primary without delay. The Acquisition and Data Processing Unit (ADPU) is responsible for data acquisition and conversion, besides the data processing. The Printed Boards Power Supply (PBPS) supplies the

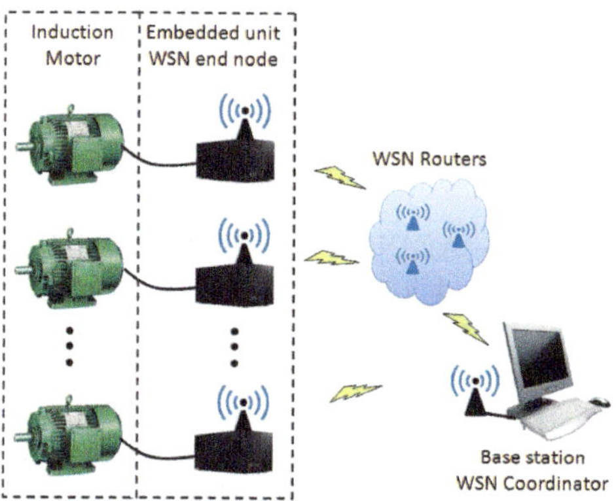

Fig. 13 Embedded system integrated into the WSN [10]

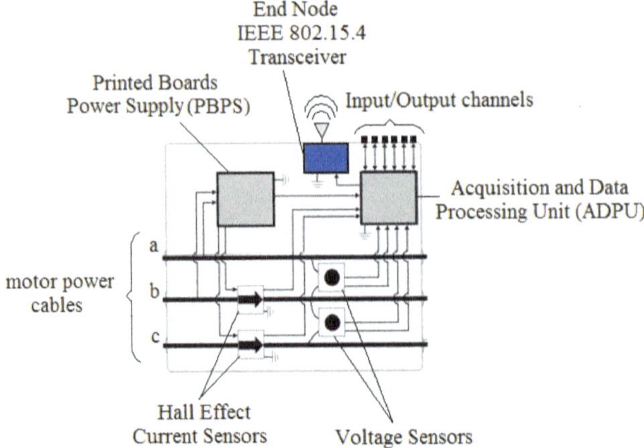

Fig. 14 Block Diagram of the Embedded System [10]

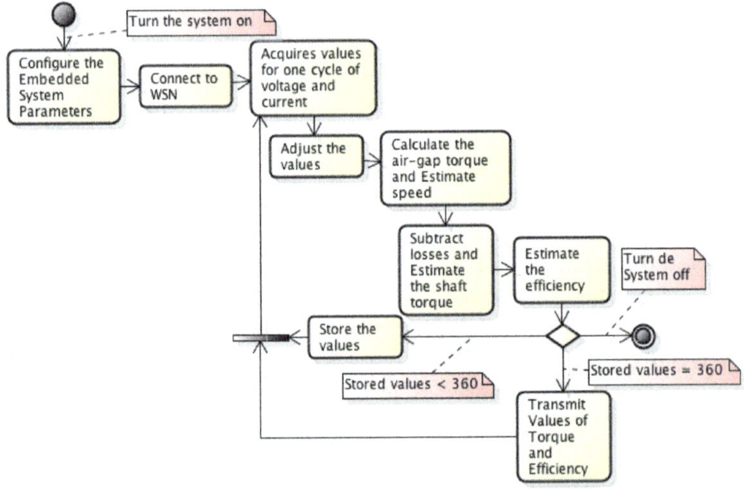

Fig. 15 Activity Diagram

current and voltage for the sensors, the IEEE 802.15.4 transceiver, and the ADPU.

The main element of the ADPU is a dsPIC33FJ64GP706, which is a digital signal controller designed for applications that require high processing capacity. It has two integrated analog/digital converters (ADC), which perform simultaneous acquisition of the voltage and current sensors. The Input/Output channels can be used for user interface, and possible connections to auxiliary sensors and actuators. The values of torque and motor efficiency are transmitted using the IEEE 802.15.4 Transceiver. We have used an MRF24J40 transceiver, designed by *Microchip* [TM]. The connection between the transceiver and the dsPIC is accomplished using a Serial Peripheral Interface Bus (SPI).

The internal operation of the embedded system is illustrated by the activity diagram shown in Figure 15. When the system starts, the embedded system parameters are configured. These parameters include the wireless network settings (e.g., address, channel), and the ADC settings. To obtain good accuracy from a simple numerical integration method, such as trapezoidal (used to implement the algorithm), then it should be used a sample rate greater than 2 kHz [52]. In our system, we set the ADC to operate with 3 kHz and 10 bits of resolution.

After the first step, the system connects to the WSN. The embedded system only begins to acquire and process data after successfully connecting to a coordinator operating in the same channel. Then, the system gets into the acquisition loop, processing and transmitting data, which is repeated until the system shuts down. The voltage and current values, after acquired, must be adjusted to reflect the real values measured from the sensors. After that, the algorithm is executed to compute the air-gap torque, according to Equation 10.7. After that, the losses are removed, and the shaft torque is estimated according to Equation 10.9. Using the shaft torque values, the system estimates the motor speed and efficiency.

The embedded systems were configured to calculate a set of 360 values (2 bytes each) of torque and efficiency, and then transmit these values aggregated into 20 packets with 72 bytes of payload each. The time necessary to acquire the signals and calculate the 360 values of torque and efficiency is about 11 seconds (6 seconds to acquire 360 cycles of current and voltage, and 5 seconds to perform the calculations). Thus, the system transmits data in burst mode, spending only about 8% of the time transmitting data, at a rate of 20 packets per second (about 14 kbps, including control overhead).

10.4.3   Experimental Results

Experimental Workbench

In order to validate the method and the instrument proposed, a workbench is used to apply a sinusoidal torque on the motor shaft, which values are adjustable and well known. Figure 16 shows its sketch, which consists in a 550-W induction motor, with nominal rotation speed of 1680 RPM, coupled to a reducer that provides an output speed of around 15 RPM. A steel disc was fitted on the output shaft, symmetrically coupled with a metal bar. At one end of this bar can be inserted masses, according to the desired torque value.

At the time the motor starts, the system (reducer shaft , steel disc, and steel bar) rotates and the masses placed at the end of the steel bar impose a variable torque on the motor, and consequently on the shaft of the reducer. The resultant torque has sinusoidal shape with values dependent on

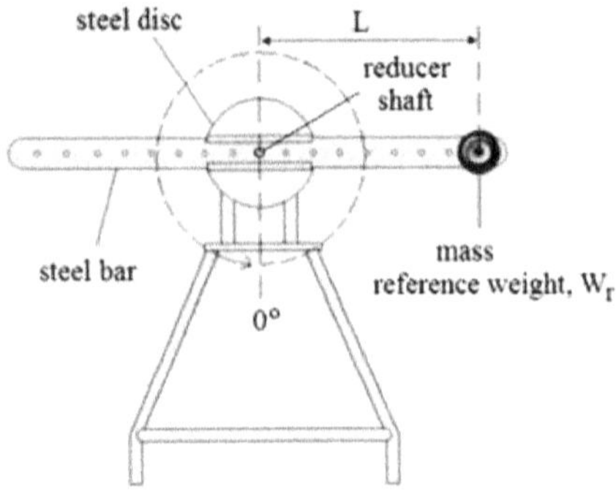

Fig. 16 Workbench employed for system analysis

the weights and the position in which the bar is. The mathematical model of the reducer shaft torque ($T_reducer$) is obtained by the dynamic equilibrium analysis. Through the sum of the moments around the center of the reducer shaft is obtained:

$$\sum (M_0)_{ext} = \sum (M_0)_{inertia} \tag{10.14}$$

In which $(M_0)_{ext}$ are the external moments and $(M_0)_{inertia}$ are the inertia moments components. Applying the Equation 10.14 for the system on Figure 16, we obtained the Equation 10.15.

$$T_{reducer} = LW_r sin(\theta) + \alpha(I_d + I_b + L^2 m) \tag{10.15}$$

In which L is the distance between the masses center and the reducer shaft, W_r is the reference weight, m is the reference mass, θ is the angular position of the bar, and α is the angular acceleration on the reducer output. The moment of inertia relative to the disk and bar are the variables I_d and I_b, respectively. The model of torque transformation between the low-speed side and the high-speed side is obtained according to the Equation 10.16.

$$T_{ref} = \frac{T_{reducer}\omega_{reducer}}{\omega_r} + J_{reducer}\alpha \tag{10.16}$$

In which $\omega_{reducer}$ is the reducer angular speed, $J_{reducer}$ is the reducer intertia with respect to the high-speed side, and α is the rotor angular acceleration. The workbench is instrumented, by using Hall Effect sensors and magnets, to measure θ, w_r and α, used in torque and efficiency equations. The $J_{reducer}$ is obtained from the manufacturer, and tests are performed to provide the torque as a function of the load.

In Equation 10.15 we can see that, in the first quarter cicle, $T_{reducer}$ ranges from zero to its maximum value. Since the workbench efficiency model is determined by the substitution of T_{shaft} by T_{ref} in Equation 10.13, the efficiency curve of the motor can be obtained for all of their operating range, using an appropriate value of W_r.

Analysis of the Embedded System Estimated Values.

Figure 17 shows the workbench employed to analyze the system and its components. The embedded system was placed close to the motor to acquire the voltage and current data. Torque and efficiency are calculated by the ADPU (Acquisition and Data Processing Unit), and then, transmitted through the WSN using the IEEE 802.15.14 transceiver. Both torque and efficiency are received at the monitoring base station, where they can be visualizaed and stored.

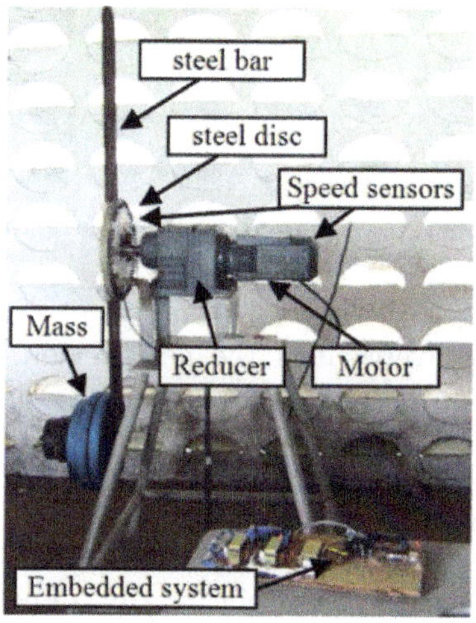

Fig. 17 Experimental setup for the torque and efficiency analysis [10]

Figure 18 shows the estimated torque curves received at the monitoring base station T_{shaft}, calculated in ADPU using Equation 10.9, and the reference torque obtained from the workbench dynamic model (T_{ref}) , obtained from Equation 10.16. As described in the previous section, the torque in this workbench is sinusoidal, however the positive cycle is the one used in this application, so during this time the workbench system works as a motor, while in the negative cycle the same system works as a generator.

The curves in Figure 18 were obtained for the first half cycle of the steel bar (Figure 16), from $0°to180°$), using two differente masses. The curves comply with the dynamic model of the workbench that consists of a sinusoid, corresponding to the first part of Equation 10.15, and modulated by the acceleration components, regarding the second part of Equations 10.15 and 10.16.

As shown in the curve regarding the reference weight with mass equal to 10kg, the estimated toque follows the reference torque, and it captures the workbench vibration. The relative error between the two curves is less the 2 % . For the curve with a reference weight of a 40 kg mass, we observed that

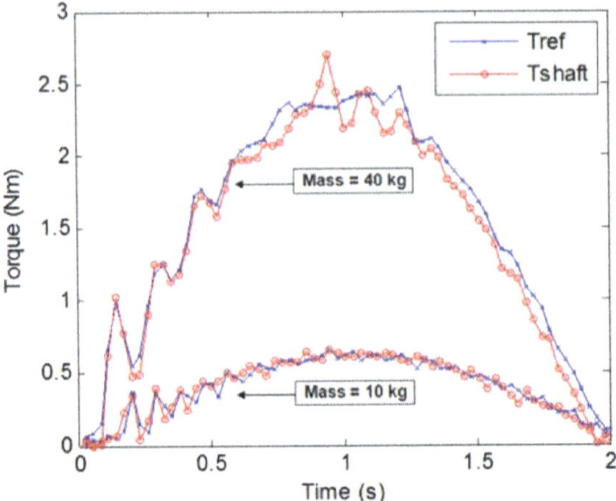

Fig. 18 Comparison between estimated and reference torque measurements for two masses

the torque amplitudes diverge near the peak region, when the workbench presents an increase in vibration amplitude.

To the system's efficiency, differents reference weights are used for computing the peak reference torque and the peak estimated torque. The corresponding speed and power input values were also used in the calculation. Thus, the reference and estimated efficiencies were calculated using Equation 10.10, by replacing T_{shaft} for the reference torque T_{ref} (see Equation 10.16) , and the estimated torque (see Equation 10.9), respectively.

In the Figure 19 is shown the reference curve and the estimated values by the embedded system. On the X-axis there is the engine load range, between 0% and 85% of nominal power, the maximum error did not exceed 2%. Although the use of the embedded processing and wireless transmition, this result corrobates to other works that use the AGT efficiency method [35] [36].

For other operating ranges, it was also possible to obtain relatively accurate efficiency with respect to the reference torque, even in the presence of strong workbench vibrations, which occur with greater intensity in the region near the nominal load.

The motor speed is estimated through a linear approximation using the AGT method. In Figure 20 the measured speed, from magnets and hall effect sensors, and the estimated speed for the two masses. The maximum error observed is 0.26% for the reference weight with the 10 kg mass, and 0.4% for the reference weight with the 40 kg mass.

A software was developed to run in the monitoring base station. The system allows to view the values obtained from all embedded systems connected to the WSN in real time. Figure 21 shows both torque and efficiency curves received in real time at the monitoring base.

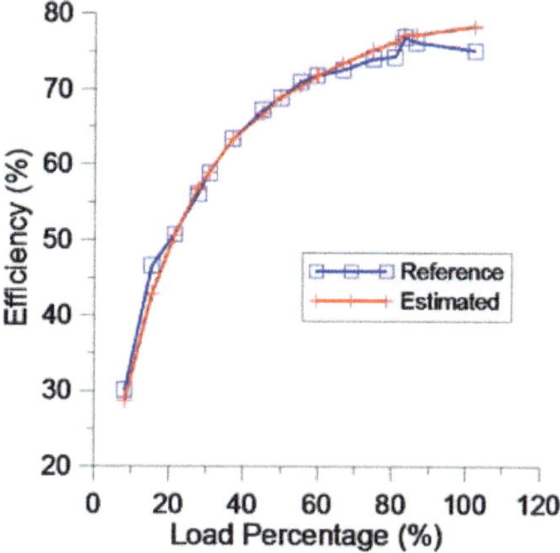

Fig. 19 Comparison of estimated and reference efficiencies versus load [10]

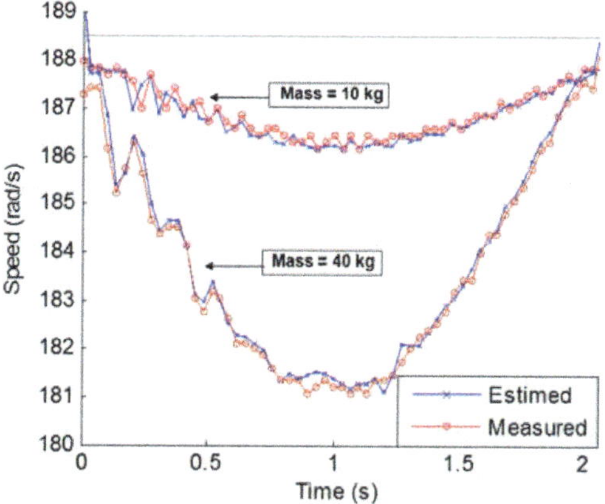

Fig. 20 Comparison of estimated and measured motor speed [10]

10.4.4 Methodology of WSN Performance Evaluation

It is very important to conduct performance studies of wireless systems in industrial environment, mainly due to the lack of reliability inherent to wireless networks. Therefore, this paper presents a study on the performance of the proposed WSN in order to observe its limitations and provide recommendations when developing new solutions for achieving better performance of such systems.

The communication performance among end nodes and the coordinator was evaluated, while performing spectrum analysis in the surrounding environment. Packet Error Rate was the chosen performance metric. The spectrum was divided into channels, according to the IEEE 802.15.4 specification.

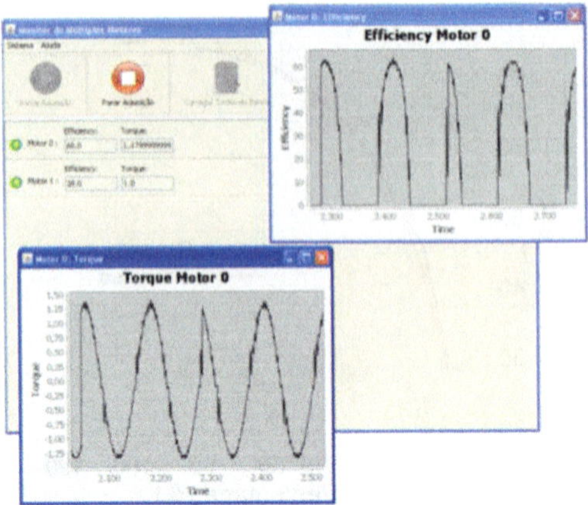

Fig. 21 Base monitoring system [10]

A comparison was made between the power values induced for each channel and the corresponding performance results. Thus, the effect of inserting new interference sources in the environment was studied, by verifying its impact on the spectrum occupation and on the communication performance within each channel. Below are detailed the factors and response variables considered in the experiment:

- Primary Factors (FP)

  PF-1 - Channel: this factor is categorical and it has three levels: Channel 13, Channel 18, and Channel 24.

  PF-2 - Interference Sources: this factor is categorical and it has three levels: "Only Microwave oven on", "Only IEEE 802.11g on (channel 6)" and "Microwave oven off and IEEE 802.11g off".

- Response Variables (RV)

  RV-1 - Spectrum Occupancy (SO): this response variable is obtained from the average power induced on the spectrum range of each channel.

  RV-2 - Packet Error Rate (PER): this response variable is the number of incorrectly transmitted data packets divided by the total number of transmitted data packets.

We decided to compare SO and PER considering only three channels. This way, the time needed for the experiments was reduced and allowed us to get a good analysis of the spectrum occupation distribution and the corresponding relationship between spectral occupancy and communication performance. To increase the significance level of the findings, three replications were performed for every experiment. The confidence level used in all statistical tests was 95%.

Calculation of Spectral Occupancy

Let f_j be a frequency component of the spectrum and F be the set of all frequencies on the spectrum range considered in the experiment.

c_r is a communication channel, and C_r is the set of frequency components of the channel, where $C_r \subset F$.

The induced power on the frequency component f in the spectrum at a given instant t is denoted by $p_t(f_j)$, and the induced power on the channel c_r, denoted by $P_t(c_r)$, obtained from the average among the powers of the frequency components belonging to C_r, is given by Equation 10.17.

$$P_t(c_r) = \frac{\sum_{f_j \in C_r} p_t(f_j)}{|C_r|} \qquad (10.17)$$

The spectral occupancy on channel c_r, denoted by $SO(c_r)_w$, in a given discrete time interval Δt_w is obtained through the average values of $P_t(c_r)$, measured at each instant during the time interval Δt_w. The values of $P_t(c_r)$ considers the power induced by all devices in the frequency range of c_r, including the IEEE 802.15.4 radios and the interference sources.

Threats to Validity

The exact moment that the experiment is performed can affect the conclusions, because the spectrum occupation pattern can vary along time. In addition to that, temporal variations during the measurements can affect the results, due to uncontrolled external factors, such as temperature and humidity. However, our experiment was replicated three times, allowing observing the system behavior at different time intervals, thus avoiding restricting the conclusions to a specific measure.

Fig. 22 Industrial Environment

Industrial Environment

The experiments were conducted inside a shed, with typical characteristics of industrial environments, such as the presence of large amounts of metallic devices. Figure 22 shows the environment where the experiments were performed.

10.4.4.1   Experiment Setup

We used three nodes to test the WSN, forming a star topology with one coordinator and two end nodes. The first node, ($N1$), was set 16 meters away from the coordinator, while the second node, ($N2$), was set 13 meters away from the coordinator. Another important detail is that node $N1$ had no line of sight to the coordinator, and it was among several metal objects; while node $N2$ had a line of sight to the coordinator. The end nodes were configured to transmit with an output power of 0 dBm.

During the experiments, to verify the impact of an IEEE 802.11g network, we performed a file transfer between two IEEE 802.11g nodes connected to a base station. One node was about one meter away from the IEEE 802.15.4 end nodes, while the other was placed next to the coordinator. The IEEE 802.11g nodes were set to transmit at power level of +15 dBm.

The embedded systems were configured to transmit 2000 packets, in each replication. During the time spent to estimate and transmit all values, it was performed spectrum measurements. The values of $SO(c_r)_w$ in the channel (c_r) used for communication was calculated for this time interval (Δt_w).

Instrumentation

For spectrum power acquisition, we have used the *Airview2/EXT* [99] spectrum analyzer. For the IEEE 802.11g network deployment, a D-Link DI-524 router and two personal computers equipped with DWL-AG132 Wireless USB Adapter were employed. The microwave oven used was a *Consul* ᵀᴹCMS25ABHNA model, with a power of 700 W.

10.4.5   WSN Performance Evaluation

It is investigated the impact on spectral occupancy and PER metrics due to the insertion of interference sources. And, also, is presented a theoretical analysis to compare the approaches with local processing and without local processing.

Figures 23 and 24 show the impact of a microwave oven and an IEEE 802.11g network (operating on channel 6), respectively. The left side of the graphs show the impact on node $N1$, and the on the right side, the impact on node $N2$. The x-axis constains the channels c_r considered in the experiment, and in the y-axis, it has PER values (left axis) and the $SO(c_r)_w$ (right axis) in dBm.

One can easily see by the graphs from Figures 23 and 24 that the inclusion of new interference sources resulted in a significant performance drop, mainly for node $N1$. Overall, node $N2$ had no great performance losses compared to node $N1$. The reason is because node $N1$ was farther away from the coordinator, without having a line of sight path to it.

Analyzing the impact of the microwave oven is observed that the node $N2$ had no great drop in performance, but when operating on channel 24, the PER presented a small variance of up to 15%. The node $N1$ had a significant impact in PER, mainly when operating on channels 18 and 24. The variance was also large for these channels, reaching 80% on channel 18 and 60% on channel 24.

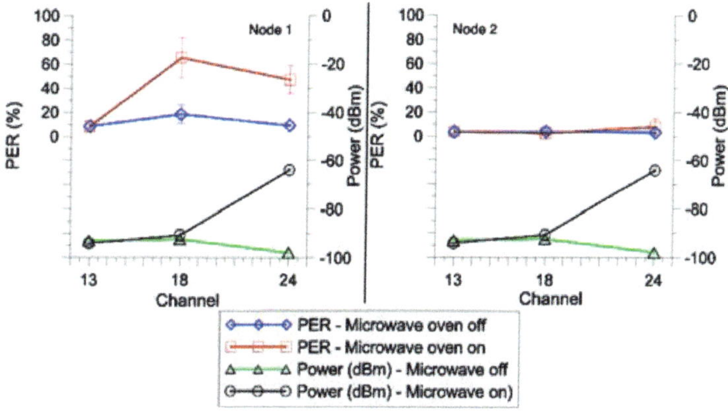

Fig. 23 Impact of a microwave oven [10]

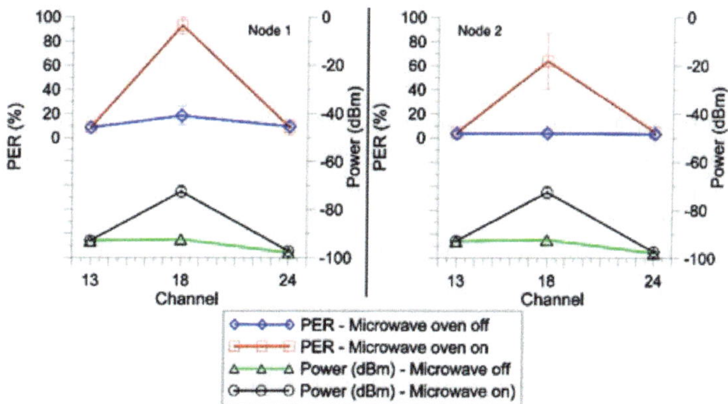

Fig. 24 Impact of an IEEE 802.11g network [10]

Concerning the impact of the IEEE 802.11g network, is observed a large correlation between SO and PER. Again, node $N1$ experienced a bigger drop in performance compared to node $N2$. For an IEEE 802.11g network operating on channel 6, the IEEE 802.15.4 nodes experienced a significant impact only when operating on channel 18. On this channel, node $N2$ presented the best performance, but with a PER variance up to 85%.

When operating on channel 13, the two end nodes kept a good quality of communication, in all scenarios under consideration.

Figure 5 in section 10.3.2 shows the induced power on the frequency components in each scenario for one of the replications performed for the channel 13. The average power induced power induced by the microwave oven can be larger than the average power induced from IEEE 802.11.g network, but the variance is also much larger.

When the microwave is turned ON, the $SO(c_r)_w$ values does not present a high variance, as seen in the Figures 23 and 24, but analyzing the frequency components individually, we see that the variance of $p_t(f_j)$ is very large during Δt_w , mainly in the more affected components.

This may explain the lower correlation between the spectral occupancy and PER for this scenario. It also explains why the interference of the IEEE 802.11g network leads to a greater performance drop both in $N1$ and $N2$, even inducing a lower average power on the channels. When the IEEE 802.11g network is turned ON, the noise level in the affected frequency range remains high all the time, with little variance. When the microwave oven is turned ON, the noise level in the affected frequency range fluctuates between very low and very high noise levels, increasing the fraction of time in which the medium remains free.

In high interference scenarios, the use of local processing becomes even more important. Each node obtains data from two current sensors and two voltage sensors with an acquisition rate of 3 kHz, where each value obtained from the sensors has 10 bits. To calculate a value of air-gap torque, a full cycle of voltage and current is needed. Therefore, to obtain one efficiency value and one torque value, 50 samples are required from each sensor. Let Q be the number of bits that must be transmitted to estimate the values, we have that $Q = 4 \times 50 \times 10 = 2000$ bits.

With local processing, to obtain one torque value and one efficiency value, it is always necessary to transmit a constant number of bits, regardless the ADC's acquisition rate and resolution. In our case, 4 bytes (32 bits) are used to store a value of torque and efficiency. Therefore, it is necessary only one packet to transmit such data. When local processing is not used, it is necessary to transmit more packets through the WSN. Whereas the packet payload is 118 bytes (944 bits), the number of packets that must be transmitted, (Q_p), is $Q_p = \lfloor \frac{Q}{944} \rfloor + 1 = 3$.

In the best case, the total number of transmitted packets is three times larger in the scenario without local processing. However, when considering an unreliable transmission medium, the difference can be greater. The transmission of a packet consists of a Bernoulli event with successful probability p, and the number of trials until the first success is defined by a Geometric distribution. In a geometric distribution, the average number of events until the first success is $\frac{1}{p}$.

Since $Q_p = 3$, when local processing is not used the probability of successfully transmit the data necessary to estimate the values is p^3, and the number of attempts before the first success is $\frac{1}{p^3}$, considering a scenario without retransmission of lost packets. As in each attempt 3 packets are transmitted, then we have an average of $\frac{3}{p^3}$ transmissions per efficiency and torque value obtained.

When local processing is not employed, packet retransmission improves performance, since only individual packets with errors need to be retransmitted. However, recurrent retransmissions increase overall delay, possibly failing to obtain the most current data. Considering that the successful probability of an acknowledgment packet is also p, then the successful probability of transmit a packet becomes p^2, and the average number of transmissions to the first hit is $\frac{3}{p^2}$. In the scenario with local processing and retransmission, the average number of transmissions until the first hit is $\frac{1}{p^2}$. Therefore, when using local processing, retransmissions increase the amount of transmitted packets.

Table 4 Transmissions

Average number of transmissions.			
Configuration	LP	LPR	NLPR	NLP
$p = 0.1$	10	100	300	3000
$p = 0.2$	5	25	75	375
$p = 0.3$	3.3	11.1	33.3	111.1
$p = 0.4$	2.5	6.25	18.75	46.87
$p = 0.5$	2	4	12	24

LP - with local processing
LPR - with local processing and retransmission
NLPR - without local processing and retransmission
NLP - without local processing

Table 4 shows the average number of transmissions required to obtain a torque value and an efficiency value, for various values of p in each scenario. From the table, we note that in high interference scenarios, there is a large reduction in the amount of data being transmitted when using local processing. When the PER is too high, it is very difficult to perform monitoring without using local processing. For example, in the scenario with an IEEE 802.11g network operating on channel 6, we see that the average PER for node $N1$ (on channel 18) was 90%. In this scenario, without local processing it would be required, on average, 300 transmissions to obtain data on the target, considering the use of retransmissions. Using local processing without retransmission, only 10 transmissions are needed in average. Node N2 presents best performance in this scenario, but the PER variance was high (up to 85%).

Moreover, as a torque and an efficiency value occupy only 4 bytes, we can aggregate multiple measures in a single packet. The embedded system was configured to transmit a set of 18 measures of torque and 18 measures of efficiency in a single packet, with a total payload of 72 bytes. Besides that, it is important to note that due to an increase in the number of transmitted packets, considering an approach without local processing, the packet error rate tends to increase.

From these results, we note that the deployment of industrial WSN still presents serious challenges related to communication reliability. In order to keep a certain level of quality of service, radios need to be aware of the environment where they are operating, adopting a dynamic spectrum allocation approach, not only at the beginning of its operation, but also during the entire period of operation. When there are changes in the distribution of spectrum usage, probably resulting from other sources of interference, the distribution of spectrum usage along the available channels, and the communication performance on each channel, will change, what may imply on switching to a less polluted channel.

Despite high PER in some scenarios, it is important to note that due to local processing capability of the embedded systems, all the data arriving at the coordinator are useful information that can be employed for decision-making. Without local processing, probably it would not have been possible to obtain useful information via the WSN, in some scenarios.

Some studies [100, 101] have proposed solutions for mitigating the interference effects in IEEE 802.15.4 networks. By combining the local processing capability, as explored in this work, with dynamic spectrum allocation techniques and techniques for mitigating the effects of interference, it may be possible to achieve a good quality of service in motor monitoring applications based on WSNs.

10.5 Conclusions

In this chapter, we assessed the use of WSN technology for the implementation of motor monitoring systems in industrial environments. Besides, we presented a WSN for monitoring torque, efficiency and speed of induction motors. All aspects of the system's implementation were discussed, including methods for estimating the parameters and challenges for the implementation and deployment of embedded systems and wireless sensor networks in industial environment.

To implement our system, the Air Gap Torque (AGT) method is used to estimate the shaft torque and motor efficiency. The computations for estimating the targeted values are done locally and then transmitted to a monitoring base unit through an IEEE 802.15.4 Wireless Sensor Network (WSN).

Experimental tests were performed to analyze the torque values obtained by the system, and then compared with torque values based on the workbench dynamic model. The estimated efficiency was compared with a reference value, presenting an error smaller than 2.0% in the range of 085% loading. This An experimental study is conducted aiming at identifying the relation between spectral occupancy and Packet Error Rate (PER) for the proposed WSN. The experiments were conducted inside a shed, with typical characteristics of industrial environments.

The study demonstrated that the addition of new interference sources can significantly affect the spectral occupancy, by also having a direct impact on the communication performance. Even with the difficulties in data transmission using the WSN in some scenarios, the system was able to provide useful monitoring information, since all processing is done locally (i.e., only the information already computed is transmitted over the network). Without local processing, it might be impossible to use the WSN technology for this particular application, considering an unreliable transmission medium. In addition to the local processing capacity, other techniques can be developed to mitigate interference in those environments, leading to better communication performance.

References

1. Hanitsch, R.: Energy Efficient Electric Motors. In: RIO 2002 - World Climate & Energy Event, pp. 6–11 (2002)
2. Kim, K., Parlos, A.G.: Induction Motor Fault Diagnosis Based on Neuropredictors and Wavelet Signal Processing. IEEE/ASME Transactions on Mechatronics 7, 201–219 (2002)
3. Lima-Filho, A.C., Belo, F.A., Gomes, R.D.: Tests prove, self-powered, wireless, pump torquemeter. Oil and Gas Journal 106, 43–48 (2008)
4. Reich, R.B.: Rotary Transformer, U.S. Patent No. 4,412,198 (1983)
5. Buchele, W.F.: Strain-Gauge Brushless Torque-Meter, U.S. Patent No. 3,881,347 (1975)
6. Meng, Z., Liu, B.: Research on Torque Real Time Monitoring System of Rotary Machine. Chin. J. Sci. Instrum. 26, 38–39 (2005)
7. Lima-Filho, A.C., Belo, F.A., Santos, J.L.A., Anjos, E.G.: Experimental and theoretical study of a telemetric dynamic torque meter. Journal of the Brazilian Society of Mechanical Sciences and Engineering 32, 241–249 (2011)
8. Hsu, J.S., Amin, A.M.A.: Torque calculations of current-source induction machines using the 1-2-0 coordinate system. IEEE Trans. on Industrial Electronics 37, 34–40 (1990)
9. Hsu, J.S.: Capacitor effects on induction motors fed by quasi rectangular current sources. IEEE Trans. on Energy Conversion 7, 509–516 (1992)

10. Lima-Filho, A.C., Gomes, R.D., Adissi, M.O., Silva, T.A.B., Belo, F.A., Spohn, M.A.: Embedded System Integrated Into a Wireless Sensor Network for Online Dynamic Torque and Efficiency Monitoring in Induction Motors. IEEE/ASME Trans. on Mechatronics 17, 404–414 (2012)

11. Lee, T.H., Low, T.S., Tseng, K.J., Lim, H.K.: An Intelligent Indirect Dynamic Torque Sensor for Permanent Magnet Brushless DC Drives. IEEE Trans. on Industrial Electronics 41, 191–200 (1994)

12. Lima-Filho, A.C., Belo, F.A., Santos, J.L.A., Anjos, E.G.: Self-Powered Telemetric Torque Meter. Journal of Dynamic Systems, Measurement, and Control 133, 1–7 (2011)

13. IEEE Standard Test Procedure for Polyphase Induction Motors and Generators, IEEE Standard 112-1996 (2004)

14. Hsu, J.S., Kueck, J.D., Olszewski, M., Casada, D.A., Otaduy, P.J., Tolbert, L.M.: Comparison of Induction Motor Field Efficiency Evaluation Methods. IEEE Trans. on Industry Applications 34, 117–125 (1998)

15. Gandhi, A., Corrigan, T., Parsa, L.: Recent Advances in Modeling and Online Detection of Stator Interturn Faults in Electrical Motors. IEEE Trans. on Industrial Electronics 58, 1564–1575 (2011)

16. Antonino-Daviu, J., Aviyente, S., Strangas, E.: Scale Invariant Feature Extraction Algorithm for the Automatic Diagnosis of Rotor Asymmetries in Induction Motors. IEEE Trans. on Industrial Informatics 9, 100–108 (2012)

17. Bouzida, A., Touhami, O., Ibtioen, R., Belouchrani, A., Fadel, M., Rezzoug, A.: Fault Diagnosis in Industrial Induction Machines Through Discrete Wavelet Transform. IEEE Trans. on Industrial Electronics 58, 4385–4395 (2011)

18. Suratsayadee, K., Himanshu, J., Lee, W., Kwan, C.: Wireless health monitoring system for vibration detection of induction Motors. In: IEEE Industrial and Commercial Power Systems Technical Conference, pp. 1–6 (2010)

19. Zhang, H., Zanchetta, P., Bradley, K.J., Gerada, C.: A Low-Intrusion Load and Efficiency Evaluation Method for In-Service Motors Using Vibration Tests With an Accelerometer. IEEE Trans. on Industry Applications 46, 1341–1349 (2010)

20. Pillay, P., Xu, Z.: Labview Implementation of Speed Detection for mains-fed motors using motor current signature analysis. IEEE Power Engineering Review 18, 47–48 (1998)

21. Ghomson, W.T., Fenger, M.: Current Signature Analysis to Detect Induction Motor Faults. IEEE Industry Applications Magazine 7, 26–34 (2001)

22. Schoen, R.R., Habetler, T.G., Kamran, F., Bartfield, R.G.: Motor Bearing Damage Detection Using Stator Current Monitoring. IEEE Transactions on Industry Applications 31, 1274–1279 (1995)

23. Riera-Guasp, M., Pineda-Sanchez, M., Perez-Cruz, J., Puche-Panadero, R., Roger-Folch, J., Antonino-Daviu, J.A.: Diagnosis of Induction Motor Faults via Gabor Analysis of the Current in Transient Regime. IEEE Transactions on Instrumentation and Measurement 61, 1583–1596 (2012)

24. Pineda-Sanchez, M., Riera-Guasp, M., Antonino-Daviu, J.A., Roger-Folch, J., Perez-Cruz, J., Puche-Panadero, R.: Diagnosis of Induction Motor Faults in the Fractional Fourier Domain. IEEE Transactions on Instrumentation and Measurement 59, 2065–2075 (2010)

25. Torkaman, H., Afjei, E., Yadegari, P.: Static, Dynamic, and Mixed Eccentricity Faults Diagnosis in Switched Reluctance Motors Using Transient Finite Element Method and Experiments. IEEE Transactions on Magnetics 48, 2254–2264 (2012)
26. Silva, A., Povinelli, R.J., Demerdash, N.A.O.: Rotor Bar Fault Monitoring Method Based on Analysis of Air-Gap Torques of Induction Motors. IEEE Transactions on Industrial Informatics 9, 2274–2283 (2013)
27. Gyftakis, K.N., Spyropoulos, D.V., Kappatou, J.C., Mitronikas, E.D.: A Novel Approach for Broken Bar Fault Diagnosis in Induction Motors Through Torque Monitoring. IEEE Transactions on Energy Conversion 28, 267–277 (2013)
28. Filippetti, F., Franceschini, G., Tassoni, C., Vas, P.: Recent developments of induction motor drives fault diagnosis using AI techniques. IEEE Transactions on Industrial Electronics 47, 994–1004 (2000)
29. Su, H., Chong, K.-T.: Induction Machine Condition Monitoring Using Neural Network Modeling. IEEE Transactions on Industrial Electronics 54, 241–249 (2007)
30. Seshadrinath, J., Singh, B., Panigrahi, B.K.: Investigation of Vibration Signatures for Multiple Fault Diagnosis in Variable Frequency Drives Using Complex Wavelets. IEEE Transactions on Power Electronics 29, 936–945 (2014)
31. Betta, G., Liguori, C., Paolillo, A., Pietrosanto, A.: A DSP-based FFT-analyzer for the fault diagnosis of rotating machine based on vibration analysis. IEEE Transactions on Instrumentation and Measurement 51, 1316–1322 (2002)
32. Washington State University-Energy Program, In-Service Motor Testing, research report E99-040 (1999)
33. Kueck, J.D., Olszewski, M., Casada, D.A., Hsu, J.S., Otaduy, P.J., Tolbert, L.M.: Assessment of methods for estimating motor efficiency and load under field conditions. Lockheed Martin Energy Research Corp., Oak Ridge, TN, Rep. ORNL/TM-13165 (1996)
34. Hsu, J.S., Scoggins, B.P.: Field test of motor efficiency and load changes through air-gap torque. IEEE Trans. on Energy Conversion 10, 471–477 (1995)
35. Lu, B., Habetler, T.G., Harley, R.G.: A Nonintrusive and In-Service Motor-Efficiency Estimation Method Using Air-Gap Torque With Considerations of Condition Monitoring. IEEE Trans. on Industry Applications 44, 1666–1674 (2008)
36. Lu, B., Gungor, V.C.: Online and Remote Motor Energy Monitoring and Fault Diagnostics Using Wireless Sensor Networks. IEEE Trans. on Industrial Electronics 56, 4651–4659 (2009)
37. Baronti, P., Pillai, P., Chook, V.W.C., Chessa, S., Gotta, A., Hu, Y.F.: Wireless sensor networks: A survey on the state of the art and the 802.15.4 and ZigBee Standards. Computer Communications 30, 1655–1695 (2007)
38. Salvadori, F., Campos, M., Sausen, P.S., de Camargo, R.F., Gehrke, C.S., Rech, C., Spohn, M.A., Oliveira, A.C.: Monitoring in Industrial Systems Using Wireless Sensor Network With Dynamic Power Management. IEEE Trans. on Instrumentation and Measurement 58, 3104–3111 (2009)
39. Willig, A.: Recent and Emerging Topics in Wireless Industrial Communications: A Selection. IEEE Trans. on Industrial Informatics 4, 102–124 (2008)

40. Gungor, V.C., Hancke, G.P.: Industrial Wireless Sensor Networks: Challenges, Design Principles, and Technical Approaches. IEEE Trans. on Industrial Electronics 56, 4258–4265 (2009)
41. Hou, L., Bergmann, N.W.: Novel Industrial Wireless Sensor Networks for Machine Condition Monitoring and Fault Diagnosis. IEEE Trans. on Instrumentation and Measurement 61, 2787–2798 (2012)
42. Hou, L., Bergmann, N.W.: Induction motor condition monitoring using industrial wireless sensor networks. In: 2010 Sixth International Conference on Intelligent Sensors, Sensor Networks and Information Processing, pp. 49–54 (2010)
43. Hu, J.: The Application of Wireless Sensor Networks to In-Service Motor Monitoring and Energy Management. In: First International Conference on Intelligent Networks and Intelligent Systems, pp. 165–169 (2008)
44. Hu, J.: In-service motor monitoring and energy management system based on wireless sensor networks. In: International Conference on Electrical Machines and Systems, pp. 823–826 (2008)
45. Esfahani, E.T., Wang, S., Sundararajan, V.: Multisensor Wireless System for Eccentricity and Bearing Fault Detection in Induction Motors. IEEE/ASME Trans. on Mechatronics, 1–9 (2013)
46. Wei, H., Chen, Y., Tan, J., Wang, T.: Sambot: A Self-Assembly Modular Robot System. IEEE/ASME Trans. on Mechatronics 16, 745–757 (2011)
47. Takahashi, J., Yamaguchi, T., Sekiyama, K., Fukuda, T.: Communication Timing Control and Topology Reconfiguration of a Sink-Free Meshed Sensor Network With Mobile Robots. IEEE/ASME Trans. on Mechatronics 14, 187–197 (2009)
48. Santos, J.L.A., Araujo, R.C.C., Lima-Filho, A.C., Belo, F.A., Lima, J.A.G.: Telemetric System for Monitoring and Automation of Railroad Networks. Transportation Planning and Technology 34, 593–603 (2011)
49. Nagornyy, A., Wallace, A.K., Jouanne, A.V.: Stray Load Losses Efficiency Connections. IEEE Industry App. Magazine 10, 62–69 (2004)
50. Meng, Z., Liu, B.: Research on the Laser Doppler Torque Sensor. Journal of Physics: Conference Series 48, 202–206 (2006)
51. National Electrical Manufacturers Association (NEMA), Standards Publication ANSI/NEMA MG 1, Motors and Generators, Rosslyn, VA, USA (2009)
52. Lu, B., Habetler, T.G., Harley, R.G.: A survey of efficiency-estimation methods for in-service induction Motors. IEEE Trans. Industrial Applications 42, 924–933 (2006)
53. Gulez, K., Adam, A.A., Pastaci, H.: A novel direct Torque Control Algorithm for IPMSM With Minimum Harmonics and Torque Ripples. IEEE/ASME Trans. on Mechatronics 12, 223–227 (2007)
54. Ozturk, S.B., Toliyat, H.A.: Direct Torque and Indirect Flux Control of Brushless DC Motor. IEEE/ASME Trans. on Mechatronics 16, 351–360 (2011)
55. Nguyen, Q.D., Ueno, S.: Modeling and Control of Salient-Pole Permanent Magnet Axial-Gap Self-Bearing Motor. IEEE/ASME Trans. on Mechatronics 16, 518–526 (2011)
56. Akyildiz, I.F., Vuran, M.C.: Wireless Sensor Networks, 1st edn. Wiley (2010)
57. Verdone, R., Dardari, D., Mazzini, G., Conti, A.: Wireless Sensor and Actuator Networks: Technologies, Analises and Design, 1st edn. Elsevier (2007)
58. Fette, B., Aiello, R., Chandra, P., Dobkin, D., Bensky, D., Miron, D., Lide, D., Dowla, F., Olexa, R.: RF & Wireless Technologies: Know It All. Elsevier (2007)

59. Gomes, R.D., Adissi, M.O., Lima-Filho, A.C., Spohn, M.A., Belo, F.A.: On the Impact of Local Processing for Motor Monitoring Systems in Industrial Environments Using Wireless Sensor Networks. Int. Journal of Distributed Sensor Networks 2013, 1–14 (2013)
60. Willig, A., Matheus, K., Wolisz, A.: Wireless Technology in Industrial Networks. Proceedings of the IEEE 93, 1130–1151 (2005)
61. Tanghe, E., Joseph, W., Verloock, L., Martens, L., Capoen, H., Herwegen, K.V., Vantomme, W.: The Industrial Indoor Channel: Large-Scale and Temporal Fading at 900, 2400, and 5200 MHz. IEEE Trans. on Wireless Communications 7, 2740–2751 (2008)
62. Stenumgaard, P., Chilo, J., Angskog, P.: Challenges and Conditions for Wireless Machine-to-Machine Communications in Industrial Environments. IEEE Communications Magazine 51, 187–192 (2013)
63. Ferrer-Coll, J., Angskog, P., Chilo, J., Stenumgaard, P.: Characterisation of highly absorbent and highly reflective radio wave propagation environments in industrial applications. IET Communications 6, 2404–2412 (2012)
64. Angskog, P., Karlsson, C., Coll, J.F., Chilo, J., Stenumgaard, P.: Sources of disturbances on wireless communication in industrial and factory environments. In: Asia-Pacific Symposium on Electromagnetic Compatibility, pp. 281–284 (2010)
65. Guo, W., Healy, W.M., Zhou, M.: An Experimental Study of Interference Impacts on ZigBee-based Wireless Communication Inside Buildings. In: Proceedings of the 2010 IEEE International Conference on Mechatronics and Automation, pp. 1982–1987 (2010)
66. Gomes, R.D., Spohn, M.A., Lima-Filho, A.C., Anjos, E.G., Belo, F.A.: Correlation between Spectral Occupancy and Packet Error Rate in IEEE 802.15.4-based Industrial Wireless Sensor Networks. IEEE Latin America Transactions 10, 1312–1318 (2012)
67. Sikora, A., Groza, V.: Coexistence of IEEE 802.15.4 with other Systems in the 2.4 GHz ISM Band. In: Instrumentation and Measurement Technology Conference, pp. 1786–1791 (2005)
68. Akyildiz, I.F., Lee, W., Vuran, M.C., Mohanty, S.: Next generation/dynamic spectrum access/cognitive radio wireless networks: a survey. Computer Networks Journal 50, 2127–2159 (2006)
69. Karl, H., Willig, A.: Protocols and Architectures for Wireless Sensor Networks, 1st edn. Wiley (2005)
70. Cao, L., Jiang, W., Zangh, Z.: Network wireless Meter Reading System Based on ZigBee Technology. In: IEEE Control and Decision Conference (2008)
71. Lee, J.: Performance Evaluation of IEEE 802.15.4 for Low-rate Wireless Personal Area Networks. IEEE Trans. on Consumer Electronics 52, 742–749 (2006)
72. Cheong, P., Chang, K., Lai, Y., Ho, S., Sou, I., Tam, K.: A zigbee-based wireless sensor network node for ultraviolet detection of flame. IEEE Trans. on Industrial Electronics 58, 5271–5277 (2011)
73. Hanzálek, Z., Jurcík, P.: Energy Efficient Scheduling for Cluster-Tree Wireless Sensor Networks With Time-Bounded Data Flows: Application to IEEE 802.15.4/ZigBee. IEEE Trans. on Industrial Informatics 6, 438–450 (2010)
74. Caione, C., Brunelli, D., Benini, L.: Distributed Compressive Sampling for Lifetime Optimization in Dense Wireless Sensor Networks. IEEE Trans. on Industrial Informatics 8, 30–40 (2012)

75. Ascariz, J.M.R., Boquete, L.: System for measuring power supply parameters with zigbee connectivity. In: IEEE Instrumentations Measurement Technology Conference, pp. 1–5 (2010)
76. Microchip - MiWi Protocol (2013), http://www.microchip.com/miwi/
77. Song, J., Han, S., Mok, A.K., Chen, D.: Wirelesshart: Applying wireless technology in real-time industrial process control. In: IEEE Real-Time and Embedded Technology and Applications Symposium, pp. 377–386 (2008)
78. Petersen, S., Carlsen, S.: Performance evaluation of wirelesshart for factory automation. In: IEEE Conference on Emerging Technologies & Factory Automation, pp. 1–9 (2009)
79. Frey, J., Lennvall, T.: Wireless Sensor Networks for Automation. In: Zurawski, R. (ed.) Networked Embedded Systems, vol. 27, pp. 1–27. CRC Press (2009)
80. ISA100, Wireless Systems for Automation (2013), http://www.isa.org/isa100
81. Flammini, A., Marioli, D., Sisinni, E., Taroni, A.: Design and Implementation of a Wireless Fieldbus for Plastic Machineries. IEEE Trans. on Industrial Electronics 56, 747–755 (2009)
82. Noergaard, T.: Embedded Systems Architecture: A Comprehensive Guide for Engineers and Programmers, 1st edn. Elsevier (2005)
83. Wolf, W.: Computers as Components: Principles of Embedded Computing System Design, 1st edn. Elsevier (2008)
84. Simon, D.E.: An Embedded Software Primer, 1st edn. Pearson (1999)
85. Boano, C.A., Tsiftes, N., Voigt, T., Brown, J., Roedig, U.: The Impact of Temperature on Outdoor Industrial Sensornet Applications. IEEE Trans. on Industrial Informatics 6, 451–459 (2010)
86. Bello, L., Toscano, E.: Coexistence Issues of Multiple Co-located IEEE 802.15.4/Zigbee Networks Running on Adjacent Radio Channels in Industrial Environments. IEEE Trans. on Industrial Informatics 5, 157–167 (2009)
87. Tang, L., Wang, K., Huang, Y., Gu, F.: Channel Characterization and Link Quality Assessment of IEEE 802.15.4-compliant Radio for Factory Environments. IEEE Trans. on Industrial Informatics 3, 99–110 (2007)
88. Gungor, V.C., Lu, B., Hancke, G.P.: Opportunities and challenges of wireless sensor networks in smart grid. IEEE Trans. on Industrial Electronics 57, 3557–3564 (2010)
89. Akerberg, J., Gidlund, M., Bjorkman, M.: Future Research Challenges in Wireless Wensor and Actuator Networks Targeting Industrial Automation. In: IEEE International Conference on Industrial Informatics, pp. 410–415 (2011)
90. Hu, J., Wu, B.: New Integration Algorithms for Estimating Motor Flux Over a Wide Speed Range. IEEE Trans. on Power Electronics 13, 969–977 (1998)
91. Zerbo, M., Sicard, P., Ba-razzouk, A.: Accurate Adaptative Integration Algorithms for Induction Machine Drive Over a Wide Speed Range. In: IEEE International Conference on Electric Machines and Drives, pp. 1082–1088 (2005)
92. Lee, S.B., Habetler, T.G., Harley, R.G., Gritter, D.J.: An Evaluation of Model-Based Stator Resistance Estimation for Induction Motor Stator Winding Temperature Monitoring. IEEE Transactions on Energy Conversion 17, 7–15 (2002)
93. Habetler, T.G., Profumo, F., Griva, G., Pastorelli, M., Bettini, A.: Stator Resistance Tuning in a Stator-Flux Field-Oriented Drive Using an Instantaneous Hybrid Flux Estimator. IEEE Trans. Power Electronics 13, 125–133 (1998)
94. Jacobina, C.B., Filho, J.E.C., Lima, A.M.N.: On-line Estimation of the Stator Resistance of Induction Machines based on Zero Sequence Model. IEEE Trans. on Power Electronics 15, 346–353 (2000)

95. Bharadwaj, R.M., Parlos, A.G.: Neural Speed Filtering for Induction Motors with Anomalies and Incipient Faults. IEEE/ASME Trans. on Mechatronics. 9, 679–688 (2004)
96. Ishida, M., Iwata, K.: Steady-state characteristics of a torque and speed control system of an induction motor utilizing rotor slot harmonics for slip frequency sensing. IEEE Trans. on Power Electronics 2, 257–263 (1987)
97. Ferrah, A., Bradley, K., Asher, G.: Sensorless speed detection of inverter fed induction motors using rotor slot harmonics and fast Fourier transform. In: IEEE-PESC Conf. Rec., vol. 1, pp. 279–286 (1992)
98. Hurst, K.D., Habetler, T.G.: Sensorless speed measurement using current harmonic spectral estimation in induction machine drivers. IEEE Trans. on Power Electronics 11, 66–73 (1996)
99. Airview (2013), http://www.ubnt.com/airview
100. Huang, J., Xing, G., Zhou, G., Zhou, R.: Beyond Co-existence: Exploiting WiFi White Space for ZigBee Performance Assurance. In: IEEE International Conference on Network Protocols, pp. 305–314 (2010)
101. Liang, C., Priyantha, N., Liu, J., Terzis, A.: Surviving Wi-Fi Interference in Low Power ZigBee Networks. In: ACM Conference on Embedded Networked Sensor Systems, pp. 309–322 (2010)

Chapter 11
Advanced Monitoring System on Debris Flow Hazards

Y.-M. Huang, Y.-M. Fang, and T.-Y. Chou

Abstract. The debris flow had become a common natural hazard in Taiwan for past years. The Taiwan government had contributed a lot effort on debris flow hazard mitigation. The monitoring system of debris flow had played an important role in the debris flow disaster response. In this chapter, the debris flow monitoring system in Taiwan was well described, including the monitoring sensors and instruments, communication methods, data collection, and power system. Different types of monitoring stations were also introduced, including the advanced monitoring device, the portable unit. A case of debris flow in Shenmu, Taiwan, was addressed to explain the debris flow monitoring system and the results. With the advantages from the intelligent debris flow monitoring system, the debris flow disaster response has been enhanced in terms of efficiency. The advanced monitoring system has been a great benefit to the public regarding debris flow warning and hazard prevention.

11.1   Introduction

Locating at the Pacific-Rim Seismic Zone, Taiwan has geology of numerous faults, steep slopelands and shallow soil layers. Geographically located in the path of monsoons and typhoons in the West Pacific, Taiwan has frequently faced typhoons, which bring abundant rainfall, and earthquakes. All these natural phenomena are highly likely to cause debris flows in susceptible areas.

The debris flow has become a common hazard in Taiwan in the last decade. The disaster of debris flow has usually caused serious damage to the infrastructures and people's property, especially at the mountain areas. In order to

Y.-M. Huang · Y.-M. Fang · T.-Y. Chou
GIS Research Center, Feng Chia University,
Taichung, Taiwan

© Springer International Publishing Switzerland 2015
H. Leung and S.C. Mukhopadhyay (eds.), *Intelligent Environmental Sensing,*
Smart Sensors, Measurement and Instrumentation 13, DOI: 10.1007/978-3-319-12892-4_11

monitor the debris flows and prevent hazards, the Soil and Water Conservation Bureau (SWCB), the Taiwan authority responsible for debris flow mitigation, had established debris flow monitoring stations since 2002. Currently there are 19 debris flow monitoring stations (Figure 1) in Taiwan. Most of the stations are located at the central part of Taiwan.

Instruments of rain gauge, water level meter, wire sensor, soil water moisture sensor and CCD camera are used for monitoring debris flows. Each station has a data center to receive and transmit debris flow information from the site to the emergency operation center (EOC). The debris flow monitoring system has successfully assisted the SWCB and the public in response to the debris flow hazards in the past years. The monitoring stations have played an important role in debris flow monitoring in Taiwan. However, because of the restrictions on budget and time scale, it is not reasonable to establish monitoring stations at each location vulnerable to the debris flow.

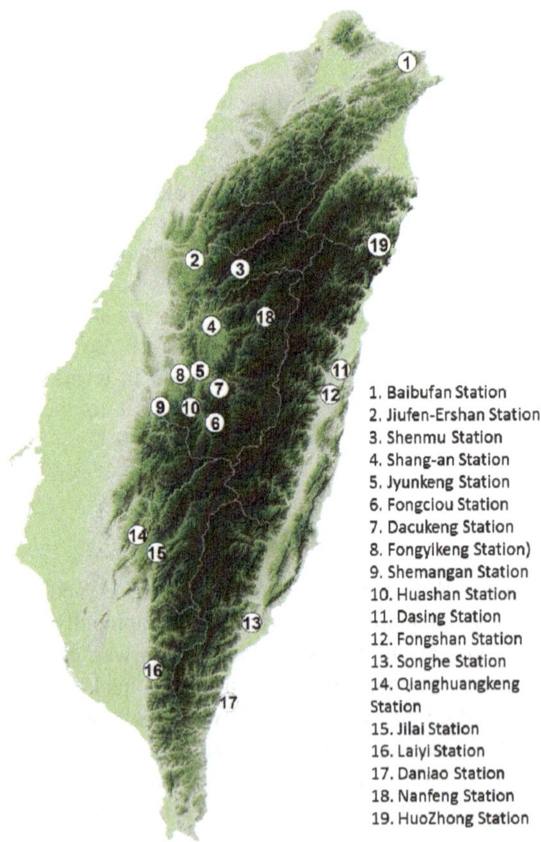

Fig. 1 Debris flow monitoring stations in Taiwan

Fig. 2 Mobile station and portable unit

Thus, the alternatives of mobile stations and portable units (Figure 2) had been developed to extend the capability of debris flow monitoring system. A concept of basin-wide monitoring network [1] had been proposed to enhance the disaster operations for debris flow hazards. All three types of monitoring stations: on-site, mobile and portable ones, work together to construct the debris flow monitoring network (Figure 3). The main purposes of monitoring are to observe and collect debris flow data for studies and provide assistance during disaster response in the rainy seasons.

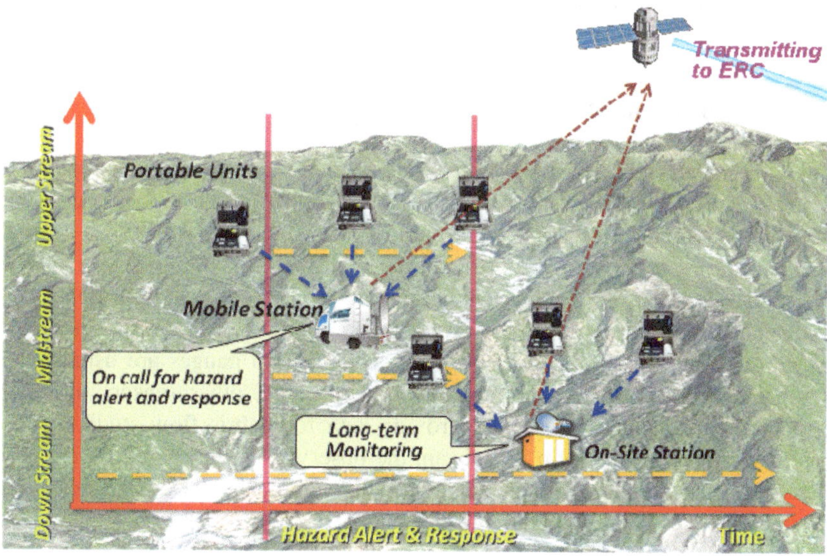

Fig. 3 Basin-wide monitoring network

11.2   Debris Flow

11.2.1 The Causes

Debris flow is a natural phenomenon with triggering conditions of abundant deposits, sufficient water and steep slope. The development of a debris flow consists of the source (producing section), transportation (flowing section) and deposition (settlement section). The source area is often on relatively steep areas at the upstream of valleys. Water currents sharply erode riverbed and side banks, causing landslide to produce a large amount of debris. When water mixes with debris to a certain concentration, certain flow velocity and erosion energy to the riverbed will be created at appropriate grades. In addition, the sediments and woody debris travelling at the front end of a debris flow can also cause considerable damage. A debris flow at this stage is known as in the transportation section. When the debris flow reaches relatively wider river channel or moderate slopes at the downstream, it slows down and becomes gradually dried, with gradually deposition of sediment. Alluvial fan is formed at the opening of a valley or river mouth and is known as the settlement area.

In the past, debris flows were considered as a natural phenomenon as they generally occurred in sparsely populated areas. However, rapid economic development has caused an increase in population and industries in mountainous areas. As a result, debris flows have threatened to people, villages and communities, farms, and public facilities. The serious impact of debris flows has been considered as a disaster to the public.

11.2.2 Observation Work

Because the causes of a debris flow include sufficient water, proper slope, and abundant source, the observation work for debris flow needs to consider what sensors can provide necessary information about the debris flow causes. Compared to the slope and debris source, the rainfall, the source of water, is the easiest to measure and most important data in the monitoring system. Rainfall has been considered an important triggering factor to debris flows [2,3,4]. Therefore, when constructing a debris flow monitoring system, rain gauges are usually the first sensor to be added. In addition to the rain gauge, other sensors like soil moisture sensor and geophone, also provide data for debris flow study. In the debris flow monitoring system in Taiwan, the observation work is classified into three types: direct and indirect, and supporting approaches. Detail description about the sensors is addressed in the next section.

Direct Approach
The direct approach in debris flow monitoring system refers to the instruments and sensors that response when a debris flow actually occurs. These instruments and sensors are wire sensors and geophones. Wire sensors are simply steel wires

acrossing the river with signal transducers at ends. When a debris flow occurs, the flowing mass of debris will break the wire and results in a signal being sent out. Geophones are sensors installed underground to detect ground surface vibration which comes from the movement of debris in the channel.

Indirect Approach

The indirect approach is the application of rainfall data. Most of researches about debris flow had been contributed in developing the relationship between the debris flow occurrence and rainfall. Rainfall is easy to measure and if a reasonable prediction model is using rainfall as a variable, it can provide a time window that can greatly improve the disaster warning and evacuation. Therefore, the monitoring station usually equips with rain gauges to provide important rain distribution with time.

Supporting Approach

The supporting approach includes sensors of CCD camera, soil moisture, water level, and flow speed. CCD provides the real-time images and streaming video for reference when confirmation of a debris flow is needed. The soil moisture sensor, water level meter, and flow speed meter provide local environmental data for further analysis about debris flow.

11.3   Debris Flow Monitoring System

The debris flow monitoring system consists of three major parts: sensors and instruments, communication, and power unit. The sensors and instruments are the most important part of a debris flow monitoring system. Each sensor or instrument measures data from the site, and pass the information to the operation center through the communication network. Different communication options are considered in the system. All of the aforementioned sensors, instruments, data collection, and communication options rely on stable and reliable power supply. The power design and equipment is last key part of a debris flow monitoring system. The following sections describe the details.

11.3.1 Sensors and Instruments

Wire sensor

Wire sensor is simply a steel wire installed crossing the river channel. A signal transducer is set at one end of the wires. Wire sensor will be broken by the front of debris flow and the circuit will be disconnected, sending out the "broken" signal to the operation center. The locations chosen for the wire sensor are usually the spillway of Sabo dam, slit dam, and submerged dam (Figure 4). The elevation of wire sensor is determined based on the design water level and the debris flow histories. Usually a set of two wires are employed on the same section to effectively capture the debris flow movement (Figure 5).

Fig. 4 Setup of wire sensors

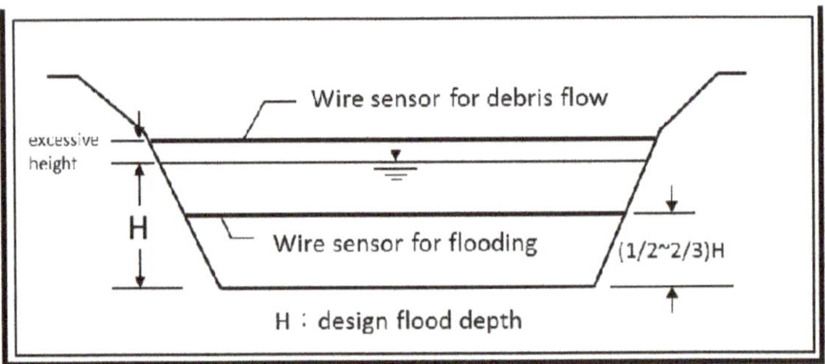

Fig. 5 Installation of wire sensors

Geophone

Debris flows consist of a flowing body of mixed mud, sand, and gravel under the action of gravity. Therefore, when debris flows occur, in addition to the huge low frequency noise, slight vibration of the earth's surface can be felt. This kind of vibration wave, which is transmitted by the earth's crust, is called debris flow underground sound. When the debris flow occurs, the ditch bed undergoes corrosion, attrition, and collision, thus producing vibration wave. The wave transmission in air is only a small portion of the energy and most of the energy is transmitted in the rock stratum of the ditch bed. The primary reason behind the ground vibration is the big gravels striking the bed especially the front-end of debris flows. The front end of debris flows always appears wave-shaped and includes big gravels. The size and density of gravels in rear end are smaller. Therefore, the collision force of front end is stronger.

Based on the results of the relevant research, Chinese scholars adopted piezoelectric geophone (1983-1985) and detected 18 debris-flow underground sounds in Jiangjiagou. Among them, the frequency distribution of the underground sounds of sticky debris flows is 25 to 80Hz, and the significant frequency is around 50Hz. Itakura et al. (1997) [5] detected that most of the frequency of debris flows underground sounds of Nojiri River was between 20 to 60Hz. Besides, according to the indoor and outdoor experimental results [6,7], irrespective of the gravel diameter, the major distribution of acoustic frequency produced by the collision of soil and gravel is between 20 to 80Hz, and most of them are in the range of 40 to 60 Hz. Therefore, the significant frequency of debris flows flow is around 40 to 60Hz. Besides, Huang et al. [8] used stone to strike the river bed of Ye River in Fengqiu area and observed that the frequency of earthquake sounds is between 10 to 150Hz. The wave velocity is around 1000 m/s. Lahusen [9] from U.S. Geological Survey also found that the underground sound frequency of debris flows was between 10 to 300Hz and the significant frequency was between 30 to 80Hz.

Fig. 6 Setup of the geophone

In order to obtain the actual characteristic frequency of debris flows, the geophone (eg., Geo Space GS-20DX type; see Figure 6) is used to capture the underground sound from the 3 axes (along the river, transverse to the river, and the vertical direction, see Figure 7). To record the debris flow underground sound the acquisition frequency is set to 500Hz.

Rain gauge

The main purpose of the rain gauge (see Figure 8) is to measure the real-time rainfall around the monitoring station. The rainfall data will be analysed and compared with the warning level of rainfall. If the measurement reaches the warning level, the monitoring system will notify the operation center and the correspondent procedure will be followed to evacuate the people if necessary.

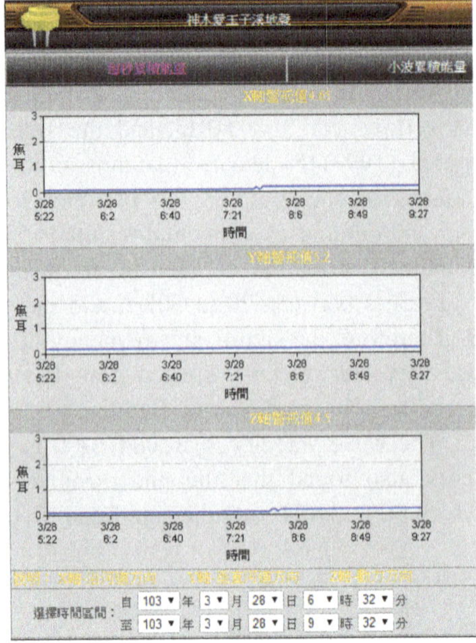

Fig. 7 Records of the accumulated energy per second of underground sounds

Fig. 8 Setup of the rain gauge

The rainfall factors used for debris flow analysis include the rainfall intensity, accumulated rainfall, the antecedent rainfall, and the period of rainfall [10,11]. Based on the practical experience in Taiwan, the rainfall-based warning system of debris flow adopts the combination of rainfall intensity, accumulated rainfall, and the period of rainfall to set up the warning value at each monitored location. In order to achieve the desired warning performance, the rainfall data is suggested to record every 10 minutes in the monitoring system.

CCD Camera

The image capture equipment in the monitoring system employs color and black-and-white mode to capture the best image. These two modes can be switched automatically by the CCD camera. The light compensation device (including visible light and infrared light) is necessary to capture clear image (Figure 9). Through the video server, the analog images are transfered into digital ones and stored on-site. The digital image is then transmitted to the operation center through the satellite communication or ADSL Internet. The local environment and conditions can been displayed directly on the monitoring system (Figure 10).

Fig. 9 CCD camera and infrared spotlight

Soil moisture sensor

Soil moisture sensor is installed underground at the monitoring site to measure the change of water content along slopes (Figure 11). The major purpose of installing soil moisture sensor is to provide useful data for studies about the debris source. Usually the debris source comes from the landslides along the slopes at the upstream area of rivers. With the historical data of soil moisture and the corresponding rainfall, the correlation of occurrence of debris flow, rainfall, and soil moisture can be further studied. In the monitoring system, it is suggested to install the soil moisture sensors at the upstream area where the landslide could occur.

Fig. 10 Illustration of real-time image transmitted from monitoring station

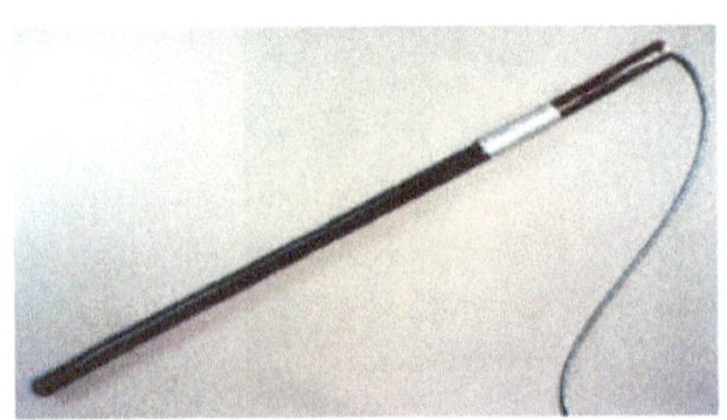

(a) Example of soil moisture sensor

(b) Installation of soil moist sensor

Fig. 11 Soil moisture sensor

Water Level Meter

In addition to measure the debris flow-related data (e.g., the rainfall and wire sensor signal), the monitoring system also measures the water level of streams. The major purpose of measuring the water level is to estimate the flood or debris flow volume which can be considered as a scale index or impact reference of an event. In the monitoring system, the water level meters are suggested to be installed at the up- and mid-stream sections of the site (Figure 12).

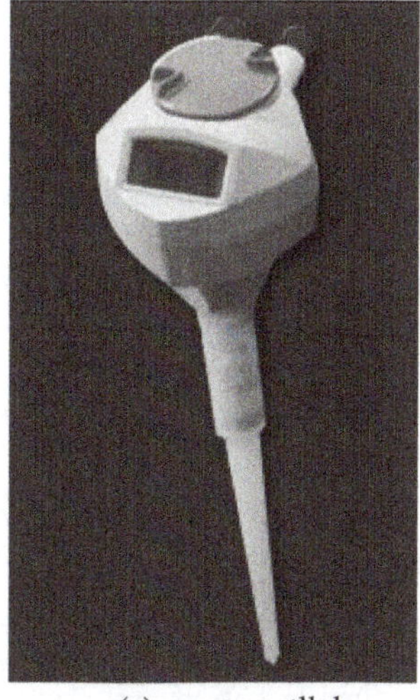

(a) sensor outllok

(b) on-site setup

Fig. 12 Water level and the setup

Flow Speed Meter

The flow speed meter is used with the water level meter to estimate the volume of flood or debris flow. The flow speed meter is usually installed above the river channel (Figure 13) at the position that the sensor can direct detect the water flow. A truss crossing the river is necessary when

(a) speed meter

(b) Installation

Fig. 13 Flow speed meter

The Layout of Sensors:

The layout of sensors considers the location and environment of the monitored debris flow. Usually the sensors of direct approach are arranged on the upper and mid-stream sections. The layout of wire sensors and geophones is usually in a pattern to capture changes at upper and mid-stream. The idea of pairing wire sensors and geophones is to provide a time window of warning before a debris flow damages the downstream area.

Fig. 14 The sensor layout of Jufen-ershan monitoring station

Other sensors, like rain gauge and CCD camera, are placed near the monitoring station or at locations where can overview the torrent. Figure 14 shows an example of sensor layout at a monitoring station of Jufen-ershan in the middle Taiwan.

11.3.2 Data Acquisition and Communication System

Data acquisition
The monitoring system requires a data acquisition device or system to collect and store the observation information from sensors and instruments (Figure 15). The data acquisition devices can be set on the site and in the operation center. The suggested data format and frequency of sensors are shown in Table 1.

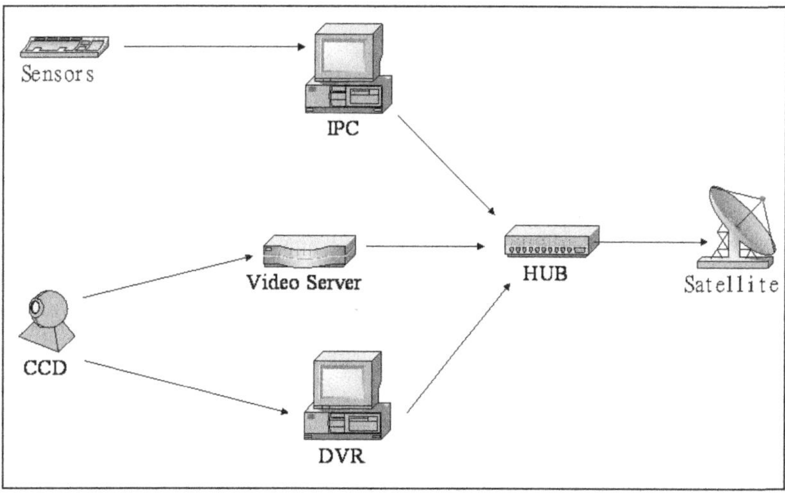

Fig. 15 Setup of the data acquisition center

Table 1 Capture frequency and storage format of each monitoring sensor

Type of sensor	Capture frequency	Storage format	Extension name	Remark
CCD camera	30 images/s (Share for the station)	movie	VGX	All CCD of the station share the same resources (e.g. there are 2 sets of CCD and each CCD records in 15 images / s), and is stored in the information center.
	around 10s per image	ortrait	JPG	Transmit to back-end database instantly
Rain gauge	1min	text file	Txt	Accumulation of rainfall per min
Geophone	500Hz/axis	binary file	D00	
Ultrasonic level transmitter	1Hz	binary file	D01	Ultrasonic level transmitter and pore pressure transducer gauge share the same capture software
Pore pressure transducer gauge				
Wire sensor	change of status	assess database	Mdb	

Communication System

The communication in a monitoring system is very important. Different options of communication can be used in the monitoring work. For debris flow monitoring, the communication methods can consider using satellite transmission, 3G/GPRS, and ADSL (Internet). Among these methods, the satellite transmission is highly recommended as the major communication option in the debris flow monitoring system. Because a debris flow usually occurs in the mountain under bad weather (e.g., heavy rain), the 3G/GPRS signal or ADSL may not stable and reliable given the tough conditions. From the practical experience, the satellite transmission is more reliable during a debris flow event. Other methods of 3G/GPRS and ADSL can be used as backup options. Therefore, it is suggested to employ more than one communication option in the monitoring system. Figure 16 illustrates the satellite transmission.

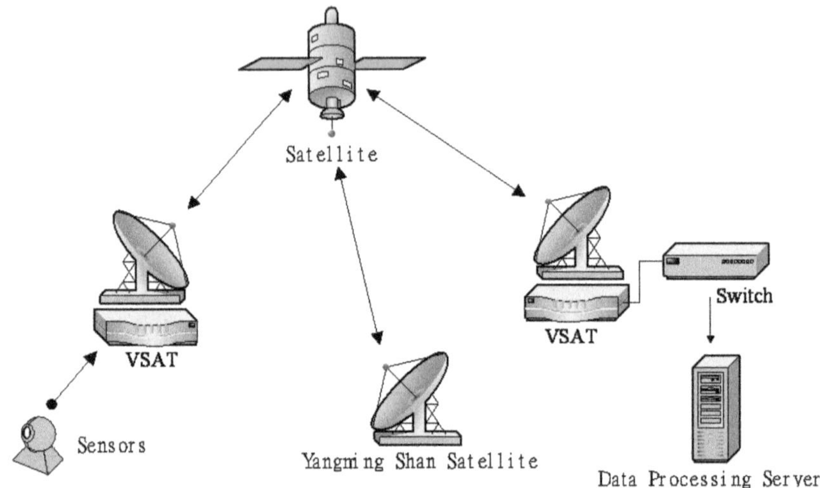

Fig. 16 Structure of the satellite transmission

11.3.3 Monitoring Stations

On-site Stations

The on-site monitoring stations are usually constructed at locations where are highly susceptible to the debris flow. A monitoring station consists of monitoring sensors and instruments, communication equipment, data acquisition devices, and power system. To provide continuous observation from the site, the monitoring station is required to operate non-stoppable under any weather conditions. The power system is the key to achieve the goal. In a monitoring station, a set of high performance batteries and a power generator are prepared as a backup option. In normal days, the power is supplied from local electricity company. If the power outage happens for some reasons, the monitoring operation still goes on with

Fig. 17 Typical monitoring station design

power supply from the batteries and power generator. In the case of Taiwan, the monitoring station can run by itself up to 7 days. The on-site monitoring station is a small scale of debris flow monitoring system at a region. Each monitoring stations can connected through communication to establish a large monitoring network.

In Taiwan, the government stated to build on-site monitoring station since 2002. The locations of stations were determined by an evaluation system. The evaluation considered factors of environmental vulnerability (e.g., slope of stream and landslide area), the impact (e.g., population and infrastructure), and case histories of debris flow. The screening process was applied to each potential torrent of debris flow, and the priority of monitoring was determined. Because of the limited budget from the government, monitoring stations cannot be built on each potential location of debris flow. Currently there are 18 on-site monitoring stations in Taiwan. Most of them are located in the middle Taiwan. Figure 17 shows a typical monitoring station outlook.

Mobile Stations

Because of the budget concern, an alternative to the on-site monitoring station is needed. A mobile station, which is a truck capable of debris flow monitoring, was developed. The mobile station equips same monitoring sensors and devices as an on-site monitoring station. The major difference is that the mobile station can be deployed at locations of interest, and it is an event-driven application. Not like the on-site monitoring station, the mobile station is called on duty when a typhoon is approaching. The deployment of mobile stations is usually determined by estimation based on the rain forecast and prediction of typhoon path.

Since 2003, the SWCB has worked with Feng Chia University to develop mobile stations. Up to now, three mobile stations have been completed and on service (Figure 18).

(a) No 1.

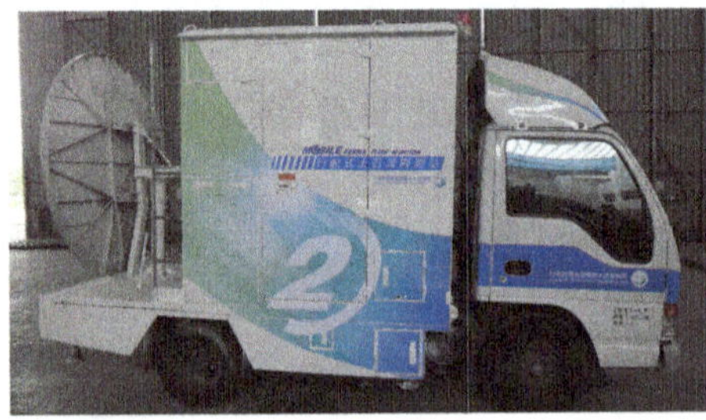

(b) No 2.

(c) No. 3.

Fig. 18 Mobile stations

Portable Units

To accomplish the idea of basin-wide monitoring network, the Taiwan government had worked with the research team of Feng Chia University to develop a compact monitoring device. This device can work with the current on-site and mobile stations to construct a networking covering the remote areas in the upstream, mid- and downstream areas for the debris flow basin. In this section, the development of the new device, called portable unit, will be described in the following sections.

Design Concept

The portable unit needs to match the requirements of easy-to-carry and basic monitoring functions. Thus, the development of portable unit adopts the light-weight, compact size, and self-operation design. In addition, to fulfill the requirement of mobility during an event, the monitoring functions of a portable unit need to start automatically when getting on site. The one-switch design was included to enable monitoring functions, such that anyone can use it and the operation becomes as easy as three steps: carry on, turn it on, and leave it. Communication and power are another two key design points for portable unit. Three communication methods were used and high-efficiency batteries were installed to power the system for three days.

The portable unit was also designed in three types: basic, advanced, and premium types. The basic model includes a rain gauge and a soil moisture sensor. The advanced one has an additional geophone sensor, and the premium type includes a set of CCD camera. Different types of portable units will be used depending on the site condition and the requirement of data collection. These models of portable units provide flexibility in deployment during debris flow monitoring. Fig. 3 shows the outlook of an advanced portable unit.

Overall, the portable unit features basic monitoring functions, compact size, light weight, self-operation, and flexible deployment. More description about the portable unit will be addressed in the following sections.

Instruments

A portable unit can equip 3 monitoring sensors and one CCD camera. Figure 19 shows the instruments installed in a portable unit. The sensors include rain gauge, soil moisture sensor, and geophone. Rain gauge and soil moisture senor are basic monitoring instruments in all three portable models. Geophone is only available in advanced and premium model. The CCD camera is only for premium model, in which extra battery sets are added to make up the additional power consumption from the camera. All these instruments are light-weighted and considerably small in size. Each portable unit can record and transmit data of rainfall and soil moisture, the two most important factors about the debris flow. Additional information is available when using advanced and premium models.

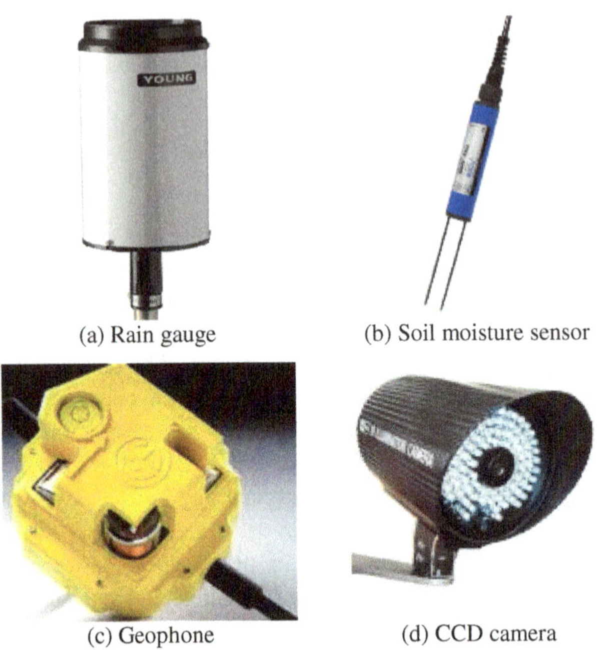

<table>
<tr><td>(a) Rain gauge</td><td>(b) Soil moisture sensor</td></tr>
<tr><td>(c) Geophone</td><td>(d) CCD camera</td></tr>
</table>

Fig. 19 Instruments in the portable unit

Hardware

The hardware of a portable unit includes the case, the computer, batteries, and a radio transmitter. The case of potable unit is in military-spec: strong, tough, water proof, and weather-resistant. The case body has rollers and grips for carrying. A touch-screen portable PC is installed running control system to receive and transmit monitoring data. The power of a portable unit comes from batteries. Two high efficiency batteries of 50 Ah are used to power the unit for three days. The thin film solar panel is also used in the unit to charge the battery when the condition permits. Figure 20 shows the thin film solar cells.

Fig. 20 The membrane solar panel

The last key hardware is a radio transmitter (FM) which is responsible for data communication between the portable units and monitoring stations. The communication methods are summarized in Table 2 and Figure 21 illustrates the transmission options.

Table 2 Communication options of portable unit

Condition	Option	Priority
within the range of ISP service	Use GPRS/3G/3.5G for images and data	1
within 1-2km distance of mobile to monitoring stations	Use 2.4GHz for images, FM for data	2
within 3-5km distance of mobile to monitoring stations	Use FM for data	3

Software
In each unit, system modules of data receiving, data transmission, and data display are implemented to control the monitoring functions. A portable unit captures data per minute, and saves a data set per day, which contains 1440 entries. Each entry of data records rainfall, soil moisture, and geophone data.

The transmission module works to send data to the nearby monitoring stations or to the operation center with predefined frequency and communication option. The module will send all data every 10 minutes. 3G wireless communication will be used when the service is available in the area, and the data will be sent directly to the server in the operation center. If 3G wireless is not available, the FM radio will be used to communicate with monitoring stations and transmit data. The data can be displayed on the screen at the site through the portable PC. Figure 22 shows the data checking screen, and Figure 23 is the portable unit on duty.

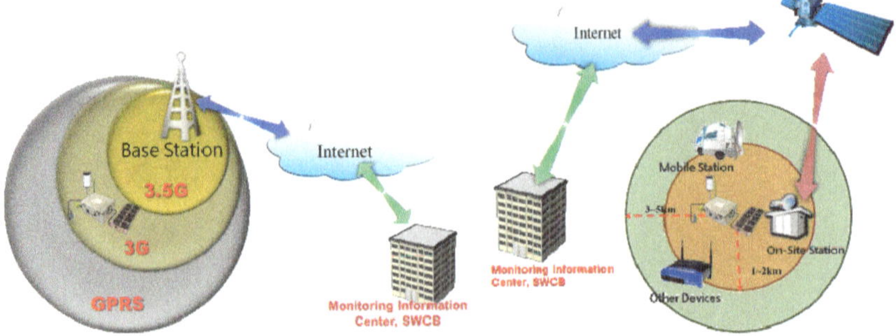

Fig. 21 Illustration of communication options

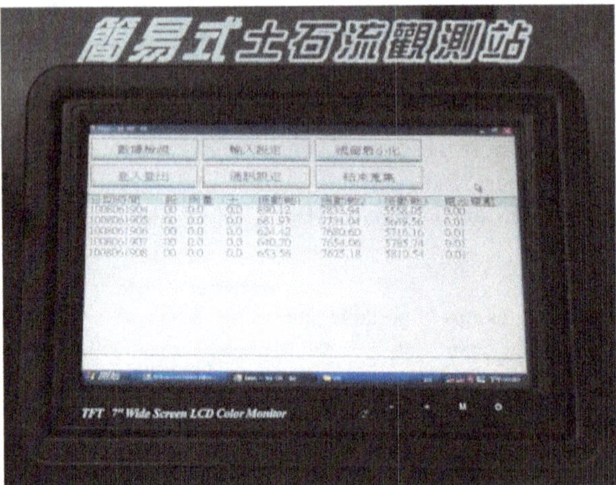

Fig. 22 Checking data status

Fig. 23 Operation of portable unit

11.3.4 Debris Flow Warning

Jan et al [4] and Lee [11] had pointed out the most empirical prediction models of debris flow relied on the rainfall parameters. A good literature review [11] had summarized five rainfall-based methods used for debris flow warning: I-R model, I-T model, R-T model, I-Po model and else, where rainfall intensity (I), accumulated rainfall (R), rainfall duration (T) and antecedent rainfall (Po) were used as rainfall variables in the models. In the previous five models, the I-R and R-T models are the most common methods in debris flow study.

The current rainfall warning method for debris flow in Taiwan was developed using an index of *RTI* which is the product of hourly rainfall (I) and effective accumulated rainfall (R) [10]. The expression of *RTI* is shown below.

$$RTI = I \times R \tag{1}$$

and

$$R = R_0 + \sum_{i=1}^{7} \alpha^i R_i = \sum_{i=0}^{7} \alpha^i R_i \tag{2}$$

where Ro is the pre-debris flow rainfall of an event, Ri is the daily rainfall of previous ith day before current event and α (=0.8) is the weighting factor of daily rainfall. Figure 24 illustrates the rainfall definitions in the equations.

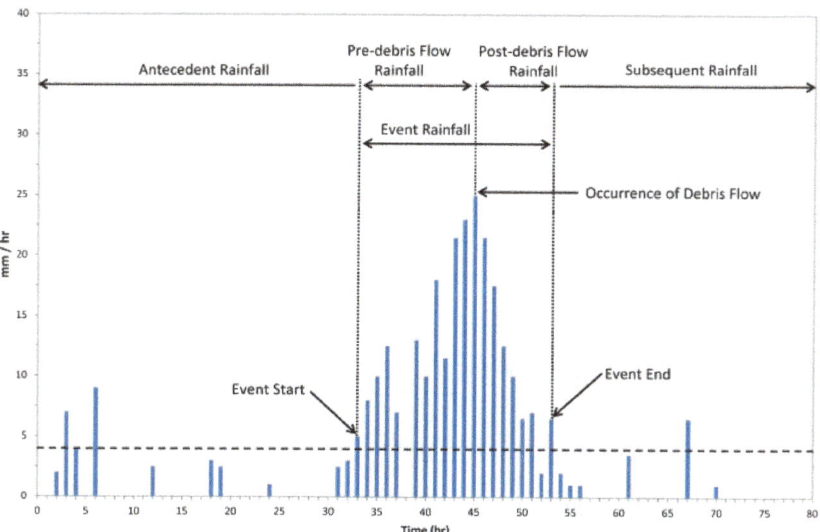

Fig. 24 Schematic diagram of the definition of a rainfall event. [10]

Based on the *RTI* estimation, the rainfall warning of debris flow at each potential location was determined by historical rainfall data and debris flow events. Table 3 summarizes the rainfall warning of debris flow in Taiwan.

11.4 Case Study: Shenmu Area in Taiwan

11.4.1 Environment and Monitoring System in Shenmu Area

The Shenmu area is located at the central Taiwan and a debris flow monitoring station had been established in 2002. The local village is adjacent to the confluence of three streams: Aiyuzi Stream (DF226), Huosa Stream (DF227) and Chushuei Stream (DF199). These streams are classified as potential debris flow torrents with high risk [12,13]. Table 4 summarizes the environmental characteristics around the Shenmu Station. Figure 25 shows the topographic map of Shenmu. Table 5 and Figure 26 indicate the area and locations of landslides

Table 3 The rainfall warning of debris flow in Taiwan (from SWCB, Taiwan)

縣市	鄉鎮(潛流數)	警戒值(mm)
宜蘭縣(143)	三星鄉(5)	600
	大同鄉(41)	550*
	冬山鄉(13)	600
	南澳鄉(10)	450*
	員山鄉(15)	550
	頭城鎮(23)	550
	礁溪鄉(16)	500
	蘇澳鎮(20)	500
基隆市(34)	七堵區(9)	550
	中山區(3)	550
	中正區(3)	550
	仁愛區(2)	600
	安樂區(6)	600
	信義區(5)	600
	暖暖區(6)	550
新北市(221)	八里區(10)	550
	三芝區(4)	500
	三峽區(23)	500
	土城區(4)	550
	中和區(1)	550
	五股區(9)	500
	平溪區(7)	550
	石門區(2)	500
	石碇區(9)	500
	汐止區(9)	500
	坪林區(9)	550
	金山區(8)	500
	泰山區(13)	500
	烏來區(3)	550
	貢寮區(7)	600
	淡水區(4)	500
	深坑區(6)	500
	新店區(20)	500
	新莊區(9)	550
	瑞芳區(25)	500
	萬里區(12)	550
	樹林區(5)	600
	雙溪區(21)	550
	鶯歌區(1)	500
台北市(50)	士林區(7)	500
	中山區(1)	550
	內湖區(12)	500
	文山區(3)	500
	北投區(17)	500
	信義區(5)	600
	南港區(5)	600
桃園縣(51)	大溪鎮(9)	550
	桃園市(2)	550
	復興鄉(30)	350
	龜山鄉(10)	550
新竹縣(76)	五峰鄉(15)	300
	北埔鄉(2)	550
	尖石鄉(26)	300

縣市	鄉鎮(潛流數)	警戒值(mm)
新竹縣(76)	竹東鎮(2)	500
	芎林鄉(4)	550
	峨眉鄉(3)	600
	新埔鎮(1)	500
	橫山鄉(8)	500
	關西鎮(15)	500
苗栗縣(79)	三灣鄉(1)	500
	大湖鄉(16)	500
	公館鄉(4)	500
	竹南鎮(1)	550
	卓蘭鎮(7)	350
	南庄鄉(15)	500
	苑裡鎮(1)	550
	泰安鄉(20)	400
	通霄鎮(3)	550
	獅潭鄉(4)	500
	銅鑼鄉(7)	450
台中市(107)	太平區(9)	400
	外埔區(2)	550
	沙鹿區(1)	600
	和平區(42)	300
	東勢區(20)	300
	新社區(21)	350
	潭子區(1)	550
	霧峰區(8)	500
	北屯區(3)	450
南投縣(249)	中寮鄉(7)	350
	仁愛鄉(34)	250
	水里鄉(32)	250
	名間鄉(2)	500
	竹山鎮(6)	350
	信義鄉(48)	250
	埔里鎮(48)	300
	草屯鎮(7)	450
	國姓鄉(36)	300
	魚池鄉(7)	350
	鹿谷鄉(21)	300
	集集鎮(1)	500
彰化縣(9)	二水鄉(6)	500
	田中鎮(2)	550
	社頭鄉(1)	600
雲林縣(12)	古坑鄉(12)	300
嘉義縣(80)	大埔鄉(4)	450
	中埔鄉(8)	400
	竹崎鄉(22)	350
	阿里山鄉(22)	250
	梅山鄉(14)	300
	番路鄉(10)	450
台南市(48)	六甲區(1)	550
	玉井區(1)	500
	白河區(11)	500
	東山區(16)	350
	南化區(11)	350

縣市	鄉鎮(潛流數)	警戒值(mm)
台南市(48)	楠西區(7)	450
	龍崎區(1)	550
	內門區(3)	550
高雄市(110)	六龜區(31)	300
	田寮區(1)	600
	甲仙區(17)	350
	杉林區(4)	450
	那瑪夏區(14)	250
	岡山區(1)	600
	阿蓮區(1)	600
	美濃區(9)	500
	茂林區(3)	400
	桃源區(16)	250
	旗山區(7)	500
	鼓山區(3)	550
屏東縣(70)	三地門鄉(7)	400
	牡丹鄉(9)	600
	來義鄉(11)	350
	枋山鄉(1)	550
	春日鄉(4)	550
	泰武鄉(5)	550
	高樹鄉(4)	550
	獅子鄉(17)	500
	萬巒鄉(1)	600
	滿州鄉(2)	400
	瑪家鄉(7)	400
	霧台鄉(2)	350
台東縣(165)	大武鄉(21)	450
	太麻里鄉(16)	450
	台東市(5)	550
	成功鎮(9)	500
	池上鄉(2)	600
	卑南鄉(38)	500
	延平鄉(10)	500
	東河鄉(20)	550
	金峰鄉(7)	400
	長濱鄉(4)	500
	海端鄉(18)	600
	鹿野鄉(3)	600
	達仁鄉(8)	450
	關山鎮(4)	600
花蓮縣(167)	玉里鎮(24)	600
	光復鄉(18)	400
	吉安鄉(7)	550
	秀林鄉(28)	450*
	卓溪鄉(15)	600
	花蓮市(2)	600
	富里鄉(8)	600
	瑞穗鄉(9)	500
	萬榮鄉(11)	500
	壽豐鄉(20)	550
	鳳林鎮(9)	450
	豐濱鄉(16)	400

along these three streams. It should be noted that the landslide area (the green-yellow blocks) in Figure 26 was recognized by the satellite image taken after the Typhoon Morakot in 2009 and overlaid with the aerial photo of Shenmu. The landslide area extended and increased after the Typhoon Morakot because of its extremely heavy and record-high rainfall in the area.

Table 4 Environment of Shenmu Station

Location	Shenmu Village, Nantou County	Debris Flow No.	DF199, DF227, DF226
Catchment	Zhuoshui River	Streams	Chusuei, Huosa, Aiyuzi
Debris Flow Warning Rainfall	250 mm	Hazard Type	Channelized debris flow
Monitored Length	5.518 km	Catchment Area	7,216.45 ha (Shenmu)
Geology	neogene sedimentary rock	Slope at Source	30~50°
Landslide area	Large, 1%≦ landslide ratio ≦5%	Sediment	Average debris material size: 3"-12"
Vegetation	Natural woods, medium sparse	Damaged by	debris, overflow
Engineering Practice	None	Priority of Mitigation	High
Station Elevation	1,187 m	Coordinate (TWD97)	X: 235367 Y: 2602749
Protected Targets	**Residents** > 5 households	**Facility** school	**Transportation** roads, bridges

Table 5. The landslide area in Shenmu (after 2009)

Debris Flow No.	Stream	Length (km)	Catchment Area (ha)	Landslide Area (ha)
DF199	Chusuei Stream	7.16	861.56	33.29
DF227	Huosa Stream	17.66	2,620	149.32
DF226	Aiyuzi Stream	3.30	400.64	99.85

The monitoring practice of Shemu Station includes instruments and sensors like rain gauge, soil moisture sensor, geophone, wire sensor and CCD camera. Figure 27 shows the monitoring layout and the location of Shenmu Station (the data center in the figure). The station continuously collects the observation data of rainfall, soil moisture and geophone signal and transmits data back to the EOC servers. The data is used for alert in response to debris flow disaster and further analysis in advanced studies.

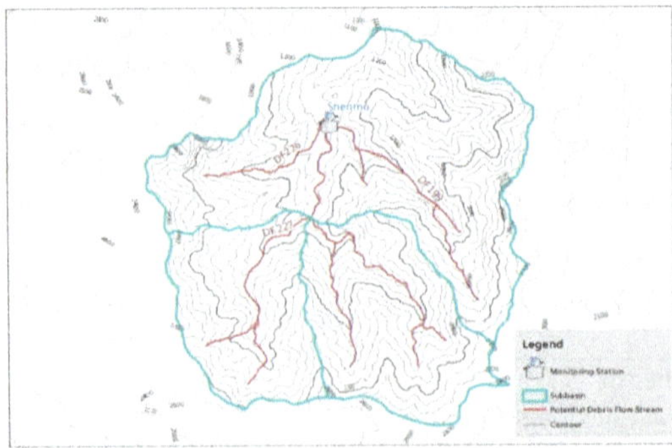

Fig. 25 The topographic map of Shenmu area

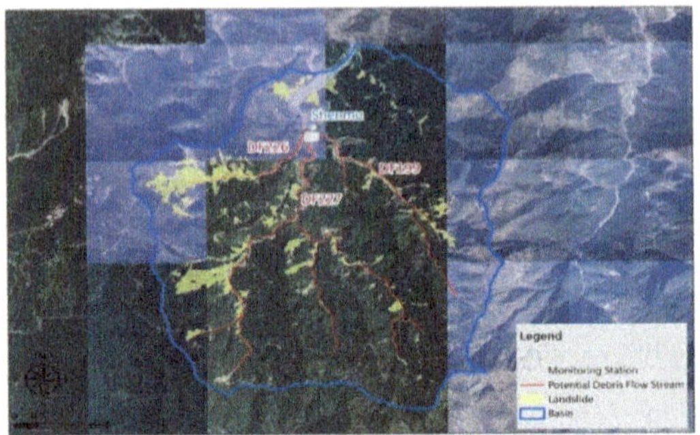

Fig. 26 The landslide areas of Shenmu site (image taken in 2009 after Typhoon Morakot)

Fig. 27 The monitoring layout of Shenmu Station

Among these sensors, the rain gauges and soil moisture sensors measure the water variation, a major cause of debris flow, in real-time manner and are usually used for warning criteria. The wire sensors and geophones function as indicators when a debris flow actually occurs. It should be noted that the data from geophones is processed using Harr wavelet transform, the low frequency range 0~62.5 Hz is used as debris flow characteristic frequency [14]. Figure 28 illustrates typical charts of observed data during an event. The CCD camera, different to other instruments, is used for real-time inspection if any warning is triggered and for debris flow image capture.

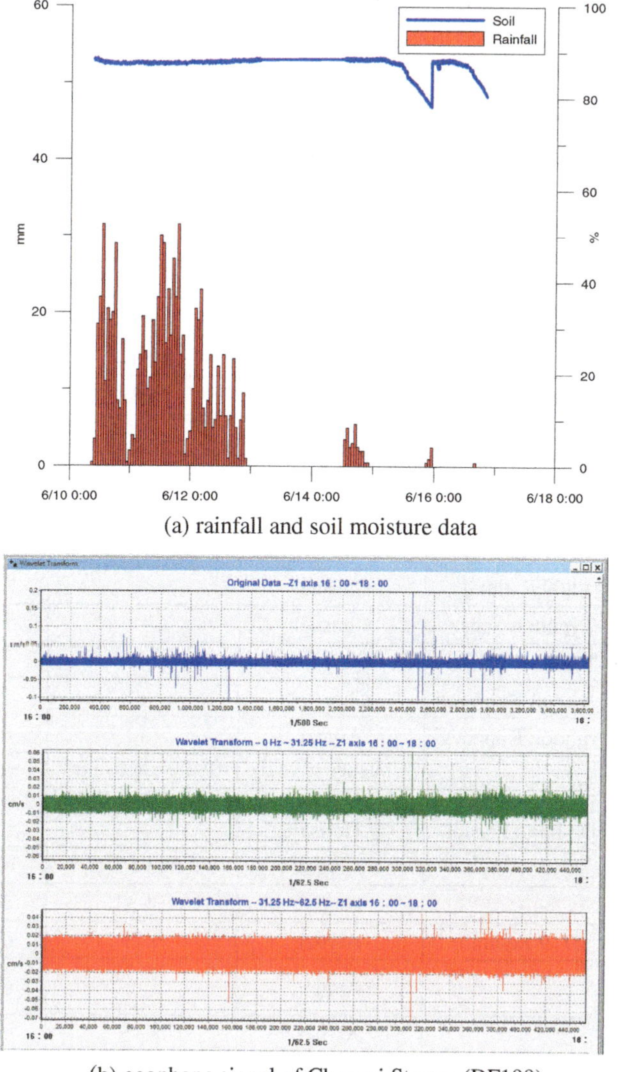

(a) rainfall and soil moisture data

(b) geophone signal of Chusuei Stream (DF199)

Fig. 28 Observations of the 0610 Heavy Rainfall event on June 10-17, 2012

11.4.2 Debris Flow Hazard History in Shenmu

Debris flow had frequently occurred at Shenmu area. As early as 1996, the Typhoon Herb had caused debris flow in Chusuei Stream. Since then, the Chusuei Stream in Shenmu had become the location of frequent debris flow disaster. In July 2001, Typhoon Toraji had brought 478 mm of maximum daily rainfall that triggered debris flow and caused severe casualties and property loss to people living at the downstream. After the Typhoon Toraji, the SWCB established a monitoring station in Shenmu in 2002. The station was designed to watch and monitor the three streams around the Shenmu village. In 2009, another catastrophic event, Typhoon Morakot, had caused serious damage to Shenmu. Typhoon Morakot had brought extremely heavy rainfall up to 1,550 mm in three days. It had caused the Aiyuzi Bridge broken due to the high-rise flood and debris.

Table 5 Debris flow hazard history of Shenmu

Date	Event	Location (stream)	Occurrence	Hazard Type
2004/5/20	-	Aiyuzi	14:53	debris flow
2004/5/21	-	Aiyuzi	16:08	debris flow
2004/5/29	-	Aiyuzi	16:19	debris flow
2004/6/11	-	Aiyuzi	16:42	debris flow
2004/7/2	Typhoon Mindulle	Aiyuzi	16:41	debris flow
2005/7/19	Typhoon Haitang	Chusuei, Aiyuzi	-	flood
2005/8/4	Typhoon Matsa	Chusuei, Aiyuzi	-	flood
2005/9/1	Typhoon Talim	Chusuei, Aiyuzi	-	flood
2006/6/9	0609 Rainfall	Chusuei, Aiyuzi	about 08:00	debris flow
2007/8/13	0809 Rainfall	Chusuei	-	flood
2007/8/18	Typhoon Sepat	Chusuei	-	flood
2007/10/6	Typhoon Krosa	Chusuei	-	flood
2008/7/17	Typhoon Kalmaegi	Chusuei	-	flood
2008/7/18	Typhoon Kalmaegi	Aiyuzi	-	flood
2009/8/8	Typhoon Morakot	Chusuei, Aiyuzi, Huosa	08:00 (landslide) 16:57 (debris flow)	landslide, debris flow
2010/9/19	Typhoon Fanapi	Huosa	-	flood
2011/11/10	-	Aiyuzi	13:17	debris flow
2011/7/13	-	Aiyuzi	14:33	debris flow
2011/7/19	0719 Rainfall	Aiyuzi	3:19	debris flow
2012/5/4	-	Aiyuzi	15:56 16:09	debris flow
2012/5/20	-	Aiyuzi	8:15	flood
2012/6/10	0610 Rainfall	Aiyuzi	10:34 15:14	debris flow
2012/6/11	0610 Rainfall	Chusuei	17:08	flood

The debris from the upper stream of Aiyuzi Stream had deposited at the bridge location. Foundation of buildings had also been scoured at the stream sides. The debris flow occurred almost every year after 2009 (Table 6). It had been noted that after the Typhoon Morakot in 2009, all the debris flow occurred at the Aiyuzi Stream. The reason of this phenomenon partly comes from the facts that Aiyuzi Stream is shorter than the other two streams and has relatively more landslide area (99.85 ha) at its upstream slopes after Typhoon Morakot. These factors contribute to the frequent occurrence of debris flow in Aiyuzi Stream. Until now, the Shenmu is still under the threat of debris flow during typhoons and heavy rainfalls. To protect the property and lives, a warning system is necessary and had been implemented in the Shenmu monitoring system.

11.4.3 Debris Flow on Nov. 10, 2011

The Aiyuzi Stream of Shenmu began to rain at about midnight on Nov. 10, 2011. The rain continued and the stream water began to increase after 13:20. The geophone at the Aiyuzi Stream signaled alert on 13:17 and soon the debris flow occurred. The debris flow was captured by the CCD camera. The stream was back to normal after about 13:30. Figure 29 and Figure 30 show the debris flow images at the upper and downstream of Aiyuzi Stream. This event was identified as a small-scale debris flow.

<table>
<tr><td>(a) 13:13</td><td>(b) 13:14</td></tr>
<tr><td>(c) 13:17</td><td>(d) 13:20</td></tr>
</table>

Fig. 29 The upstream images of Aiyuzi Stream

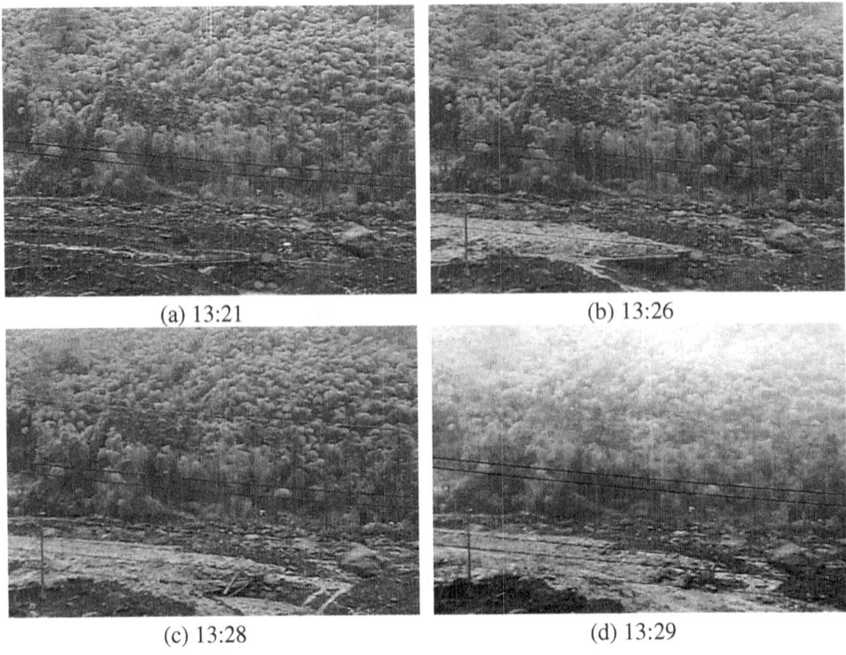

(a) 13:21   (b) 13:26

(c) 13:28   (d) 13:29

Fig. 30 The downstream images of Aiyuzi Stream

Direct Approach

Figure 31 shows the results of geophones at upper stream of Aiyuzi in the case of Nov. 11, 2011. Spikes are shown in the figure indicating the vibration from the debris flow. The warning threshold of the geophone was reached on 13:17 and the wire sensors were broken on 13:20 as shown in fig. 8. An alert message was automatically sent by the monitoring system when the geophone response reached the warning threshold. The debris flow arrived the downstream at about 13:29. In this case, the time window of warning was about 12 minutes. No casualty was reported in this event.

Indirect Approach

The warning threshold is 250 mm in Shenmu in terms of effective accumulated rainfall. In this case, the debris flow occurred when the effective accumulated rainfall was about 75 mm indicating this event was triggered by much less rainfall than the past cases. Figure 32 shows the hourly rainfall and soil moisture changes of the event. Although the indirect method of using rainfall in this case did not have the expected pre-warning to debris flow, the rainfall data was still important when considering the small-scaled and lower rainfall-triggered debris flow in Shenmu hazard history.

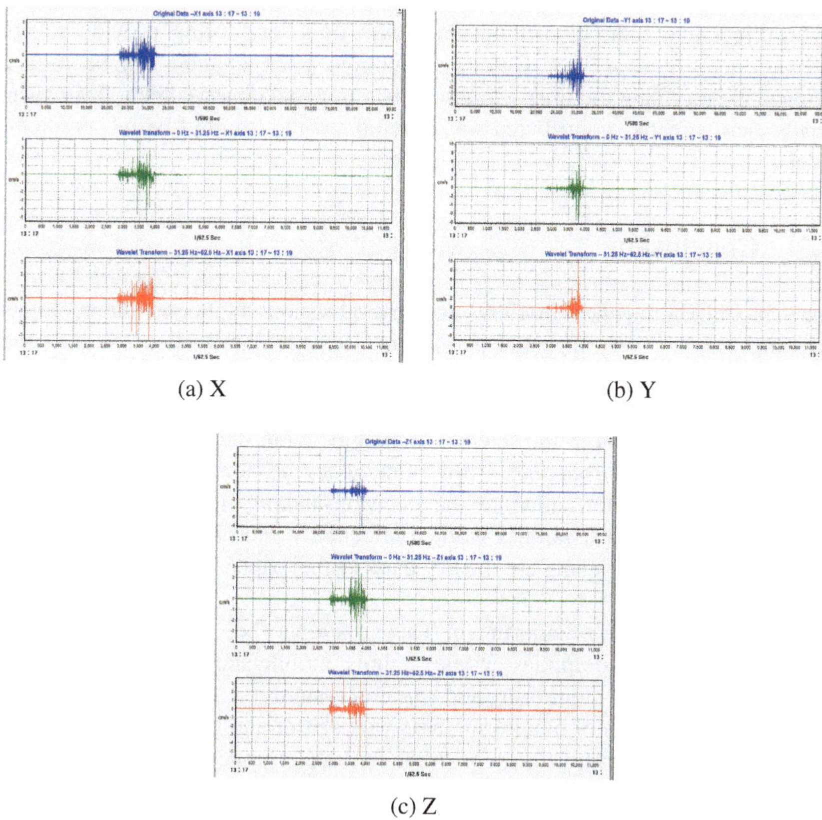

(a) X

(b) Y

(c) Z

Fig. 31 The geophone signals of Aiyuzi Stream at 13:17~13:19 on Nov. 10, 2011

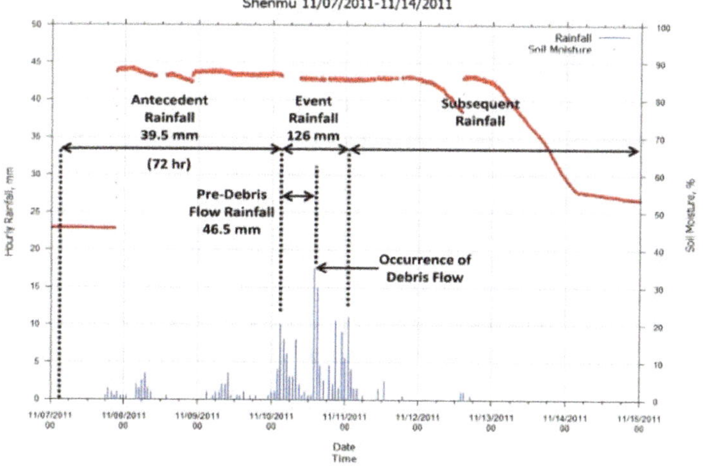

Fig. 32 The rainfall and soil moisture data of event on Nov. 10, 2011

The direct and indirect methods in debris flow monitoring provide reliable information for debris flow disaster prevention. Each method can be used as reference to determine the time window of warning and observed data from events actually has greatly contributed in developing the rainfall-based prediction models.

11.5 Conclusion

After Chi-Chi Earthquake (in 1999) and several serious typhoons, the geological status had become very unstable in Taiwan. Heavy rainfall often caused debris flows, which had greatly threatened to people's life and property. A monitoring system of debris flow was needed and had been developed using advanced sensors and technologies. Through the satellite communication, the real-time observation can be transmitted to the debris flow emergency response center in order to provide supporting information to the decision-making units. The debris flow monitoring system in Taiwan has worked effectively and brought great benefit to the general public. Following are the key parts of this paper.

(1) The debris flow monitoring in Taiwan adopts the concept of basin-wide monitoring network and applies on-site stations, mobile stations, portable units to construct a debris flow monitoring and warning system.

(2) The direct option of monitoring uses wire sensor and geophone as indicators to debris flow. The arrangement of these sensors at upper and mid-stream can provide a time window of warning. Indirect method of rain gauges is more commonly used in researches for rainfall-based model for debris flow. A reliable rainfall-model can also provide a better warning procedure for debris flow. The case described in this chapter had shown the benefit of applying both direct and indirect methods in debris flow monitoring in Shenmu.

(3) The design of portable units matches the requirement of easy-to-carry and self-operation resulting, a successful monitoring unit which can support the existing monitoring network and system. Portable units provide great potential for debris flow monitoring in terms of network flexibility and prompt response to hazards.

References

[1] Lee, B.J., et al.: Basin-Wide Monitoring Network for Debris Flow at Laonong and Qishan Rivers, Project Report. Soil and Water Conservation Bureau, 364 (2010) (in Chinese)

[2] Dai, F.C., Lee, C.F., Wang, S.J.: Analysis of rainstorm-induced slide-debris flows on natural terrain of Lantau Island, Hong Kong. Engineering Geology 51(4), 279–290 (1999)

[3] Ding, M.T., Wei, F.Q.: Distribution Characteristics of Debris Flows and Landslides in Three Rivers Parallel Area. Disaster Advances 4(3), 7–14 (2011)

[4] Jan, C.D., Lee, M.H., Huang, T.H.: Effect of Rainfall on Debris Flows in Taiwan. In: Proceedings of the International Conference on Slope Engineering, Hong Kong, vol. 2, pp. 741–751 (2003)

[5] Itakura, Y., Koga, Y., Takahama, J.I., Nowa, Y.: Acoustic detection sensor for debris flow. In: The First Int. Conf. on Debris-Flow Hazards Mitigation: Mechanics, Prediction, and Assessment, San Francisco, U.S.A., pp. 747–756 (1997)

[6] Liu, G., Li, X.: Application For Underground Sound Sensor. In: Annual Cross-Channel Mountainous Disasters and Environmental Preservation Conference, pp. 161–169 (2000)

[7] Liu, G., Li, X.: Application For Underground Sound Sensor. In: Annual Cross-Channel Mountainous Disasters and Environmental Preservation Conference, pp. 161–169 (2000)

[8] Huang, C., Ye, C., Yin, H., Wang, J.: Study of Debris Flow for Integration with a Geophone. Chinese Water and Landscape Conservation Journal 36(1), 39–53 (2005)

[9] Lahusen, R.G.: Detecting Debris Flows Using Ground Vibrations, USGS Fact Sheet, pp. 236–296 (1996)

[10] Jan, C.D., Lee, M.H.: A Debris-Flow Rainfall-Based Warning Model. Journal of Chinese Soil and Water Conservation, CSWCS 35(3), 275–285 (2004)

[11] Lee, M.H.: A Rainfall-Based Debris Flow Warning Analysis and Its Application, Ph.D. Thesis, National Cheng Kung University, Taiwan (2006) (in Chinese)

[12] Lee, B.J., et al.: The, On-site Data gathering and Monitoring Station Maintenance Program, Project Report. Soil and Water Conservation Bureau, 364 (2011) (in Chinese)

[13] Lee, B.J., et al.: The 2012 On-site Data gathering and Monitoring Station Maintenance Program, Project Report. Soil and Water Conservation Bureau, 476 (2012) (in Chinese)

[14] Fang, Y.M., et al.: Analysis of Debris Flow Underground Sound by Wavelet Transform-A Case Study of Events in Aiyuzih River. Journal of Chinese Soil and Water Conservation 39(1), 27–44 (2008)

Author Index